Effects of Environmental Organic Pollutants on Environment and Human Health: The Latest Updates

Effects of Environmental Organic Pollutants on Environment and Human Health: The Latest Updates

Editors

Zhen-Guang Yan
Zhi-Gang Li
Jinzhe Du

MDPI

Basel • Beijing • Wuhan • Barcelona • Belgrade • Novi Sad • Cluj • Manchester

Editors
Zhen-Guang Yan
State Key Laboratory of Environmental Criteria and Risk Assessment
Chinese Research Academy of Environmental Sciences
Beijing
China

Zhi-Gang Li
State Key Laboratory of Environmental Criteria and Risk Assessment
Chinese Research Academy of Environmental Sciences
Beijing
China

Jinzhe Du
School of Marine Science and Engineering
Qingdao Agricultural University
Qingdao
China

Editorial Office
MDPI
St. Alban-Anlage 66
4052 Basel, Switzerland

This is a reprint of articles from the Special Issue published online in the open access journal *Toxics* (ISSN 2305-6304) (available at: https://www.mdpi.com/journal/toxics/special_issues/39RVK1GZCN).

For citation purposes, cite each article independently as indicated on the article page online and as indicated below:

Lastname, A.A.; Lastname, B.B. Article Title. *Journal Name* **Year**, *Volume Number*, Page Range.

ISBN 978-3-7258-0834-2 (Hbk)
ISBN 978-3-7258-0833-5 (PDF)
doi.org/10.3390/books978-3-7258-0833-5

Contents

About the Editors

Zhen-Guang Yan

Dr. Zhen-Guang Yan is a research professor of ecotoxicology and water quality criteria in the State Key Laboratory of Environmental Criteria and Risk Assessment of the Chinese Research Academy of Environmental Sciences. He received his Ph.D. in biochemistry and molecular biology from Tsinghua University in 2008. He is the academic leader for the study of aquatic life criteria in the Key Laboratory. Dr. Yan serves as the committeeman for the Committee on Environmental Standards and Criteria and the Committee on Environmental and Ecological Toxicity. In 2003, he received the Excellent Young Teacher Awards of Qingdao, Shandong Province. In 2011, he received the National Excellent Youth Science and Technology Award from the Chinese Society of Toxicology. His research interests encompass ecotoxicology, water quality criteria, and risk assessment. Dr. Yan's current research activities focus on the development of water quality criteria for priority pollutants and the screening of native test organisms. He is in the process of developing technological platforms for water quality criteria, which includes constructing an environmental criteria database, screening Chinese resident aquatic species for criteria development, and establishing suitable test methods. Dr. Yan has published more than 20 SCI papers in core journals, among them *Environmental Pollution*, *Journal of Biological Chemistry*, *Biomacromolecules*, and *Science China*. He has also coauthored four academic books in collaboration with his colleagues.

Zhi-Gang Li

Zhi-Gang Li is an Associate Professor at the State Key Laboratory of Environmental Criteria and Risk Assessment, Chinese Research Academy of Environmental Sciences. In recent years, he has been researching the effect of exposure to environmental pollutants on health. In the field of the toxic mechanisms of air pollution exposure to health, he found the epigenetic modification of air pollution to contribute to metabolic disorders, circadian disorders, cardiopulmonary damage, etc. In the field of environmental epidemiology, he found the adverse effects of air pollution exposure on human health and the exposure–response relationship between them, which could provide the theoretical basis for health risk management of air pollution. He has published over 40 papers in *Environment International*, *Environmental Pollution*, *Science of the Total Environment*, etc.

Jinzhe Du

Dr. Jinzhe Du is a research scientist in marine environmental science and shellfish biomineralization in the School of Marine Science and Engineering at Qingdao Agricultural University. She received her Ph.D. in biochemistry and molecular biology from Tsinghua University in 2021. Dr. Du's current research activities focus on the impact of ocean acidification on biomineralization. She has accumulated rich experience in the molecular regulatory mechanism of shell mineralization. Dr. Du has published more than 10 SCI papers in core journals, among them *Science of the Total Environment*, *Crystal Growth and Design*, *Crystengcomm*, *Scientific Reports*, and *Proteomics*. These research results have important academic significance for the field of biomineralization.

toxics

MDPI

Editorial

Effects of Environmental Organic Pollutants on Environment and Human Health: The Latest Updates

Zhen-Guang Yan [1,*], Zhi-Gang Li [1] and Jin-Zhe Du [2]

1 State Key Laboratory of Environmental Criteria and Risk Assessment, Chinese Research Academy of Environmental Sciences, Beijing 100012, China; lizg@craes.org.cn
2 College of Marine Science and Technology, Qingdao Agricultural University, Qingdao 266109, China; djz@qau.edu.cn
* Correspondence: zgyan@craes.org.cn

1. Introduction

This editorial introduces the Special Issue "Effects of Environmental Organic Pollutants on Environment and Human Health: The Latest Updates". Environmental Organic Pollutants include volatile organic compounds (VOCs) and persistent organic pollutants (POPs). VOCs originate from motor vehicle emissions and various manufactured products, such as building materials, paints, and cleaning agents, which often pollute the atmosphere around us. POPs have an intrinsic resistance to natural degradation processes and are found in polluted water, soil, atmosphere, sediment, etc. In addition, many emerging organic pollutants are considered potentially harmful to human health [1], such as Pharmaceuticals and Personal Care Products, and thus they are prioritized in biomonitoring surveillance.

This Special Issue presents the most recent advancements in research on VOCs and POPs, encompassing investigations and risk assessments of environmental organic pollutants, studies on the toxic mechanisms of such pollutants, as well as comprehensive summaries and reviews of the research progress and historical developments within related fields.

2. An Overview of Published Articles

There are two articles focused on VOCs. Wang et al.'s (contribution 4) field observations of CO, NO, NO_2, O_3, and VOCs were conducted in Beijing, Baoding, and Shanghai. They illustrated the pollution characteristics, source analysis, and risk assessment of VOC in the three cities. The research results indicated that motor vehicle exhaust was the main source of VOCs in all three cities. Acrolein was the only substance with an average hazard quotient greater than 1, indicating a significant non-carcinogenic risk. In Beijing, 1,2-dibromoethane had an R-value of 1.1×10^{-4} and posed a definite carcinogenic risk. Singh et al. (contribution 10) used statistical analysis to determine that anthropogenic activities were considerable sources of emission of VOCs in industrial areas. During the lockdown, the major factors behind the crucial decrease in TVOC levels were the complete and partial restrictions on industrial activities, transport, and marketplace openings. Comparatively, the lifetime cancer risk (LCR) value for males and females was estimated to be higher throughout the lockdown period than in the pre-and post-lockdown periods. These findings showed that exposure to VOCs induced adverse health effects, including carcinogenic and non-carcinogenic risks.

There are 11 articles focused on POPs. The research fields of these studies include toxic mechanisms, epidemiology, and ecological risk.

Among them, there are two articles on the toxic mechanisms of exposure to POPs on health. Exposure to Bisphenol A (BPA) has led to an increased risk of obesity and nonalcoholic fatty liver diseases (NAFLDs). Li et al. (contribution 2) investigated the effects of BPA on the hepatic lipid metabolism function and its potential mechanisms in mice

Citation: Yan, Z.-G.; Li, Z.-G.; Du, J.-Z. Effects of Environmental Organic Pollutants on Environment and Human Health: The Latest Updates. *Toxics* **2024**, *12*, 231. https://doi.org/10.3390/toxics12040231

Received: 8 March 2024
Accepted: 15 March 2024
Published: 22 March 2024

through a comparison of the BPA exposure model and the BPA exposure + cessation of drug treatment model. The results showed that the mice exposed to BPA manifested NAFLD features. Importantly, BPA could significantly decrease the level of APOD protein, whereas an extremely significant increase occurred after they stopped exposure. Meanwhile, APOD over-expression suppressed TG accumulation in AML12 cells. In conclusion, the damage caused by BPA can be repaired by upregulation of APOD, and it is a potentially effective biochemical detection indicator for the treatment of obesity or NAFLDs caused by BPA exposure. The persistent pollutants in wastewater can enter the food chain and ultimately endanger human health [2]. Yang et al. (contribution 3) studied the cytotoxicity of the industrial wastewater treatment. They conducted a broad evaluation of the environmental health risks from industrial wastewater along the Yangtze River, China, using a battery of bioassays. The toxicity tests on the wastewater samples showed that the wastewater treatment processes were effective at lowering acetylcholinesterase (AChE) inhibition, HepG2 cells' cytotoxicity, the estrogenic effect in T47D-Kbluc cells, DNA damage in Euglena gracilis, and the mutagenicity of *Salmonella typhimurium.* These two studies provide a good basis for the health risk assessment of BPA and industrial wastewater, and Yang et al.'s finding also provides a scientific reference for the optimization and operation of wastewater treatment processes.

Furthermore, in this Special Issue, Niu et al. (contribution 11) systematically summarize contemporary findings from epidemiological surveys, and they explored the mechanistic correlation between exposure to emerging pollutants (including endocrine disruptors, perfluorinated compounds, microplastics, and antibiotics) and blood glucose dysregulation. Their work provides a basic reference for further research on the complex interaction between new pollutants and diabetes.

It is clear that a poor ecological environment will damage our health. Ecological risks are closely related to health risks. In this Special Issue, there are eight articles that focus on the ecological risk of organic matter. Wang et al. (contribution 1) researched the nine pesticide pollutants included in the "List of New Key Pollutants for Control (2023 Edition)" issued by the Chinese government. They analyzed the environmental exposure to pesticide pollutants in sediments along the coast of China and derived baseline standards for sediment quality using the balanced distribution method. They also conducted a multi-level ecological risk assessment of pesticides in sediment. The risk quotient assessment showed that endosulfan and DDT posed medium environmental risks to the Chinese coastal sediment environment, and PCBs posed medium risks in some bays of the East China Sea. The semi-probabilistic optimized evaluation and the joint probability curve (JPC) assessments all showed that endosulfan and DDT pose a certain degree of risk to the environment. Hong et al. (contribution 5) analyzed the spatial distribution characteristics of nine alkylphenols (APs) in the Yongding River and Beiyun River. The differences in the concentrations and spatial distribution patterns of the nine APs were systematically evaluated using principal component analysis (PCA). The results demonstrated that the APs were widely present in both rivers, and the pollution risks associated with the APs were more severe in the Yongding River than in the Beiyun River. This study provides theoretical data support and a basis for AP pollution risk evaluation in Yongding River and Beiyun River. Triclosan (TCS), a commonly used antibacterial preservative, has been demonstrated to have high toxicological potential, and it adversely affects water bodies [3]. Lu et al. (contribution 6) addressed the adverse effects of TCS on freshwater microalgae (E. gracilis), including morphological alterations, reduced photosynthesis, and oxidative stress. They showed that the main toxic mechanisms of TCS exposure for E. gracilis were changes in ROS and antioxidant enzyme activities, which stimulated algal cell damage, and the inhibition of the TCA cycle metabolic system, including carbon metabolism, nitrogen metabolism, and the D-glutamine and D-glutamate metabolism pathways controlled by the downregulation of DEGs, which were further manifested as oxidative stress and photosynthesis inhibition effects. Wang et al. (contribution 7) investigated the Yellow River Estuary region and found that a total of 34 antibiotics, including macrolides, sulfonamides, quinolones, tetracyclines, and chloramphenicol, were pollutants. The results show that antibiotics were widely present in the water bodies of the Yellow River Estuary, with 14 antibiotics detected to varying degrees, including a high detection rate for lincomycin hydrochloride. Farming wastewater and domestic sewage were the primary sources of antibiotics in the Yellow River Estuary. This study provides beneficial information for the assessment of the ecological risk presented by antibiotics in the Yellow River Estuary water bodies and a scientific basis for future antibiotic pollution control in the Yellow River Basin. Hou et al. (contribution 9) conducted a comprehensive field investigation of Baiyangdian Lake and assessed the ecological risk of PAEs, which can provide data support and a theoretical basis for the formulation of water quality standards and the future prevention and control of PAE pollution. The Water Quality Criteria (WQC) for the protection of aquatic organisms mainly focus on the maximum threshold for pollutants that do not have harmful effects on aquatic organisms. Feng et al. (contribution 13) systematically discussed an overview of water biological conservation, its theoretical methods, and its research progress and detailed the key scientific issues that need to be considered in WQC research. Combined with the specific characteristics of emerging pollutants, some new ideas and directions for future research on the WQC protection of aquatic organisms were proposed. Liu et al. (contribution 12) comprehensively reviewed the development process of WQC in China, focusing on the methodological progress and challenges in selecting priority pollutants, biological screening tests, and standardizing ecotoxicity testing protocols. They also provided critical assessments of the necessary minimum data requirements for quality assurance measures, data validation techniques, and ALC assessment. Moreover, in Men et al.'s (contribution 8) study, a non-experimental approach was used to calculate the RfD values, which explored the potential correlation between toxicity and physicochemical characteristics and the chemical structure of pesticides. The molecular descriptors of contaminants were calculated using T.E.S.T software from the EPA, and a prediction model was developed using a stepwise multiple linear regression (MLR) approach. Approximately 95% and 85% of the data points differed by less than ten-fold and five-fold between the predicted values and true values, respectively, which improved the efficiency of the RfD calculation. The model prediction values have certain reference values in the absence of experimental data, which is beneficial to the advancement of contaminant health risk assessment.

3. Conclusions

The papers published in this Special Issue include investigations and the risk assessments of organic pollutants in multiple regions, providing a scientific basis for governments and businesses to formulate environmental policies and technological improvements. The toxicity mechanism research provides a foundation and reference for subsequent research. The review paper systematically summarizes epidemiological investigation findings and delves into the mechanism of correlation between exposure to emerging pollutants and blood glucose imbalance. It comprehensively reviews the development process of WQC in China and systematically provides an overview of WQC, its theoretical methods, and research progress for aquatic organism protection. This Special Issue provides important references for subsequent research.

Conflicts of Interest: The author declares no conflicts of interest.

List of Contributions

1. Wang, J.; Zhao, Q.; Gao, F.; Wang, Z.; Li, M.; Li, H.; Wang, Y. Ecological Risk Assessment of Organochlorine Pesticides and Polychlorinated Biphenyls in Coastal Sediments in China. *Toxics* **2024**, *12*, 114. https://doi.org/10.3390/toxics12020114
2. Li, C.; Shen, N.; Yang, S.; Wang, H.-L. Effects of BPA Exposure and Recovery on the Expression of Genes Involved in the Hepatic Lipid Metabolism in Male Mice. *Toxics* **2023**, *11*, 775. https://doi.org/10.3390/toxics11090775

3. Yang, B.; Cui, H.; Gao, J.; Cao, J.; Klobučar, G.; Li, M. Using a Battery of Bioassays to Assess the Toxicity of Wastewater Treatment Plant Effluents in Industrial Parks. *Toxics* **2023**, *11*, 702. https://doi.org/10.3390/toxics11080702
4. Wang, Z.; Zhang, P.; Pan, L.; Qian, Y.; Li, Z.; Li, X.; Guo, C.; Zhu, X.; Xie, Y.; Wei, Y. Ambient Volatile Organic Compound Characterization, Source Apportionment, and Risk Assessment in Three Megacities of China in 2019. *Toxics* **2023**, *11*, 651. https://doi.org/10.3390/toxics11080651
5. Hong, Y.; Chen, M.; Zhu, Z.; Liao, W.; Feng, C.; Yan, Z.; Qiao, Y.; Mei, Y.; Xu, D. The Distribution Characteristics and Ecological Risks of Alkylphenols and the Relationships between Alkylphenols and Different Types of Land Use. *Toxics* **2023**, *11*, 579. https://doi.org/10.3390/toxics11070579
6. Lu, T.; Zhang, T.; Yang, W.; Yang, B.; Cao, J.; Yang, Y.; Li, M. Molecular Toxicity Mechanism Induced by the Antibacterial Agent Triclosan in Freshwater *Euglena gracilis* Based on the Transcriptome. *Toxics* **2023**, *11*, 414. https://doi.org/10.3390/toxics11050414
7. Wang, J.; Yan, Z.; Qiao, Y.; Liu, D.; Feng, C.; Bai, Y. Distribution and Characterization of Typical Antibiotics in Water Bodies of the Yellow River Estuary and Their Ecological Risks. *Toxics* **2023**, *11*, 400. https://doi.org/10.3390/toxics11050400
8. Men, S.-H.; Xie, X.; Zhao, X.; Zhou, Q.; Chen, J.-Y.; Jiao, C.-Y.; Yan, Z.-G. The Application of Reference Dose Prediction Model to Human Health Water Quality Criteria and Risk Assessment. *Toxics* **2023**, *11*, 318. https://doi.org/10.3390/toxics11040318
9. Hou, Y.; Tu, M.; Li, C.; Liu, X.; Wang, J.; Wei, C.; Zheng, X.; Wu, Y. Risk Assessment of Phthalate Esters in Baiyangdian Lake and Typical Rivers in China. *Toxics* **2023**, *11*, 180. https://doi.org/10.3390/toxics11020180
10. Singh, B.P.; Sohrab, S.S.; Athar, M.; Alandijany, T.A.; Kumari, S.; Nair, A.; Kumari, S.; Mehra, K.; Chowdhary, K.; Rahman, S.; et al. Substantial Changes in Selected Volatile Organic Compounds (VOCs) and Associations with Health Risk Assessments in Industrial Areas during the COVID-19 Pandemic. *Toxics* **2023**, *11*, 165. https://doi.org/10.3390/toxics11020165
11. Niu, H.; Xu, M.; Tu, P.; Xu, Y.; Li, X.; Xing, M.; Chen, Z.; Wang, X.; Lou, X.; Wu, L.; et al. Emerging Contaminants: An Emerging Risk Factor for Diabetes Mellitus. *Toxics* **2024**, *12*, 47. https://doi.org/10.3390/toxics12010047
12. Liu, C.; Geng, Z.; Xu, J.; Li, Q.; Zhang, H.; Pan, J. Advancements, Challenges, and Future Directions in Aquatic Life Criteria Research in China. *Toxics* **2023**, *11*, 862. https://doi.org/10.3390/toxics11100862
13. Feng, C.; Huang, W.; Qiao, Y.; Liu, D.; Li, H. Research Progress and New Ideas on the Theory and Methodology of Water Quality Criteria for the Protection of Aquatic Organisms. *Toxics* **2023**, *11*, 557. https://doi.org/10.3390/toxics11070557

References

1. Li, M.-R.; Men, S.-H.; Wang, Z.-Y.; Liu, C.; Zhou, G.-R.; Yan, Z.-G. The application of human-derived cell lines in neurotoxicity studies of environmental pollutants. *Sci. Total Environ.* **2024**, *912*, 168839. [CrossRef] [PubMed]
2. Hakak, S.; Khan, W.-Z.; Gilkar, G.-A. Industrial wastewater management using blockchain technology: Architecture, requirements, and future directions. *IEEE Internet Things* **2020**, *3*, 38–43. [CrossRef]
3. Machado, M.-D.; Soares, E.-V. Toxicological effects induced by the biocide triclosan on Pseudokirchneriella subcapitata. *Aquat. Toxicol.* **2021**, *230*, 105706. [CrossRef] [PubMed]

toxics

MDPI

Article

Ecological Risk Assessment of Organochlorine Pesticides and Polychlorinated Biphenyls in Coastal Sediments in China

Jie Wang [1,2,†], Qi Zhao [3,†], Fu Gao [4,†], Ziye Wang [2], Mingrui Li [2], Haiming Li [1,*] and Yizhe Wang [2,*]

1 College of Marine and Environmental Sciences, Tianjin University of Science and Technology, Tianjin 300457, China; wangjiew2021@163.com
2 State Key Laboratory of Environmental Criteria and Risk Assessment, Chinese Research Academy of Environmental Sciences, Beijing 100012, China; wzy972953@163.com (Z.W.); limingrui212@mails.ucas.ac.cn (M.L.)
3 Bayingoleng Ecological Environment Monitoring Station of Weiwu 'er Autonomous District, Xinjiang 841000, China; mingkongqi2023@163.com
4 Technical Centre for Soil, Agriculture and Rural Ecology and Environment, Ministry of Ecology and Environment, Beijing 100012, China; gaofugaofu@126.com
* Correspondence: lhm@tust.edu.cn (H.L.); wangyz@craes.org.cn (Y.W.)
† These authors contributed equally to this work.

Abstract: Although the ecological risk of emerging contaminants is currently a research hotspot in China and abroad, few studies have investigated the ecological risk of pesticide pollutants in Chinese coastal sediments. In this study, nine pesticide pollutants included in the "List of New Key Pollutants for Control (2023 Edition)" issued by the Chinese government were used as the research objects, and the environmental exposure of pesticide pollutants in China's coastal sediments was analyzed. The baseline sediment quality criteria were deduced using the balanced distribution method, and a multi-level ecological risk assessment of pesticides in sediment was performed. The results showed that the nine pesticide pollutants were widespread in Chinese coastal sediments, with concentrations ranging from 0.01 ng·g^{-1} to 330 ng·g^{-1}. The risk quotient assessment showed that endosulfan and DDT posed medium environmental risks to the Chinese coastal sediment environment, and PCBs posed medium risks in some bays of the East China Sea. The semi-probabilistic, optimized semi-probability evaluation and joint probability curve (JPC) assessments all show that endosulfan and DDT pose a certain degree of risk to the environment.

Keywords: coastal; sediment; pesticides; ecological risk assessment

Citation: Wang, J.; Zhao, Q.; Gao, F.; Wang, Z.; Li, M.; Li, H.; Wang, Y. Ecological Risk Assessment of Organochlorine Pesticides and Polychlorinated Biphenyls in Coastal Sediments in China. *Toxics* **2024**, *12*, 114. https://doi.org/10.3390/toxics12020114

Academic Editor: Josef Velíšek

Received: 22 November 2023
Revised: 22 January 2024
Accepted: 23 January 2024
Published: 29 January 2024

1. Introduction

Persistent organic pollutants (POPs) have characteristics such as biological toxicity [1–3], environmental persistence [4,5], and bioaccumulation [6,7]. They are widespread in the environment and pose great risks to human health and the environment. The Chinese government has attached great importance to the control of pollutants in recent years. In December 2022, the Chinese government released the "List of New Pollutants for Key Control (2023 Edition)". The list contains 14 chlorinated hydrocarbons including chlorinated pesticides: Chlordane, Mirex, DDTs, HCH isomers, Endosulfans, HCB, and PCBs.

Pesticide pollution is very common. For example, organochlorine and other pesticide pollutants were detected in Cirebon [8], the East China Sea [9], and Xiangshan Bay [10] in Indonesia, and the total concentration ranges were 10–120 ng·L^{-1}, 183.49–1363.77 ng·L^{-1}, and 2.88–34.72 ng·L^{-1}, respectively. Similarly, a variety of pesticide pollutants were also found in the sediments of the Ebro River Delta [11], the Vasai River in Mumbai [12], the iSimangaliso Wetland Park in South Africa [13], and the west coast of India [14], with concentration ranges of 50.8–1912 ng·g^{-1}, 597–1538 ng·g^{-1}, 26.29–283 ng·g^{-1}, and 0.39–21.16 ng·g^{-1}, respectively. In addition, PCBs and organochlorine pesticides were found in the sediments of Shantou Bay, China, with concentrations ranging from 0.54 to 55.5 $ng \cdot g^{-1}$ and 2.19 to 16.9 $ng \cdot g^{-1}$ [15]. Organochlorine pesticides were found in the sediments of areas of North Bohai Sea, China, and concentrations of HCH and DDT in sediments ranged from below detection (<LOD) to 1964.97 $ng \cdot g^{-1}$ and <LOD to 86.46 $ng \cdot g^{-1}$, respectively [16]. PCBs were found in the Jiaojiang Estuary of the East China Sea, with concentrations ranging from 4.93 $ng \cdot g^{-1}$ to 108.79 $ng \cdot g^{-1}$ [17].

The ecological risk of pesticides in the aquatic environment is a topic of widespread concern around the world. For example, Guo et al. used the risk quotient method and the probabilistic risk assessment method to conduct an ecological risk assessment of organochlorine pesticides in the surface waters of Meiliang Bay, Gonghu Bay, and Huikou Bay in Taihu Lake [18]. The results showed that DDT, endosulfan, and hexachlorocyclohexane (HCH) presented relatively high risks. Xu et al. used the risk quotient method to assess the ecological risks of 35 pesticides in seven watersheds in China [19]. The results showed that the ecological risks of each watershed were at a potential medium level, and pesticides were the main compounds that posed risk. Hong et al. conducted an ecological risk assessment of DDT in pond sediments in the Yangtze River–Huai River region of China, and they found that there was a moderate to high ecological risk [20]. The results showed that the total content of PCBs and DDT posed a moderate ecological risk [21]. Zhao et al. conducted an ecological risk assessment of organochlorine pesticides and polychlorinated biphenyls in surface sediments of Tianjin Haihe Estuary based on sediment quality guidelines. The results showed that organochlorine pesticides and PCBs have potential ecological risks [22].

In addition, the ecological risks of pesticides in the ocean have also received widespread attention. For example, Xie et al. used the risk quotient method to assess the ecological risk of pesticides in the coastal waters of the Liaodong Peninsula in China, and the results showed that atrazine and acetochlor had higher risks to aquatic organisms than other pesticides [23]. Wang et al. conducted an ecological risk assessment of organochlorine pesticides in the waters of Hangzhou Bay, China, and the results showed that the potential danger of organochlorine pesticides in sediments was worrying [24]. The above-mentioned studies mainly conducted ecological risk assessments based on acute toxicity data, the assessment methods were not uniform, and the areas were relatively scattered [25]. There have been few systematic studies on the ecological risk assessment of pesticides in sediments in China's coastal waters.

Based on previous research, this study (1) analyzed the environmental exposure of pesticides in China's coastal sediments; and (2) conducted a multi-level ecological risk assessment of pesticides in sediments based on the sediment quality criteria derived using the phase equilibrium distribution method to provide a reference for the environmental management of pollutants in China.

2. Materials and Methods

2.1. *Evaluation and Selection of Data*

2.1.1. Environmental Concentration Data of Pollutants

The nine pesticides listed in the "List of New Pollutants for Key Control (2023 Edition)" released by China, namely chlordane, mirex, hexachlorobenzene, DDT, α-HCH, β-HCH, lindane, endosulfan technical and its related isomers, and PCBs, were used as research objects; the keywords "sediment", "risk assessment", "pollutants", and "China" were used to search the Web of Science and China National Knowledge Infrastructure; and we screened them from the obtained literature on the exposure data of pesticide pollutants in China's coastal waters. A total of 3379 exposure data points from 47 documents were collected. The detection time was from January 1998 to April 2023. The survey areas were the Bohai Sea, East China Sea, Yellow Sea, and South China Sea in China's coastal waters. The survey areas included the key sea areas of China's CEC control, such as the Yangtze River Estuary, the Pearl River Estuary, and Hangzhou Bay. If the data collected in this study contained values greater than the method detection limit, the actual measured values were recorded. If the monitoring data were displayed in graphic form, the average concentrations were recorded. Given the number of studies in the literature, the mean concentration for a location was calculated using measured values if greater than the method detection limits (MDL), the 1/2 MDL if < MDL, or 0 if not detected.

2.1.2. Environmental Toxicity Information

According to the principles of the accuracy, relevance, and reliability of pollutant toxicity data proposed by the US Environmental Protection Agency [26], Klimmisch [27], Durda [28], Hobbs [29], and Moermond [30], the toxic effect data for nine pesticides were searched in the ECOTOX database (https://cfpub.epa.gov/ecotox/search.cfm, accessed on 7 April 2022). During the toxicity data screening process, the data quality was evaluated from the following aspects: (1) experimental design, including the testing method, experimental process, and the validity and quality control of the experimental results; (2) reagent purity (>90%); (3) the source of the tested organisms; (4) exposure conditions, including the applicability of the test system to the test substance and the test organism, the test concentration interval, exposure time, and biomass loading; and (5) data analysis, including the statistical methods and concentration response curves. Toxicity data were used for classification. In principle, the most sensitive effect indicators were selected, and the chronic toxicity data of no observed effect concentration (NOEC) or 10% effect concentration (EC10) were preferred, followed by the lowest observed effect concentration (LOEC) or the median effect concentration (EC50) [31].

2.2. Derivation of Sediment Quality Benchmarks Using the Phase Equilibrium Distribution Method

There are many methods for deriving sediment benchmarks. For non-ionic organic compounds, the US Environmental Protection Agency recommends derivation using the equilibrium distribution method [32]. This method assumes that benthic and overlying aquatic organisms have the same sensitivity to the same pollutant. When the pollutants in interstitial water reach the water quality criteria (WQC), the content of pollutants in the sediment is the sediment quality criteria (SQC). The sediment quality benchmark calculation formula is as follows [32]:

$$K_p = f_{oc} \times K_{oc}, \tag{1}$$

$$SQC = K_p \times WQC, \tag{2}$$

where f_{oc} is the organic carbon content, dimensionless; the K_{oc} organic carbon partition coefficient can be deduced from the octanol/water partition coefficient K_{ow}, and the unit is $L \cdot kg^{-1}$; K_p is the partition coefficient of pollutants between the sediment phase and the interstitial water phase, dimensionless; WQC is the reference value of water quality, $ng \cdot L^{-1}$; and SQC is the sediment quality standard value, $ng \cdot g^{-1}$.

2.3. Ecological Risk Assessment

Multi-level ecological risk assessment (MLERA) is a method for the comprehensive ecological risk assessment of pollutants from a low level to a high level. This study performed an MLERA based on the ecotoxicology risk assessment framework [33,34], the risk assessment technical guidance document [35], the NORMAN priority framework for substances, and previous studies [36,37].

2.3.1. First Level: Quotient Value Method

The quotient value method is the most commonly used and extensive risk assessment method. The calculation method of the risk quotient is the ratio of the average concentration of a single chemical in the sediment to the predicted no-effect concentration (PNEC). The formula is as follows [38]:

$$RQ = C/PNEC_{sediment}, \tag{3}$$

In the formula, RQ is the risk quotient, dimensionless; C is the average concentration calculated from the measured value set of a single chemical, $ng \cdot g^{-1}$; and $PNEC_{sediment}$ is the predicted no-effect concentration derived by the most sensitive toxicity data with assessment factors (Afs) of 10, 20, or 100 depending on test endpoints of NOEC or EC10, LOEC, or EC50 [31,39].

When the RQ is less than 0.1, it is considered that there is no risk; when $0.1 \leq RQ < 1$, the pollutant is considered to have low risk; when $1 \leq RQ < 10$, the pollutant is considered to have medium risk; when $RQ \geq 10$, it is considered that the pollutant has a high risk [31,40]. Although the quotient value method can initially reflect the relative risk of a pollutant, it cannot explain the actual impact of the pollutant on aquatic organisms.

2.3.2. Second Level: Semi-Probability Method

The semi-probability method compares the measured ambient concentration of an individual chemical at each sampling point with its PNEC value. Pollution concentrations above the PNEC pose a potential risk to aquatic organisms, while concentrations below the PNEC are considered to pose an insignificant risk. Therefore, frequencies exceeding the PNEC can be used to prioritize pollutants. The frequency at which a target chemical exceeds the PNEC (F) can be calculated as the number of sampling points whose concentration exceeds the PNEC divided by the total number of sampling points. The results reveal the proportion of sites showing potential risk potential. The formula is as follows [38]:

$$F = n/N \times 100\%, \tag{4}$$

where F is the frequency exceeding PNEC, dimensionless; n is the number of sampling stations whose concentration exceeds the PNEC; and N is the total number of stations [41].

2.3.3. Third Level: Semi-Probability Method for Optimization

The current RQ based on the average concentration in water may be biased by the detection frequency. When screening high-risk compounds, it is a trend to consider both the concentration and frequency, so the optimal level of risk assessment can be performed according to the NORMAN network [37,42–44]. The product of the RQ_{max} value and the frequency of PNEC exceeding the standard is the priority index (PI), which can more clearly show the pesticide pollutants that should be focused on in China's coastal sediments. The formula is as follows [38]:

$$PI = RQ_{max} \times F, \tag{5}$$

In the formula, PI is the priority index, dimensionless; RQ_{max} is the risk quotient calculated based on the maximum concentration, dimensionless; and F is the frequency at which the concentration exceeds the PNEC, dimensionless.

When the PI is less than 1, the risk of the pollutant is low and the pollutant does not pose an ecological risk to the environment.

2.3.4. Fourth Level: Joint Probability Curves

There are many subjective factors in the process of formulating PNECs, which are determined based on the effects of small concentrations reported by a limited number of studies, and the results may not be repeatable. Joint probability curve methods can remedy this deficiency through using the linear regression of two datasets to calculate the probability that a concentration will adversely affect a specific proportion (%) of a species, and classifying the risk as minimal, low, medium, or high [45–47]. The formula is as follows [38]:

$$\text{Risk product} = \text{exceedance probability} \times \text{magnitude of effect}, \tag{6}$$

Risk < 0.25% is classified as minimal risk; risk $\geq$ 0.25% and <2% is classified as low risk; risk $\geq$ 2% and <10% is classified as medium risk; and risk $\geq$ 10% is classified as high risk.

3. Results and Discussion

3.1. Distribution of Pesticides in China's Coastal Sediments

The nine pesticides in this study were from the "List of Emerging Contaminants for Key Control". Nine pesticide pollutants with relevant research data in China's coastal sediments were used as research objects to evaluate their risk levels in China's coastal sediments.

A total of 3290 exposure data points of nine target chemicals were collected in the coastal waters of China, distributed in the Bohai Sea, Yellow Sea, East China Sea, and South China Sea (Figure 1). Among them, the South China Sea had the most types of pesticide pollutants (nine types), followed by the Yellow Sea (eight types), the East China Sea (seven types), and the Bohai Sea, which had had the fewest types of pesticides (four types) (Figure 2). As shown in Figure 1, the concentration of pollutants in the Yellow Sea was the highest, and the concentration of pollutants in most sea areas ranged from 500 to 1000 ng·g^{-1} dw, with the highest concentration reaching 1188 ng·g^{-1}. It can be seen from Figure 3 that over time, the concentrations of organochlorine pesticides measured in China's coastal sediments show an increasing trend. During the five-year period from 2009 to 2013, the highest exposure concentrations of organochlorine pesticides and PCBs were measured in coastal sediments in China. Overall, the pollution of organochlorine pesticides and PCBs in China's coastal sediments is intensifying. This finding showed that there were more pesticide residues in this sea area. In addition, the concentration of pollutants in the South China Sea was relatively high. The concentration of pollutants in most sea areas was between 1 and 10 ng·g^{-1} dw, and the concentration of pollutants in some coastal areas was between 100 and 500 ng·g^{-1} dw. It can be seen from Figure 2 that the most frequently reported sea area is the South China Sea, with a total of 1238 samples reported, followed by the East China Sea, with a total of 1205 samples reported. The main reason for this is that the Yangtze River Estuary and the Pearl River Estuary are national key sea areas, and there are many research studies on pesticide pollutants. Figure 4 shows the environmental concentrations of the nine compounds. Across the country, except for mirex, other pesticide pollutants were detected at a relatively high frequency (75–100%) in China's coastal sediments. The most detected pollutant was polychlorinated biphenyls, with 880 concentration data points, which were detected in the Bohai Sea, Yellow Sea, East China Sea, and South China Sea. The pollutants α-HCH, β-HCH, and lindane, as HCH isomers, were tested the same number of times. It is worth noting that the research on some pesticide pollutants is relatively limited. For example, mirex was only reported in the South China Sea, and only 58 samples were reported. Considering its detection frequency of 72%, further research is needed.

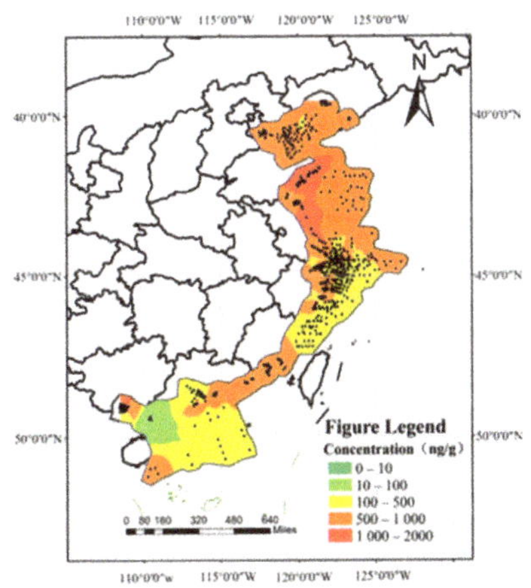

Figure 1. Concentration gradient map of pesticide pollutants in China's coastal waters.

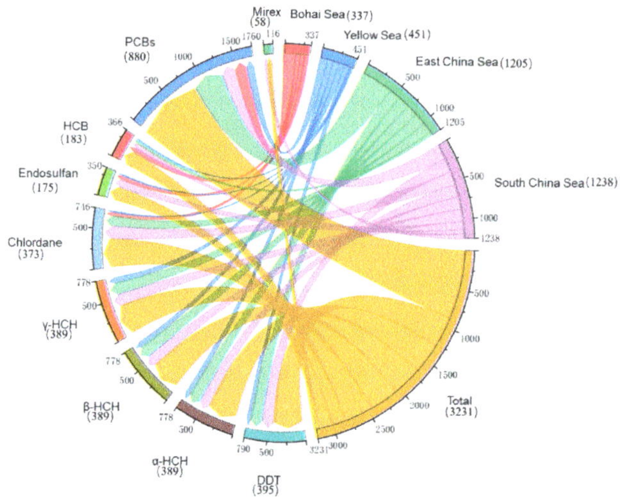

Figure 2. Total number of pesticide pollutants in China's coastal sediments.

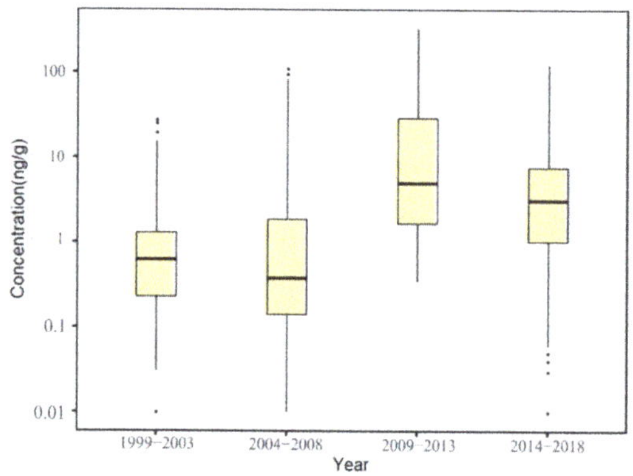

Figure 3. Temporal variation trends of pesticide pollutant concentrations in coastal sediments of China.

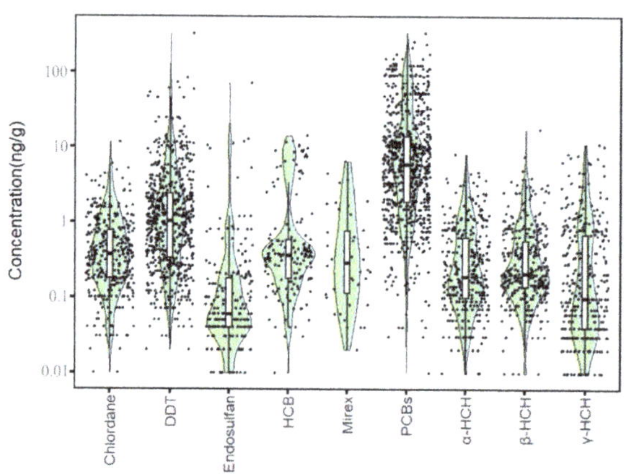

Figure 4. Concentrations of pesticide pollutants in the offshore sediments of total sea areas in China.

3.2. *Toxic Effects of Pesticide Pollutants*

In this study, available chronic toxicity data on aquatic organisms were collected for nine pesticide-type pollutants (Table 1). The results shown in Table 1 represent the most sensitive endpoints for the nine pesticides. The data obtained in this study contained individual toxicity data for nine species. Among them, there were four vertebrates, four invertebrates, and one primary producer. The threshold range of chronic toxicity endpoints in vertebrates is 0.1–32,000 ng·L^{-1}, and the threshold range of invertebrates is 1–65,000 ng·L^{-1}. The chronic toxicity endpoint for the primary producer is 12,000 ng·L^{-1}. Among the nine pesticides, endosulfan, chlordane, and DDT were extremely toxic to organisms, with chronic toxicity endpoints of 0.1 ng·L^{-1}, 1 ng·L^{-1}, and 5 ng·L^{-1}, respectively.

Table 1. Toxicity of nine pesticides to aquatic organisms.

Chemicals	Endpoint	Concentration (ng·L^{-1})	AF	WQC	$PNEC_{sediment}$
DDT	NOEC	5	10	0.5	4.9
β-HCH	NOEC	32,000	10	3200	27,646
HCB	NOEC	96.6	10	9.66	193
α-HCH	EC50	65,000	100	650	5616
Lindane	NOEC	10	10	1	8.6
Endosulfan	NOEC	0.1	10	0.01	0.03
Chlordane	LOEC	1	20	0.05	27.5
PCBs	LOEC	15	20	0.75	58.5
Mirex	EC50	12,000	100	120	841,820

Note: Toxicological data of the nine pesticides were obtained from the ECOTOX knowledge base (https://cfpub.epa.gov/ecotox/search.cfm, accessed on 7 April 2022). PNEC: predicted no-effect concentration.

3.3. *Risk Characterizations*

3.3.1. First Level: Quotient Value Method

First, the quotient value method was used to conduct the first-level risk assessment of nine chlorinated hydrocarbon pollutants. Figure 5 shows the ranking of the risk values of nine chlorinated hydrocarbon pollutants, from high to low: endosulfan, DDT, PCBs, lindane, chlordane, HCB, α-HCH, β-HCH, and mirex. Among them, the RQ values of endosulfan and DDT were between 1 and 10, indicating that endosulfan and DDT had moderate environmental risks, and the other seven pesticide pollutants had low risks. Figure 6 further shows the risk map of the top three pollutants, namely endosulfan, DDT, and PCBs, in coastal sediments in different sea areas of China. It can be seen from Figure 6 that when evaluating each sea area separately, endosulfan has a relatively high risk in the Bohai Sea and the South China Sea; DDT poses a medium risk around the Bohai Sea and the South China Sea; and PCBs have low risks near the East China Sea and South China Sea, with moderate risks in some bays of the East China Sea.

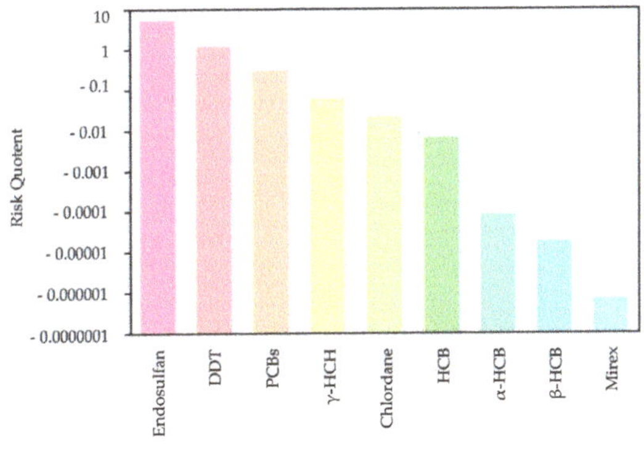

Figure 5. Sorting chart of the risk values of nine pesticide pollutants.

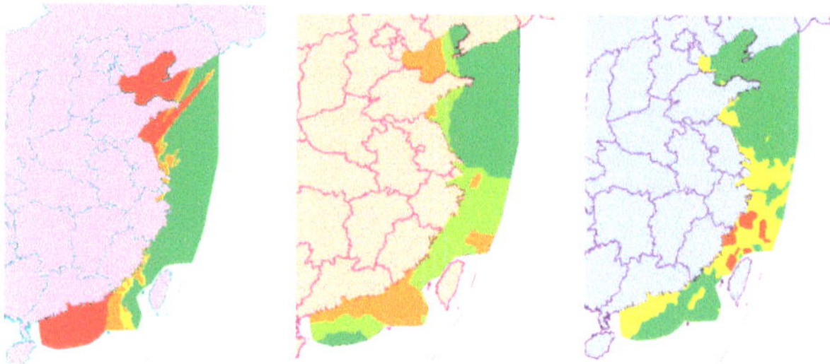

Figure 6. Risk map of three pesticide pollutants in China's coastal sediments.

3.3.2. Second Level: Semi-Probability Method

Using the average concentration to assess the risk of pollutants produces a large error, and it is impossible to accurately assess the actual exposure of pollutants in each area. Therefore, the semi-probability method was used for the second-level risk assessment to quantify the probability that the exposure concentrations of nine chlorinated hydrocarbon pollutants in China's coastal sediments would exceed the PNEC of aquatic organisms, and the percentage exceeding the PNEC value was identified. The results showed that the four chlorinated hydrocarbon pollutants of endosulfan, DDT, PCBs, and lindane had environmental risks, and they had adverse effects on some sensitive species, while the frequencies of other pollutants exceeding the PNEC were all zero and the environmental risks were low. Among them, endosulfan had the highest probability of exceeding the PNEC at 71%, followed by DDT at 25%, while lindane and PCBs each had several points where the exposure concentration exceeded the PNEC value.

The comparison of the two methods showed that the risk of the three chlorinated hydrocarbon pollutants (endosulfan, DDT, and PCBs) to the environment could be observed using the quotient value method and the semi-probability method, but the quotient value method could not evaluate the risk of lindane. For environmental risks, the semi-probability method could further accurately detect the excess of lindane at individual points and provide technical support for the next step of precise management.

3.3.3. Third Level: Semi-Probability Method for Optimization

The optimized semi-probability method was used to conduct the third-level risk assessment. The prioritization indexes (PIs) of the four chlorinated hydrocarbon pollutants of endosulfan, DDT, PCBs, and lindane were calculated, and risk characterization was performed. The PIs of four chlorinated hydrocarbon pollutants were endosulfan at a PI of 383, DDT at a PI of 29.7, PCBs at a PI of 2.6, and lindane at a PI of 3.1×10^{-2}. The results showed that the PI values of endosulfan, DDT, and PCBs were all greater than 1, and therefore posed risks to the environment, whereas the PI value of lindane was less than 1, and the risk to the environment was low. It is suggested that endosulfan and DDT be treated as the priority compounds in China's coastal sediments for key research.

The comparison revealed that for the four pesticide pollutants, although the quotient value method judged that both endosulfan and DDT had moderate risks, the difference in PI between the two obtained using the optimized semi-probability method was nearly 13 times, the main reason being that for endosulfan, the PNEC was exceeded more frequently (71%). In addition, although the commercial value method determined that lindane had no risk to the environment, and the semi-probability method indicated that lindane posed a certain risk to the environment, the evaluation using the optimized semi-probability method showed that the risk of lindane was low and it did not pose a risk to the environment. This result was attributed to the low frequency of PNEC exceedance in lindane (0.5%).

3.3.4. Joint Probability Curve Method

For the four pesticide pollutants of endosulfan, DDT, PCBs, and lindane, the joint probability curve method was used to conduct a sophisticated fourth-level risk assessment based on ecotoxicity data. The joint probability curves for each pollutant were obtained by integrating the impact of China's coastal sediment concentration distribution on the chronic toxicity data of different species, and they were used to indicate the probability of exceeding different levels of impact (Figure 7). As can be seen from Figure 7, endosulfan and DDT have a low risk of chronic effects on aquatic organisms.

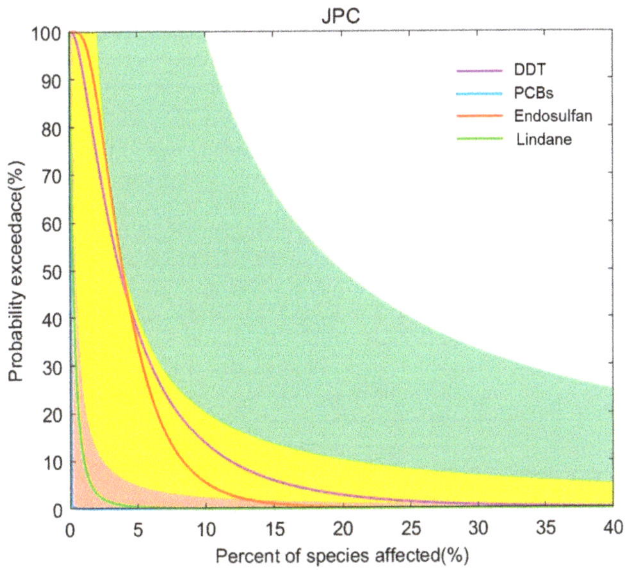

Figure 7. Joint probability curve of the toxicity of pesticide pollutants in the coastal waters of China. Risk classification: orange: minimum risk; yellow: low risk; green: medium risk; white: high risk.

3.3.5. Comparing the Risk Assessment Results Generated Using the Four Methods

In this paper, four methods were used to conduct an MLERA of pesticide pollutants in China's coastal sediments. The quotient value method is often used in most quantitative or semi-quantitative ecological risk assessments. However, this method is usually conservative in determining exposure and selecting toxicity reference values; it provides only a rough estimate of risk, and there are many uncertainties in the calculation. Moreover, the quotient value method does not take into account the differences in the exposure of individuals within the population and the differences in the chronic effects of exposed species. The semi-probability method and the optimized semi-probability method can express the probability of ecological risks posed by chemicals to aquatic organisms in the coastal areas of the country. This method conducts ecological risk assessment based mainly on the probability that the detected concentration of pollutants will exceed the PNEC. The optimized semi-probability method considers both the environmental concentration and frequency, and it evaluates the ecological risk of regional pollutants based on the actual situation, which is beneficial to the management of regional pollutants. For example, after using the semi-probability method to assess the ecological risk of lindane, it was found that the environmental concentration of lindane exceeded the PNEC value at some points, while the risk of lindane determined using the optimized semi-probability method indicated that lindane posed no significant risk (PI = 3.1×10^{-2}). However, lindane still has points exceeding the PNEC in China's coastal sediments, especially in waters where lindane poses a potential risk to aquatic organisms, and its risk should not be completely ignored. The main disadvantage of the optimized semi-probability method is that in this study, the toxicity data of the most sensitive species were used as the PNEC value, and the sediment reference value was derived using the phase equilibrium distribution method, which did not take into account the range of species present in the environment. Therefore, the predicted risk needs to be confirmed using the joint probability curve. Taking endosulfan as an example, the frequency of the PNEC exceeding the standard was 71%, and the PI was 383, but the results of the joint probability curve showed that its relative risk in sea areas was relatively low.

4. Conclusions

This study analyzed the environmental exposure of pesticide pollutants in China's offshore sediments and conducted a multi-level ecological risk assessment of pesticides in sediments. The results show the following:

(1) Nine kinds of pesticide pollutants are widespread in China's coastal sediments, with concentrations ranging from 0.01 $ng \cdot L^{-1}$ to 330 $ng \cdot L^{-1}$.

(2) The first-level quotient method assessment showed that the risk quotients of the nine chlorinated hydrocarbon pollutants were ranked in descending order as endosulfan, DDT, PCBs, lindane, chlordane, HCB, α-HCH, β-HCH, and mirex. Among these pollutants, endosulfan and DDT posed a medium environmental risk to the sediments in the coastal waters of China, and PCBs posed a medium risk in some bays of the East China Sea. The second-level semi-probability assessment showed that endosulfan, DDT, lindane, and PCBs should be considered as priority pollutants. The semi-probability evaluation results of the third-level optimization show that the focus should be on the ecological risks of endosulfan, DDT, PCBs, and lindane in China's coastal sediments, especially endosulfan and DDT. The four-level joint probability curve method assessment shows that endosulfan and DDT have a low risk of chronic effects on aquatic organisms.

(3) Based on the four evaluation methods, this study concludes that endosulfan and DDT pose environmental risks to China's coastal sediments.

Author Contributions: Conceptualization, J.W. and Q.Z.; methodology, Q.Z. and F.G.; software, J.W. and Z.W.; validation, Z.W.; formal analysis, J.W.; writing—original draft preparation, J.W.; writing—review and editing, Q.Z., Y.W. and F.G.; visualization, M.L.; supervision, H.L.; project administration, Y.W.; funding acquisition, Y.W. All authors have read and agreed to the published version of the manuscript.

Funding: This work was financially supported by the National Key Research and Development Program of China (Grant Nos. 2022YFC3703205 and 2021YFC3201005).

Data Availability Statement: Data are contained within the article.

Conflicts of Interest: The authors declare that they have no known competing financial interests or personal relationships that could have appeared to influence the work reported in this paper. The authors declare no conflicts of interest.

References

1. Brain, R.A.; Ramirez, A.J.; Fulton, B.A.; Chambliss, C.K.; Brooks, B.W. Herbicidal effects of sulfamethoxazole in Lemna gibba: Using p-aminobenzoic acid as a biomarker of effect. *Environ. Sci. Technol.* **2008**, *42*, 8965–8970. [CrossRef] [PubMed]
2. Li, M.-R.; Men, S.-H.; Wang, Z.-Y.; Liu, C.; Zhou, G.-R.; Yan, Z.-G. The application of human-derived cell lines in neurotoxicity studies of environmental pollutants. *Sci. Total Environ.* **2024**, *912*, 168839. [CrossRef] [PubMed]
3. Xu, J.-Y.; Wang, K.; Men, S.-H.; Yang, Y.; Zhou, Q.; Yan, Z.-G. QSAR-QSIIR-based prediction of bioconcentration factor using machine learning and preliminary application. *Environ. Int.* **2023**, *177*, 108003. [CrossRef] [PubMed]
4. Daughton, C.G.; Ternes, T.A. Pharmaceuticals and personal care products in the environment: Agents of subtle change? *Environ. Health Perspect.* **1999**, *107*, 907–938. [CrossRef] [PubMed]
5. Peck, A.M. Analytical methods for the determination of persistent ingredients of personal care products in environmental matrices. *Anal. Bioanal. Chem.* **2006**, *386*, 907–939. [CrossRef] [PubMed]
6. Yang, G.; Fan, M.; Zhang, G. Emerging contaminants in surface waters in China—A short review. *Environ. Res. Lett.* **2014**, *9*, 074018. [CrossRef]

7. Qian, M.-M.; Wang, Z.-Y.; Zhou, Q.; Wang, J.; Shao, Y.; Qiao, Q.; Fan, J.-T.; Yan, Z.-G. Environmental DNA unveiling the fish community structure and diversity features in the Yangtze River basin. *Environ. Res.* **2023**, *239*, 117198. [CrossRef]
8. Lan, J.; Jia, J.; Liu, A.; Yu, Z.; Zhao, Z. Pollution levels of banned and non-banned pesticides in surface sediments from the East China Sea. *Mar. Pollut. Bull.* **2019**, *139*, 332–338. [CrossRef]
9. Li, W.; Yang, H.; Gao, Q.; Pan, H.; Yang, H. Residues of organochlorine pesticides in water and suspended particulate matter from Xiangshan Bay, East China Sea. *Bull. Environ. Contam. Toxicol.* **2012**, *89*, 811–815. [CrossRef]
10. Peris, A.; Barbieri, M.V.; Postigo, C.; Rambla-Alegre, M.; Lopez de Alda, M.; Eljarrat, E. Pesticides in sediments of the Ebro River Delta cultivated area (NE Spain): Occurrence and risk assessment for aquatic organisms. *Environ. Pollut.* **2022**, *305*, 119239. [CrossRef]
11. Singare, P.U. Persistent organic pesticide residues in sediments of Vasai Creek near Mumbai: Assessment of sources and potential ecological risk. *Mar. Pollut. Bull.* **2015**, *100*, 464–475. [CrossRef] [PubMed]
12. Buah-Kwofie, A.; Humphries, M.S. The distribution of organochlorine pesticides in sediments from iSimangaliso Wetland Park: Ecological risks and implications for conservation in a biodiversity hotspot. *Environ. Pollut.* **2017**, *229*, 715–723. [CrossRef] [PubMed]
13. Sarkar, A.; Nagarajan, R.; Chaphadkar, S.; Pal, S.; Singbal, S.Y.S. Contamination of organochlorine pesticides in sediments from the Arabian Sea along the west coast of India. *Water Res.* **1997**, *31*, 195–200. [CrossRef]
14. Shi, J.; Li, P.; Li, Y.; Liu, W.; Zheng, G.J.; Xiang, L.; Huang, Z. Polychlorinated biphenyls and organochlorine pesticides in surface sediments from Shantou Bay, China: Sources, seasonal variations and inventories. *Mar. Pollut. Bull.* **2016**, *113*, 585–591. [CrossRef] [PubMed]
15. Hu, W.; Wang, T.; Khim, J.S.; Luo, W.; Jiao, W.; Lu, Y.; Naile, J.E.; Chen, C.; Zhang, X.; Giesy, J.P. HCH and DDT in Sediments from Marine and Adjacent Riverine Areas of North Bohai Sea, China. *Arch. Environ. Contam. Toxicol.* **2010**, *59*, 71–79. [CrossRef] [PubMed]
16. Zhou, S.-S.; Shao, L.-Y.; Yang, H.-Y.; Wang, C.; Liu, W.-P. Residues and sources recognition of polychlorinated biphenyls in surface sediments of Jiaojiang Estuary, East China Sea. *Mar. Pollut. Bull.* **2012**, *64*, 539–545. [CrossRef] [PubMed]
17. Guo, G.; Wu, F.; He, H.; Zhang, R.; Li, H. Ecological Risk Assessment of Organochlorine Pesticides in Surface Waters of Lake Taihu, China. *Hum. Ecol. Risk Assess.* **2012**, *19*, 840–856. [CrossRef]
18. Xu, M.; Huang, H.; Li, N.; Li, F.; Wang, D.; Luo, Q. Occurrence and ecological risk of pharmaceuticals and personal care products (PPCPs) and pesticides in typical surface watersheds, China. *Ecotoxicol. Environ. Saf.* **2019**, *175*, 289–298. [CrossRef]
19. Zhang, H.; Shan, B. Historical distribution of DDT residues in pond sediments in an intensive agricultural watershed in the Yangtze-Huaihe region, China. *J. Soils Sediments* **2014**, *14*, 980–990. [CrossRef]
20. Tang, D.; Liu, X.; He, H.; Cui, Z.; Gan, H.; Xia, Z. Distribution, sources and ecological risks of organochlorine compounds (DDTs, HCHs and PCBs) in surface sediments from the Pearl River Estuary, China. *Mar. Pollut. Bull.* **2020**, *152*, 110942. [CrossRef]
21. Zhao, L.; Hou, H.; Zhou, Y.; Xue, N.; Li, H.; Li, F. Distribution and ecological risk of polychlorinated biphenyls and organochlorine pesticides in surficial sediments from Haihe River and Haihe Estuary Area, China. *Chemosphere* **2010**, *78*, 1285–1293. [CrossRef]
22. Xie, H.; Wang, X.; Chen, J.; Li, X.; Jia, G.; Zou, Y.; Zhang, Y.; Cui, Y. Occurrence, distribution and ecological risks of antibiotics and pesticides in coastal waters around Liaodong Peninsula, China. *Sci. Total. Environ.* **2019**, *656*, 946–951. [CrossRef] [PubMed]
23. Wang, X.; Zhang, Z.; Zhang, R.; Huang, W.; Dou, W.; You, J.; Jiao, H.; Sun, A.; Chen, J.; Shi, X.; et al. Occurrence, source, and ecological risk assessment of organochlorine pesticides and polychlorinated biphenyls in the water-sediment system of Hangzhou Bay and East China Sea. *Mar. Pollut. Bull.* **2022**, *179*, 113735. [CrossRef] [PubMed]
24. Sultan, M.; Hamid, N.; Junaid, M.; Duan, J.J.; Pei, D.S. Organochlorine pesticides (OCPs) in freshwater resources of Pakistan: A review on occurrence, spatial distribution and associated human health and ecological risk assessment. *Ecotoxicol. Environ. Saf.* **2023**, *249*, 114362. [CrossRef] [PubMed]
25. U.S. EPA. *Evaluation Guidelines for Ecological Toxicity Data in the Open Literature*; Environmental Fate and Effects Division, Office of Pesticide Programs: Washington, DC, USA, 2011.
26. Klimisch, H.J.; Andreae, M.; Tillmann, U. A systematic approach for evaluating the quality of experimental toxicological and ecotoxicological data. *Regul. Toxicol. Pharmacol.* **1997**, *25*, 1–5. [CrossRef] [PubMed]
27. Durda, J.L.; Preziosi, D.V. Data Quality Evaluation of Toxicological Studies Used to Derive Ecotoxicological Benchmarks. *Hum. Ecol. Risk Assess.* **2000**, *6*, 747–765. [CrossRef]
28. Hobbs, D.A.; Warne, M.S.; Markich, S.J. Evaluation of criteria used to assess the quality of aquatic toxicity data. *Integr. Environ. Assess. Manag.* **2005**, *1*, 174–180. [CrossRef] [PubMed]
29. Moermond, C.T.; Kase, R.; Korkaric, M.; Agerstrand, M. CRED: Criteria for reporting and evaluating ecotoxicity data. *Environ. Toxicol. Chem.* **2016**, *35*, 1297–1309. [CrossRef] [PubMed]
30. Bu, Q.; Wang, B.; Huang, J.; Deng, S.; Yu, G. Pharmaceuticals and personal care products in the aquatic environment in China: A review. *J. Hazard. Mater.* **2013**, *262*, 189–211. [CrossRef]
31. Adams, W.J.; Kimerle, R.A.; Barnett, J.W. Sediment quality and aquatic life assessment. *Environ. Sci. Tech.* **2002**, *26*, 1864–1875. [CrossRef]
32. European Commission. *Technical Guidance Document on Risk Assessment*; Joint Research Centre, Institute for Health and Consumer Protection, European Chemicals Bureau: Ispra, Italy, 2003.
33. U.S. EPA. *Guidelines for Ecological Risk Assessment*; United States Environmental Protection Agency: Washington, DC, USA, 1998.

34. U.S. EPA. *Framework for Ecotoxicological Risk Assessment*; United States Environmental Protection Agency: Washington, DC, USA, 1992.
35. Zhou, S.-X.; Li, X.-F. Interfacial debonding of an orthotropic half-plane bonded to a rigid foundation. *Int. J. Solids Struct.* **2019**, *161*, 1–10. [CrossRef]
36. Desbiolles, F.; Malleret, L.; Tiliacos, C.; Wong-Wah-Chung, P.; Laffont-Schwob, I. Occurrence and ecotoxicological assessment of pharmaceuticals: Is there a risk for the Mediterranean aquatic environment? *Sci. Total. Environ.* **2018**, *639*, 1334–1348. [CrossRef] [PubMed]
37. Liu, N.; Jin, X.; Feng, C.; Wang, Z.; Wu, F.; Johnson, A.C.; Xiao, H.; Hollert, H.; Giesy, J.P. Ecological risk assessment of fifty pharmaceuticals and personal care products (PPCPs) in Chinese surface waters: A proposed multiple-level system. *Environ. Int.* **2020**, *136*, 105454. [CrossRef] [PubMed]
38. Tarazona, J.V.; Escher, B.I.; Giltrow, E.; Sumpter, J.; Knacker, T. Targeting the environmental risk assessment of pharmaceuticals: Facts and fantasies. *Integr. Environ. Assess. Manag.* **2010**, *6*, 603–613. [CrossRef] [PubMed]
39. Ågerstrand, M.; Rudén, C. Evaluation of the accuracy and consistency of the Swedish Environmental Classification and Information System for pharmaceuticals. *Environ. Sci. Technol.* **2010**, *408*, 2327–2339. [CrossRef] [PubMed]
40. von der Ohe, P.C.; Dulio, V.; Slobodnik, J.; De Deckere, E.; Kühne, R.; Ebert, R.U.; Ginebreda, A.; De Cooman, W.; Schüürmann, G.; Brack, W. A new risk assessment approach for the prioritization of 500 classical and emerging organic microcontaminants as potential river basin specific pollutants under the European Water Framework Directive. *Environ. Sci. Technol.* **2011**, *409*, 2064–2077. [CrossRef] [PubMed]
41. Zhou, S.; Di Paolo, C.; Wu, X.; Shao, Y.; Seiler, T.B.; Hollert, H. Optimization of screening-level risk assessment and priority selection of emerging pollutants—The case of pharmaceuticals in European surface waters. *Environ. Int.* **2019**, *128*, 1–10. [CrossRef]
42. Tousova, Z.; Oswald, P.; Slobodnik, J.; Blaha, L.; Muz, M.; Hu, M.; Brack, W.; Krauss, M.; Di Paolo, C.; Tarcai, Z.; et al. European demonstration program on the effect-based and chemical identification and monitoring of organic pollutants in European surface waters. *Sci. Total. Environ.* **2017**, *601–602*, 1849–1868. [CrossRef]
43. von der Ohe, P.C.; Dulio, V. *NORMAN Prioritisation Framework for Emerging Substances*; NORMAN Association: Verneuil-en-Halatte, France, 2013.
44. Moore, D.R.; Thompson, R.P.; Rodney, S.I.; Fischer, D.; Ramanarayanan, T.; Hall, T. Refined aquatic risk assessment for aldicarb in the United States. *Integr. Environ. Assess. Manag.* **2010**, *6*, 102–118. [CrossRef]
45. Moore, D.R.; Teed, R.S.; Greer, C.D.; Solomon, K.R.; Giesy, J.P. Refined avian risk assessment for chlorpyrifos in the United States. *Rev. Environ. Contam. Toxicol.* **2014**, *231*, 163–217.
46. Whitfield Aslund, M.; Breton, R.L.; Padilla, L.; Winchell, M.; Wooding, K.L.; Moore, D.R.; Teed, R.S.; Reiss, R.; Whatling, P. Ecological risk assessment for Pacific salmon exposed to dimethoate in California. *Environ. Toxicol. Chem.* **2017**, *36*, 532–543. [CrossRef]
47. Clemow, Y.H.; Manning, G.E.; Breton, R.L.; Winchell, M.F.; Padilla, L.; Rodney, S.I.; Hanzas, J.P.; Estes, T.L.; Budreski, K.; Toth, B.N.; et al. A refined ecological risk assessment for California red-legged frog, Delta smelt, and California tiger salamander exposed to malathion. *Integr. Environ. Assess. Manag.* **2018**, *14*, 224–239. [CrossRef]

toxics

Review

Emerging Contaminants: An Emerging Risk Factor for Diabetes Mellitus

Huixia Niu [1], Manjin Xu [2], Pengcheng Tu [1], Yunfeng Xu [2], Xueqing Li [1], Mingluan Xing [1], Zhijian Chen [1], Xiaofeng Wang [1], Xiaoming Lou [1], Lizhi Wu [1,*] and Shengzhi Sun [3,*]

1 Zhejiang Provincial Center for Disease Control and Prevention, 3399 Bin Sheng Road, Binjiang District, Hangzhou 310051, China; 2111101061@nbu.edu.cn (H.N.); pchtu@cdc.zj.cn (P.T.); xqli@cdc.zj.cn (X.L.); mlxing@cdc.zj.cn (M.X.); zhjchen@cdc.zj.cn (Z.C.); xfwang@cdc.zj.cn (X.W.); xmlou@cdc.zj.cn (X.L.)
2 School of Public Health, Xiamen University, Xiang'an South Road, Xiang'an District, Xiamen 361102, China; 32620192200571@stu.xmu.edu.cn (M.X.); 32620192200574@stu.xmu.edu.cn (Y.X.)
3 School of Public Health, Capital Medical University, Beijing 100069, China
* Correspondence: lzhwu@cdc.zj.cn (L.W.); shengzhisun@ccmu.edu.cn (S.S.)

Abstract: Emerging contaminants have been increasingly recognized as critical determinants in global public health outcomes. However, the intricate relationship between these contaminants and glucose metabolism remains to be fully elucidated. The paucity of comprehensive clinical data, coupled with the need for in-depth mechanistic investigations, underscores the urgency to decipher the precise molecular and cellular pathways through which these contaminants potentially mediate the initiation and progression of diabetes mellitus. A profound understanding of the epidemiological impact of these emerging contaminants, as well as the elucidation of the underlying mechanistic pathways, is indispensable for the formulation of evidence-based policy and preventive interventions. This review systematically aggregates contemporary findings from epidemiological investigations and delves into the mechanistic correlates that tether exposure to emerging contaminants, including endocrine disruptors, perfluorinated compounds, microplastics, and antibiotics, to glycemic dysregulation. A nuanced exploration is undertaken focusing on potential dietary sources and the consequential role of the gut microbiome in their toxic effects. This review endeavors to provide a foundational reference for future investigations into the complex interplay between emerging contaminants and diabetes mellitus.

Keywords: emerging contaminants diabetes mellitus; gut microbiota; dietary exposure; emerging pollutants

Citation: Niu, H.; Xu, M.; Tu, P.; Xu, Y.; Li, X.; Xing, M.; Chen, Z.; Wang, X.; Lou, X.; Wu, L.; et al. Emerging Contaminants: An Emerging Risk Factor for Diabetes Mellitus. *Toxics* **2024**, *12*, 47. https://doi.org/10.3390/toxics12010047

Academic Editors: Zhen-Guang Yan, Zhi-Gang Li and Jinzhe Du

Received: 21 November 2023
Revised: 15 December 2023
Accepted: 17 December 2023
Published: 8 January 2024

1. Introduction

According to the International Diabetes Federation, the global number of individuals with diabetes reached 425 million in 2017 [1]. Projections suggest that by 2045, the global diabetic population will escalate to 783 million. Diabetes is understood to be a polygenic hereditary disorder with a pronounced genetic predisposition. Approximately 60% of type II diabetes patients have a familial history of the disease, showcasing a notable familial aggregation [2]. Current genome-wide association studies have pinpointed over 80 susceptibility loci associated with diabetes [3]. Beyond genetic predispositions, environmental factors exert an undeniable influence on diabetes outcomes [4], with emerging pollutants increasingly taken into consideration [5].

With the advent of modern industrial developments and the introduction of novel chemical compounds, the spectrum of environmental pollutants has expanded. Many of these emerging contaminants possess chemical and toxicological properties that remain insufficiently characterized. These pollutants emanate from diverse sources, are numerous in type, and their inherent resistance to degradation leads to their ubiquitous presence and accumulation in environmental matrices [6,7]. Despite their often low concentrations, the biotoxicity of these pollutants, coupled with their persistence and bioaccumulative potential, presents potential detrimental effects on the environment and biota [8]. As the understanding of the environmental and health impacts of chemical substances deepens, and as environmental detection technologies evolve, an increasing number of these pollutants are being identified. However, there is a noticeable absence of regulatory frameworks overseeing their presence and impact, leading to their collective designation as emerging contaminants (ECs) or emerging pollutants (EPs). It is widely acknowledged that these contaminants, even at low concentrations, can enter biological systems through ingestion, inhalation, or dermal contact [9], with ingestion being the predominant exposure route. Once inside an organism, they accumulate in various tissues and organs [10,11], influencing metabolic pathways, including glucose metabolism. For instance, when mice were exposed to 1.25 mg/kg/d PFOA for 28 days, an elevation in blood glucose levels was observed, accompanied by a reduction in hepatic glucose and glycogen content [12]. Moreover, some studies suggest that certain emerging pollutants might influence host glucose metabolism by altering the composition and function of the gut microbiome. For example, Huang et al. found that when mice fed a high-fat diet were exposed to polystyrene microplastics, there was a marked reduction in the richness and diversity of their gut microbiota, an increase in the relative abundance of Gram-negative bacteria, as well as elevated levels of insulin resistance and pro-inflammatory cytokines [13].

The phenomenon of increasing attention to the impacts of emerging environmental pollutants on the incidence, progression, and complications of diabetes has been the subject of extensive research. A survey of various databases reveals that articles have been published reviewing the changes in glucose metabolism induced by these emerging pollutants [14,15]. However, these articles have primarily focused on the impact of individual pollutants on diabetes and its complications, with a comprehensive review of all emerging pollutants, their presence in the environment, biological exposure pathways, and effects on glucose metabolism remaining unexplored. In addition, studies on the effects of exposure to emerging contaminants that cause disturbances in glucose metabolism through alterations in the structure and composition of the gut microbiota have not been reviewed. Therefore, this review summarizes the effects on glucose metabolism after exposure to the four emerging contaminants and the possible mechanisms, especially through the alteration of glucose metabolism after affecting the gut microbiota, by presenting the four new contaminants. The objective is to furnish a foundation for future research on the influence of these pollutants on diabetes, to enhance public awareness of their potential hazards, and to provide empirical support and scientific underpinning for the management of emerging pollutants and relevant environmental policies.

2. Emerging Contaminants (ECs)

Emerging contaminants (ECs) refer to environmental pollutants detectable in the environment and natural ecosystems, posing significant health and environmental risks to both humans and ecological systems. Yet, they remain either unregulated by laws and standards or inadequately addressed [16,17]. When these contaminants are introduced into the environment, their concentrations tend to be low, often rendering their short-term toxic effects unnoticeable. However, their bioaccumulative nature and resistance to degradation result in persistent accumulation. By the time they are detectable, these pollutants have already accrued and posed long-term hazards [18]. The ECs in the environment carry potential dangers to the ecosystem and all living organisms, including humans, such as chronic toxicity, genetic harm, endocrine-disrupting effects, and the "tri-effects" (carcinogenic, teratogenic, and mutagenic effects) [19]. These characteristics have prompted scientists to pay increasing attention to emerging pollutants in the environment.

Surprisingly, over 3000 types of emerging pollutants have been identified globally. Almost every country has detected the presence of these [20]. Their widespread sources and diversity are alarming. China, being a significant producer and consumer of various chemicals, faces challenges as the large-scale production, misuse, and improper management of these chemical substances introduce them into the environment, intensifying environmental pollution concerns [21]. The difficulty in monitoring and the lack of adequate regulatory measures for these contaminants, combined with their persistent and bioaccumulative nature, make them challenging to manage once they enter the environment [18,22,23].

Pharmaceutical factories, plastics, artificial sweeteners, plasticizers, illicit drugs, cleaning products, cosmetics, personal care products, beverages, and packaging are primary sources of these pollutants [24]. They primarily enter biological systems via dietary exposure. Many studies hint that animal-derived food sources are major contributors to many endocrine-disrupting agents. For instance, residues of polychlorinated biphenyl congeners in Crassostrea tulipa (oysters) and Anadara senilis (mussels) were detected at concentrations of 2.95–11.41 mg/kg wet weight and 5.55–6.37 mg/kg wet weight, respectively [25]. The likelihood of exposure to these biphenyls in bivalves is high, with median concentrations exceeding FDA action levels [26]. Wang et al. [27] detected 11 types of perfluorinated compounds in consumer products such as pork tenderloin, pork heart, pork liver, pork kidney, chicken breast, and chicken liver. Every sample contained these compounds, with pork liver having the highest average content of 3.438 ng/g, followed by pork kidney (0.508 ng/g). Additionally, researchers found microplastics in various salts and bottled waters humans consume, with the highest concentrations in sea salt (550–681 particles/kg) and bottled water showing average concentrations of 10.4 particles/L and 325 particles/L for microplastics with sizes >100 μm and <100 μm, respectively [28,29].

Emerging contaminants closely linked with daily human life can be categorized as biological (e.g., resistant genes, algal toxins), chemical (e.g., novel pesticides, endocrine disruptors, flame retardants, antibiotics, perfluorinated compounds), and physical (e.g., microplastics, nanomaterials) [22,30,31]. Based on publicly available regional or site-specific monitoring data, typical emerging pollutants in China mainly include endocrine disruptors, perfluorinated compounds, microplastics, and antibiotics, all of which are causing severe pollution issues in the air, water, and soil environments [22,30]. Numerous published articles have indicated that the presence of emerging pollutants in the environment increases the risk of diabetes in exposed populations and accelerates the onset and progression of the disease. Moreover, it is well known that the mechanisms of type I diabetes and type II diabetes are different. Type I diabetes is mostly related to genetic factors, while for type II diabetes, lifestyle and exposure to emerging contaminants seem to be more important. This article focuses on these four emerging pollutants, briefly discussing their influence on the onset and progression of diabetes, hoping to offer a scientific foundation for the management and prevention of emerging contaminants (Figure 1).

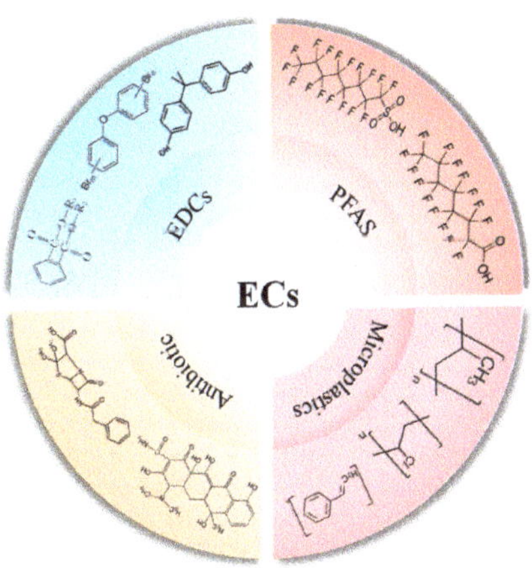

Figure 1. Typical emerging contaminants and representative compounds.

3. Endocrine-Disrupting Chemicals (EDCs)

Endocrine-disrupting chemicals (EDCs) are defined as "an exogenous substance or mixture that alters function(s) of the endocrine system and consequently causes adverse health effects in an intact organism, or its progeny, or (sub) populations" [32]. The most prevalent EDCs include persistent organic pollutants (POPs), phenolic compounds, insecticides, and flame retardants, among others. Due to the frequent use of endocrine-disrupting chemicals (EDCs) in daily life, they are ubiquitously present in the environment. The detection methods for EDCs in the environment primarily include mass spectroscopy, chromatography-based methods, and advanced sensing approaches [33–35], such as electrochemical and colorimetric methods. Owing to their rapid, portable, sensitive, and eco-friendly characteristics, sensing approaches are frequently employed to detect EDCs in environmental and food production systems. Some EDCs are lipophilic, and once ingested by humans through the food chain, they accumulate in adipose tissue [36]. An epidemiological study by Miquel Porta discovered that over 85% of study participants had detectable levels of EDCs such as polychlorinated biphenyls (PCBs), DDE, and DDT in their bloodstream. Additionally, the concentration of EDCs in adipose tissue ranged between 92 ng/g and 399 ng/g [37]. Upon entering organisms, EDCs interfere with the natural synthesis, secretion, and elimination of hormones, leading to endocrine imbalances. Such disturbances can result in endocrine disorders, including obesity and diabetes [38] (Table 1).

3.1. Persistent Organic Contaminants (POPs)

Persistent organic contaminants (POPs) are toxic chemicals that are highly resistant to environmental degradation and metabolic degradation [39]. Most POPs are hydrophobic and can accumulate continuously in the fat of animals and humans [40], causing significant biological toxicity, such as developmental defects, metabolic diseases, cancer, and even death [41]. The primary POPs in the environment include phthalates (PAEs), polybrominated diphenyl ethers (PBDEs), polychlorinated biphenyls, and dioxins.

Phthalates (PAEs) are frequently introduced into the environment as plasticizers used in various plastic products [42]. Given China's role as a major producer of PAEs and its lack of effective pollution control measures, the PAE pollution level in Chinese waters is exceptionally high [43]. Moreover, recent studies have shown that when PAEs enter an organism, they can disrupt glucose metabolism and affect blood glucose levels. A case–control study of volunteers from China found that environmental PAEs can be metabolized within the human body, and their metabolites can be excreted in urine. There is a significant positive correlation between these metabolites and fasting blood glucose and glycated hemoglobin levels, thereby interfering with normal glucose metabolism and affecting the development of type II diabetes [44]. Additionally, metabolic pathways closely related to glucose metabolism, such as galactose metabolism, amino acid metabolism, riboflavin metabolism, and pyridoxal metabolism, can also be affected by PAEs, changing their metabolic marker levels. For instance, in a case–control study of volunteers in Tianjin, China, researchers have observed that the metabolic products involved in galactose metabolism in the serum of patients with type II diabetes were significantly elevated and showed a positive correlation with serum PAE levels. Other metabolites involved in amino acid metabolism, riboflavin metabolism, and pyridoxal metabolism also showed significant changes [45]. Similarly to type II diabetes, gestational diabetes also showed a significant correlation with PAE metabolites [46,47]. While there are a growing number of population studies on the relationship between PAEs and diabetes, research on their specific mechanisms is still relatively limited. When rat insulinoma (INS-1) cells were exposed to dibutyl phthalate (DBP), Yang found that exposure at concentrations of 60 μmol/L and 120 μmol/L led to increased cell apoptosis, significant reductions in mitochondrial membrane potential, increased cellular oxidative stress levels, and decreased superoxide dismutase levels. It is speculated that DBP might reduce INS-1 cell insulin synthesis and secretion through mitochondrial apoptosis and oxidative stress pathways [48].

Polybrominated diphenyl ethers (PBDEs) are brominated flame retardants. Since the 1960s, they have been used as flame retardants in commercial and household products (44). Like PAEs, PBDEs continuously accumulate in organisms, affecting their glucose and lipid metabolism. Data from a nested case–control study indicate a significant correlation between brominated biphenyl ethers (BDEs) and gestational diabetes [49]. Furthermore, BDE-153, BDE-154, and BDE-183 all have odds ratios >1, showing a significant positive and inverted U-shaped correlation with diabetes. Similarly, Ongono and colleagues found in a cohort study that dietary exposure to hexabromocyclododecane is positively correlated with type II diabetes and dietary exposure to PBDEs has a positive non-linear relationship with type II diabetes [50]. Liu used mice orally exposed to BDE-153 to explore the potential mechanisms by which brominated flame retardants might affect glucose metabolism. He found that mouse insulin levels showed a dose-dependent relationship with BDE-153 and the expression of PPARγ and AMPKα was disturbed. It is speculated that BDE-153 might interfere with the expression of adipokines and insulin secretion by affecting the expression of PPARγ and AMPKα, leading to metabolic dysregulation [51].

In addition to PAEs and PBDEs, there are numerous other POPs present in the environment, including polychlorinated biphenyls (PCBs) and dioxins. Organisms living within such environments, especially those on a high-fat diet or at risk for diabetes, can experience altered glucose metabolism when exposed to these POPs. Ibrahim et al. [52] fed C57BL/6J mice a high-fat diet containing POPs and observed that, compared to the unexposed group, both the high-fat-diet mice and those on a Western diet exhibited exacerbated manifestations of insulin resistance, visceral obesity, and abnormalities in glucose tolerance. Additionally, it was found that mice in the low-exposure group exhibited better insulin sensitivity and glucose tolerance than those in the high-exposure group. Furthermore, when C57BL/6J mice on a high-fat diet or those with diabetes were exposed to PCBs, they displayed glucose metabolic abnormalities characterized by glucose intolerance, increased gluconeogenesis, elevated tricarboxylic acid cycle flux, hyperinsulinemia, and intensified systemic insulin resistance [41,53]. Interestingly, a study by Nicki A. Baker revealed that while exposure to PCBs induced lipid inflammation and glucose and insulin tolerance impairment in mice on a low-fat diet, the glycemic equilibrium in obese mice remained unaffected unless they underwent weight reduction [54,55].

The continuous accumulation of dioxins, specifically 2,3,7,8-tetrachlorodibenzo-p-dioxin (TCDD), in the environment and their impact on glucose metabolism is gradually gaining attention from scientists. When mice were exposed to environmental TCDD at a dose of 20 ng/kg, the hyperglycemia induced by a high-fat diet and the reduction in plasma insulin levels induced by glucose were intensified. These mice exhibited significantly elevated blood sugar levels, substantial changes in islet endocrine and metabolic pathways, and increased expression of mRNAs encoding the sodium–glucose transporter 1 and glucose transporter 2 in the intestines [56,57]. Kurita's research yielded similar conclusions: post-TCDD exposure, there was a notable decrease in plasma insulin concentrations in mice, with insulin secretion levels significantly reduced [58]. Furthermore, researchers discovered that when high-fat-diet mice undergoing mating, pregnancy, or lactation were injected with 20 ng/kg TCDD twice a week, these mice experienced accelerated weight gain, faster onset of hyperglycemia, reduced islet levels, and islet shrinkage [59].

The review identified that phthalates, polybrominated diphenyl ethers, polychlorinated biphenyls, and other persistent organic pollutants are widely present in the environment and can enter biological organisms, including humans, through various pathways. Studies have found a significant positive correlation between the concentration of persistent organic pollutants in human serum or urine and fasting blood glucose levels. This phenomenon has been further substantiated in animal and cell experiments, with discussions on potential mechanisms.

3.2. Bisphenol A (BPA) and Its Structural Analogs

Bisphenol A (BPA) and its structural analogs (BPS, BPF, BPAF) are synthetically produced and ubiquitously present in the environment, serving as endocrine disruptors. The omnipresence of BPA compounds means that human exposure to this contaminant is inevitable [60]. Once introduced to the human body, BPA can instigate a myriad of detrimental effects, disrupting metabolic processes including lipid metabolism and glycemic regulation [61]. To date, numerous epidemiological studies have scrutinized the association between urinary concentrations of BPA and its metabolites and diabetes, not only pinpointing a robust correlation between BPA and type II diabetes but also identifying it as a risk factor for gestational diabetes. Beyond this, to elucidate the specific role of BPA in the onset and progression of diabetes and to understand its potential mechanisms leading to the disease, numerous animal experiments have been conducted. For instance, when mice on a standard diet were exposed to 5, 50, 500, and 5000 μgBPA/kg for 8 months, they manifested clear hyperglycemia and hypercholesterolemia. Male mice exposed to 5000 μg/kg/d BPA displayed pronounced insulin resistance [62]. Similarly, in mice fed a high-fat diet, research indicated that BPA aggravates the pre-diabetic symptoms induced by such a diet. Intriguingly, while male mice exhibited only impaired glucose tolerance, female mice also demonstrated increased weight and elevated serum insulin levels, among other symptoms [63]. Regarding the glucose intolerance induced by BPA, a study by Moon et al. posits that it might be attributed to altered serum adipocytokine levels and skeletal muscle phosphorylation, subsequently inducing glucose tolerance abnormalities [64].

Bisphenol A, as an endocrine disruptor, is prevalent in the environment. Numerous population studies have also discovered its association with the development of diabetes, and animal experiments have been conducted to investigate the potential mechanisms of bisphenol A in disrupting glucose metabolism.

Table 1. Summary of studies related to the effects of POPs on glucose metabolism.

Class	Study Subject	Research Models	Chemical	Toxicity	Reference
Population Studies	General population (250 T2DM and 250 controls)	Case–control study	PAEs	• T2DM is significantly positively associated with urinary PAE metabolite concentrations. • Urinary mono (2-ethylhexyl) phthalate is significantly and positively correlated with fasting blood glucose levels. • Exposure to PAEs is associated with T2DM, fasting blood glucose, and glycosylated hemoglobin levels. • Association of exposure to PAEs with T2DM, fasting glucose, and glycosylated hemoglobin may vary among sex, BMI, or age.	[44]
	General population (60 T2DM and 60 controls)	Case–control study	mPAEs, bisphenols	• Positive association between urinary mPAEs and T2DM. • mPAE exposure may contribute to increased risk of T2DM by interfering with galactose metabolism, amino acid metabolism, and pyrimidine metabolism.	[45]
	First trimester (T1) pregnant women (60 GDM and 90 IGT)	Case–control study	PAEs	• PAE metabolites are significantly associated with glucose intolerance and are more strongly associated in some races.	[46]
	First trimester (T1) pregnant women (169 GDM and 169 controls)	Case–control study	PAEs	• Urinary MOP, MBZP, MEOHP, and MECPP concentrations in early pregnancy are significantly associated with GDM. • Urinary MEOHP concentration was significantly and positively correlated with GDM.	[47]
	First trimester (T1) pregnant women (439 pregnant women)	Nested case–control study	PBDEs	• The OR of GDM was significantly positively correlated with the levels of BDE-153 and BDE-183, with an inverted U-shaped correlation with GDM. • BDE-153 and -154 were significantly positively correlated with fasting blood glucose, 1 h and 2 h postprandial blood glucose.	[49]
	General population (71,415 women)	Prospective cohort study	Brominated flame retardants (BFRs)	• Positive association between dietary HBCD exposure and T2D risk. • Positive but non-linear association between dietary exposure to PBDEs and T2D risk.	[50]

Table 1. *Cont.*

Class	Study Subject	Research Models	Chemical	Toxicity	Reference
Cell Experiments	Rat insulinoma (INS-1) cells	Experimental research	DBP	• Cellular insulin synthesis and secretion were significantly reduced. • Increased apoptosis rate and significant decrease in mitochondrial membrane potential. • As the DBP exposure measure increased, the level of oxidative stress increased and the level of antioxidant index decreased.	[48]
Animal Experiments	Adult male C57BL/6J mice	Experimental research	BDE-153	• Induces disorders of glycolipid metabolism in mice. • Insulin staining positivity increased in a dose-dependent manner. • BDE-153 may interfere with adipokine expression and insulin secretion by affecting PPARγ and AMPKα expression, leading to disorders of glucolipid metabolism.	[51]
	High-fat-diet male mice (C57BL/6J)	Experimental research	POPs	• Leading to insulin resistance, visceral obesity, abnormal glucose tolerance. • Impaired phosphorylation of insulin secondary Akt and muscle glucose uptake capacity. • Better insulin sensitivity and glucose tolerance in mice fed diets with reduced POP concentrations.	[52]
	High-fat-diet male mice (C57BL/6J)	Experimental research	TSP, D_2O, PCB126	• PCB126-induced inflammatory response (including higher levels of serum cytokines and adipose-validated gene expression) observed in high-fat-diet mice. • Deterioration of impaired glucose homeostasis characterized by abnormal glucose tolerance, gluconeogenesis, and increased tricarboxylic acid cycle fluxes.	[41]
	High-fat-diet female mice (C57BL/6J)	Experimental research	PCBs	• Produces hyperinsulinemia and exacerbates whole-body insulin resistance in obese mice.	[53]

Table 1. *Cont.*

Class	Study Subject	Research Models	Chemical	Toxicity	Reference
Animal Experiments	Low-fat-diet or high-fat-diet female mice (C57BL/6J)	Experimental research	PCB-77, PCB-126	• Persistently impaired glucose and insulin tolerance in low-fat-diet mice. • Increased expression of TNF-α in fat in PCB-77-treated mice (may contribute to disruption of glucose homeostasis). • Glycemic homeostasis was not affected in obese mice in the exposed group but was damaged after weight loss.	[54]
	Low-fat-diet or high-fat-diet male mice	Experimental research	PCB-77	• Shows weight gain, lipid inflammation, and impaired glucose tolerance. • Increased amount of tumor necrosis factor-α mRNA and impaired glucose homeostasis in PCB-77-exposed mice after weight loss.	[55]
	High-fat-diet mice (C57BL/6J)	Experimental research	TCDD	• Accelerated high-fat-diet-induced hyperglycemia and glucose-induced decrease in plasma insulin levels in female mice. • Slight increase in male islet area. • Abnormal changes in endocrine and metabolic pathways in female pancreatic islets.	[56]
	Male C57BL/6J mice and DBA/2J mice	Experimental research	TCDD	• Altered villous structure and nucleoplasmic ratio of intestinal epithelial cells. • Significantly increased blood glucose levels in mice. • Increased expression of mRNAs encoding sodium–glucose transporter 1 and glucose transporter 2 in the intestine.	[57]
	Male C57BL/6J mice, AhR mice, islets from C57BL/6J mice	Experimental research	TCDD	• Significantly reduced plasma insulin concentrations in C57BL/6J mice at 60 and 120 min after glucose stimulation. • Significant decrease in insulin secretion levels.	[58]

Table 1. *Cont.*

Class	Study Subject	Research Models	Chemical	Toxicity	Reference
Animal Experiments	High-fat-diet female mice (C57BL/6J)	Experimental research	TCDD	• Accelerated weight gain in high-fat-diet mice after exposure to TCDD. • Faster onset of hyperglycemia. • Decreased plasma insulin levels induced by glucose and shrinkage of pancreatic islets. • Low-dose TCDD exposure during pregnancy has long-term adverse effects on metabolic adaptation in HFD-fed offspring mice.	[59]
	Male CD1 mice	Experimental research	BPA	• Bisphenol A exposure for 8 months causes hyperglycemia and hypercholesterolemia. • The 5000 mg/kg/d BPA exposure group of mice had significantly decreased glucose tolerance.	[62]
	High-fat-diet mice (C57BL/6J)	Experimental research	BPA	• Aggravating high-fat-diet-induced pre-diabetes symptoms. • Female mice exhibit weight gain, elevated serum insulin levels, impaired glucose tolerance. • Male mice show only impaired glucose tolerance.	[63]
	High-fat-diet male mice (C57BL/6J)	Experimental research	BPA	• Induces an increase in glucose tolerance. • Reducing skeletal muscle phosphorylation by altering serum adipocytokine levels may be one of the mechanisms by which BPA induces abnormal glucose tolerance.	[64]

4. Per- and Polyfluoroalkyl Substances (PFASs)

Per- and polyfluoroalkyl substances (PFASs) typically consist of carbon chains ranging from 4 to 14 carbons, complemented by a few functional groups [65]. Due to their intrinsic properties such as thermal stability, hydrophobicity, and oleophobicity, PFASs have found extensive applications in industrial production and consumer goods [66], for instance, non-stick cookware, grease-resistant food packaging, and personal care products [67]. Per- and polyfluoroalkyl substances (PFASs), as novel pollutants, have not been thoroughly researched. Due to their widespread application in consumer products, PFASs are omnipresent in the environment, posing potential threats to both the environment and humans. Perfluorooctane sulfonate (PFOS) and perfluorooctanoic acid (PFOA), the most frequently detected PFASs [68], persist in the environment because of the stability of their carbon–fluorine bonds [69]. The current method employed by the United States Environmental Protection Agency for detecting PFASs in the environment relies on combinations of liquid chromatography and mass spectroscopy [70]. However, due to the high costs and the need for trained specialized laboratory personnel, recent research efforts have focused on developing rapid, portable, and low-cost detection methods. The precise toxicological profile of PFASs remains elusive, but mounting research underscores the potential health risks they pose, inclusive of metabolic disturbances (Table 2).

4.1. Perfluorooctane Sulfonate (PFOS)

PFOS, a degradation product among many PFASs, emerges as one of the most scrutinized compounds in the PFAS family. Scientists have ascertained that PFOS not only remains persistent in the environment but also has a notably long half-life of approximately 5.4 years in human serum once ingested, with serum concentrations averaging around 0.05 μg/mL [11,71]. Recently, the connection between PFOS and diabetes has gained research traction.

A growing body of epidemiological evidence associates increased PFAS serum concentrations in humans with elevated fasting glucose, fasting insulin levels, changes in the insulin homeostasis model, and enhanced cellular functions. This compound has been identified not only as a risk factor for gestational diabetes but also as an augmenting agent for type II diabetes susceptibility [72–74]. Observations from human population studies are progressively being corroborated by animal experiments. For instance, Sant et al. [75] exposed zebrafish embryos in the blastula stage to 16, 32, 64 Mm PFOS, noting congenital anomalies mirroring the increased risk factors for human diabetes. The embryos and larvae exhibited perturbed pancreatic growth, pancreatic islet malformations, and a U-shaped dose–response relationship with respect to islet size and PFOS exposure. Qin et al. [76] discovered through in vivo and in vitro studies that PFOS exposure stimulates the free-fatty-acid-regulated membrane receptor G protein-coupled receptor 40 in pancreatic β-cells, thereby heightening intracellular calcium levels and insulin secretion. Moreover, insulin secretion was augmented in a concentration-dependent manner upon acute PFOS exposure, with a marked increase observed at concentrations exceeding 50 μM [77]. Intriguingly, Duan et al. [78] yielded contrary findings, indicating that prolonged PFOS exposure (48 h) inhibits glucose-stimulated insulin secretion. Furthermore, in specific cohorts such as pregnant and lactating mice, research has showcased the biological effects of elevated fasting glucose and insulin levels in both F1 juvenile and adult mice due to PFOS exposure. However, insulin resistance and glucose intolerance anomalies were conspicuously observed only in adult mice. Notably, a high-fat diet exacerbated these effects [79].

4.2. Perfluorooctanoic Acid (PFOA)

Perfluorooctanoic acid (PFOA) is frequently employed as an emulsifier in the production of polytetrafluoroethylene and fluorinated polymers. Ambient concentrations of PFOA in the air typically range from 0.07–0.9 ng/m^3 [80], but can spike to 0.12–0.91 μg/m^3 in the vicinity of fluoropolymer-manufacturing plants [81]. PFOA emissions during the manufacturing process are carried by the wind to adjacent agricultural areas, where they settle in the topsoil layer, eventually seeping downward to the water table [82]. Once introduced into organisms via environmental exposure, PFOA accumulates over time. Ehresmanet reported human serum PFOA concentrations spanning from the detection limit (5 or 10 ng/mL) to 7320 ng/mL [83]. Owing to PFOA's crucial role in metabolic processes and its newfound potential to influence human glucose metabolism, an escalating number of researchers are probing its implications for diabetes and its hypothesized operational mechanisms.

Numerous epidemiological studies have identified a correlation between serum PFOA levels and the proinsulin-to-insulin ratio, after adjusting for confounding factors. Notably, diabetic subjects exhibit significantly elevated lnPFOA levels compared to their non-diabetic counterparts, and these levels can presage the onset of diabetes [84,85]. Analogous findings have emerged from animal studies concerning PFOA and diabetes. For instance, when Zheng and colleagues administered a dose of 1.25 mg/kg/d of PFOA to mice via gavage for 28 days, the mice in the exposed group, although unchanged in weight, manifested conspicuously elevated fasting blood glucose levels, coupled with decreased hepatic glycogen and glucose content [12]. Similarly, Yan et al. observed heightened insulin sensitivity and glucose tolerance in mice exposed to 5 mg/kg/d PFOA. This was attributed to suppressed hepatic gluconeogenesis, leading to diminished liver glycogen synthesis [86]. While mounting research is spotlighting the influence of PFOA on blood glucose levels, the underpinning mechanisms remain only partially elucidated. In an investigation by He [87] on the potential impact of PFOA on the functionality of pancreatic β-cells in mice, it was discerned that at a dose of 500μM, PFOA stimulates β-cell apoptosis. Moreover, even lower doses of PFOA resulted in diminished insulin secretion upon glucose stimulation and a pronounced upregulation of endoplasmic-reticulum-stress-related gene expression.

The review finds that exposure to perfluorinated compounds leads to increased fasting blood glucose levels and disrupted glucose metabolism in mice, along with alterations in the morphology, size, and length of the islets, thereby impacting insulin secretion. However, the underlying mechanisms of these effects have yet to be fully elucidated.

Table 2. Summary of studies related to the effects of PFAS on glucose metabolism.

Class	Study Population	Research Models	Chemical	Toxicity	Reference
Population Studies	General population (474 adolescents and 969 Adults)	Cross-sectional study	PFNA, PFOS, PFOA, etc.	• Hyperglycemia is associated with elevated PFNA concentrations. • Elevated serum PFNA concentrations are negatively associated with the prevalence of metabolic syndrome. • Elevated serum PFOS concentrations correlate with elevated insulin, assessment of insulin resistance in a homeostasis model, and cellular function.	[72]
	General population (1045 adults)	Cross-sectional study	PFOS and PFOA	• Fasting blood glucose, fasting insulin, and pancreatic β-cell function correlate with increased concentrations of Br-PFOS. • Long-term exposure to PFAS isomers is associated with impaired glucose homeostasis and may increase the prevalence of type II diabetes mellitus among Chinese adults.	[73]
	Pregnant women (171 GDM and 169 controls)	Cross-sectional study	PFOS, PFOA, PFNA, etc.	• Concentrations of PFAS congeners positively correlate with fasting glucose, 1 h and 2 h blood glucose after glucose tolerance test, and sustained glucose results on glycated hemoglobin. • Risk of gestational diabetes and blood glucose levels increase significantly with increasing concentrations of PFOS mixtures.	[74]
	General population (1016 men and women aged 70 years)	Prospective cohort study	PFOA, PFNA	• After adjusting for confounders, PFNA and PFOA showed a significant non-linear relationship with diabetes mellitus. • PFOA is associated with the insulinogen/insulin ratio but not with markers of insulin resistance.	[84]
	General population (100 participants)	Prospective cohort study	PFOA, PFOS	• Log-transformed PFOA and log-transformed PFOS were significantly higher in diabetic patients than in non-diabetic patients. • LnPFOA significantly predicts diabetes.	[85]

Table 2. *Cont.*

Class	Study Population	Research Models	Chemical	Toxicity	Reference
Cell Experiments	GPR40-KO C57BL/6, C57BL/6 mice and mouse islet β-cells	Experimental research	PFOS	• Stimulation of insulin secretion and intracellular calcium levels by activation of the GPR40, an important free-fatty-acid-regulated membrane receptor on islet β-cells.	[76]
	Beta-TC-6 pancreatic cells	Experimental research	PFOS	• Acute exposure to PFOS stimulates insulin secretion and increases intracellular calcium ion concentrations. • PFOS stimulates insulin secretion at least in part by GPR40.	[77]
	Mouse pancreatic β-cells	Experimental research	PFOS	• Continuous exposure to PFOS inhibits glucose-stimulated insulin secretion but has no significant effect on insulin gene expression. • SIRT1 activators and UCP2 inhibitors partially reverse PFOS-induced impairment of insulin secretion.	[78]
	Mouse pancreatic β-cell line (MIN6 cells)	Experimental research	PFOA	• Time- and dose-dependent inhibition of cell viability. • A high dose (500 μM) of PFOA promoted β-cell apoptosis, and a low dose (300 μM) had no effect on cell survival. • Low doses also reduce glucose-stimulated insulin secretion. • Endoplasmic-reticulum-stress-related proteins were significantly increased, and inhibition of TRIM3 expression significantly protected MIN6 cells from PFOA-induced defective insulin secretion and apoptosis by ameliorating endoplasmic reticulum stress.	[87]
Animal Experiments	Zebrafish (*Danio rerio*) embryos	Experimental research	PFOS	• Abnormal islet morphology. • Islet size and pancreas length showed a U-shaped dose–response relationship with PFOS. • Embryonic PFOS exposure can disrupt pancreatic organogenesis in a manner that mimics human genetic defects and predisposes individuals to diabetes mellitus.	[75]
	Pregnant mice and offspring	Experimental research	PFOS	• Elevated fasting glucose and insulin levels in F1 offspring juveniles and adults. • Insulin resistance and abnormal glucose tolerance were only evident in F1 adults.	[79]

Table 2. *Cont.*

Class	Study Population	Research Models	Chemical	Toxicity	Reference
Animal Experiments	Adult male Balb/c mice	Experimental research	PFOA	• No significant change in body weight in the exposed group of mice. • Fasting blood glucose levels were elevated and glycogen and glucose levels in the liver were reduced. • Increased glucose production capacity.	[12]
	Male Balb/c	Experimental research	PFOA	• Induced higher insulin sensitivity and glucose tolerance. • Reduced hepatic glycogen synthesis, which may be due to inhibition of gluconeogenesis. • Changes in levels of centrally circulating proteins (including proteins that may be associated with diabetes and liver disease).	[86]

5. Microplastics

Microplastics refer to plastic particles with a diameter of ≤5 mm. Traditional methods for detecting microplastics in the environment include visual identification or microscopic observation, Fourier transform infrared spectroscopy, thermal pyrolysis, and Raman spectroscopy. However, the diverse sources and compositions of environmental microplastics, along with the presence of numerous impurities, render these conventional detection methods inadequate for comprehensive microplastic detection [88]. With increasing interest in the toxicity of microplastics and advancements in detection technology, more sensitive and high-performance detection techniques are being developed, such as a variety of remote sensing techniques including polarized light optical microscopy (PLM), atomic force microscopy, and hybrid combinations of these techniques [89]. Humans and other organisms are exposed to environmental microplastics through ingestion, inhalation, and dermal contact [90]. Furthermore, oral exposure has been reported as the primary route of microplastic exposure. Kumar et al. [91], in their review, mention seafood, beer, table salt, bottled mineral water, and milk as the main pathways for microplastics to enter the human body. Once internalized, microplastics accumulate within tissues and organs, leading to histopathological alterations and cytotoxic responses [6,92]. For instance, a study by Lu et al. on zebrafish found that exposure to microplastics first leads to accumulation in liver tissues, causing inflammation and lipid accumulation, and disrupting lipid and energy metabolism, leading to metabolic changes [93]. Cortés et al.'s cellular experiments also found that microplastics induce the production of a significant amount of ROS in Caco-2 cells, thereby increasing cytotoxicity [94]. The same conclusion was reached in cell experiments with T98G and HeLa [95]. Additionally, immune responses and changes in the structure and composition of the gut microbiota induced by microplastic exposure have also been increasingly identified [91]. It is well-established that gut microbiota dysbiosis, inflammatory reactions, oxidative stress, and changes in innate immune responses—all consequences of microplastic exposure—are major pathophysiological factors for insulin resistance. Consequently, scientists posit a strong link between microplastic exposure and insulin resistance, necessitating comprehensive research and elucidation. However, current investigations in this area remain limited (Table 3).

Studies have identified correlations between changes in blood glucose levels and insulin resistance caused by microplastic exposure, specifically noting connections to gut microbiota disruption, inflammation, and oxidative stress. Huang et al. [13] exposed mice on a high-fat diet to polystyrene microplastics of sizes 5, 50, 100, and 200 μm. The mice displayed insulin resistance accompanied by elevated levels of plasma lipopolysaccharides and pro-inflammatory cytokines (tumor necrosis factor and interleukin-1β). A reduction in gut microbiota richness and diversity was also observed, particularly with an increased relative abundance of Gram-negative bacteria. Based on these findings, scientists hypothesize that insulin resistance triggered by microplastics might be due to tissue accumulation and microbiota-induced inflammatory responses, thereby inhibiting the insulin signaling pathway. Takuro Okamura also demonstrated that mice exposed to microplastics showed elevated blood glucose levels and deposition of microplastics in the gut mucosa, resulting in an increase in intrinsic inflammatory cells and a reduction in anti-inflammatory cells [96]. Additionally, the insulin resistance and elevated blood glucose levels induced by microplastics might be associated with high levels of reactive oxygen species (ROS) in mice exposed to 5 mg/kg and 15 mg/kg, which potentially disrupt the PI3K/Art pathway related to glucose metabolism [97]. In another study, beyond increasing oxidative stress, glucose tolerance, and insulin resistance, 30 mg/kg/d microplastic exposure also led to decreased phosphorylation levels of AKT and GSK3β [98]. AKT agonists can effectively alleviate oxidative stress, elevated blood glucose levels, and insulin resistance, suggesting that part of the diabetes mechanism induced by microplastics might be related to AKT/GSK3β phosphorylation. Furthermore, research has identified that reduced cortisol levels in mice exposed to 55 μg/d microplastics might interfere with insulin secretion, thereby inducing insulin resistance [99]. A review of microplastics reveals that current research on the relationship between microplastics and glucose metabolism is relatively scarce. Studies have found that exposure to microplastics can affect an organism's glucose metabolism and the development of diabetes through inflammatory responses, oxidative stress, and disruption of the composition and structure of the gut microbiota.

Table 3. Summary of studies related to the effects of MPs on glucose metabolism.

Class	Study Population	Microplastics	Toxicity	Reference
Animal Experiments	Five-week-old high -fat-diet male mice (Mus musculus, ICR)	Polystyrene (5, 50, 100, and 200 µm)	• Mice in both the normal and high-fat-diet groups exhibited insulin resistance with elevated plasma lipopolysaccharide and pro-inflammatory cytokines (tumor necrosis factor and interleukin-1β). • Significant decrease in the abundance and diversity of intestinal microbiota and increase in the relative abundance of Gram-negative bacteria (*Pseudobacillaceae* and *Enterobacteriaceae*). • Small particles of microplastics (5 µm) aggregated in the liver, kidney, and blood vessels in mice. • Inhibition of insulin signaling pathway in liver (inhibited IRS1 and reduced PI3K expression). • The mechanism may stimulate inflammatory responses and inhibit insulin signaling pathways by regulating intestinal microbiota and polystyrene accumulation in tissues.	[13]
	High-fat-diet male mice (C57BL/6)	Polystyrene (0.45–0.53 µm)	• High-fat-diet mice exposed to polystyrene microplastics had higher blood glucose, lipid concentrations, and non-alcoholic fatty liver disease scores and more inflammatory cells and fewer anti-inflammatory cells in the lamina propria than those that were not. • High-fat diet mice exposed to polystyrene microplastics had significantly higher expression of genes associated with inflammation, long-chain fatty acid transporter proteins, and Na+ glucose cotransporter proteins than those that were not. • *Desulfovibrio* genes are significantly enriched in the intestines.	[96]
	Mice	Polystyrene nanoplastics	• Oral exposure to nanoplastics can cause organ damage, mainly in liver function and lipid metabolism. • Chronic exposure significantly increased blood glucose levels and ROS levels but did not affect plasma insulin secretion. • High levels of ROS interfere with the PI3K/Art pathway, leading to insulin resistance and elevated blood glucose.	[97]
	High-fat-diet male mice (C57BL/6)	Polystyrene (80 nm)	• Exposure to 30 mg/kg/d of microplastics alone significantly increased blood glucose, glucose tolerance, and insulin resistance. • Exposure of high-fat-diet mice to polystyrene microplastics significantly exacerbates oxidative stress, glucose tolerance, and insulin resistance and induces liver and pancreas damage. • Polystyrene microplastics exacerbate type II diabetes by an underlying mechanism in part related to AKT/GSPK3β phosphorylation (associated with ROS).	[98]
	ICR mice	Polystyrene (1 µm)	• Mice exposed to microplastics for a week suffered significant liver damage and oxidative stress, disturbed liver–gut axis, and increased risk of insulin resistance. • Oxidative stress occurred in the liver. • When exposed to microplastics for two weeks, lower cortisol in the liver may interfere with insulin secretion and induce insulin resistance. • Disturbed gut microbiota. • Fasting blood glucose, fasting insulin, and HOMA-IR levels were significantly elevated after microplastics exposure.	[99]

6. Antibiotics

Besides the aforementioned emerging pollutants, antibiotics have also been identified as a significant new class of pollutants, extensively present in the environment and water bodies. By the late 1990s, antibiotics had become widely used in medicine and established as pillars of modern medical practice. The consumption of antibiotics has seen a steady increase, with global consumption growing by 39% between 2000 to 2015. Particularly, antibiotic consumption in low-income countries surged by 77% during the same period [100]. In 2011, the global human utilization of antibiotics was estimated at 70 billion, equivalent to an annual consumption rate of 10 per individual [101]. Subsequently, these antibiotics, or their metabolites, enter the environment through human and animal urine and feces, ultimately persisting in soil and aquatic environments [102]. Current detection of antibiotics mainly relies on instrumental analysis, which is highly sensitive. However, due to high costs and laborious pre-treatment, traditional instrumental analysis methods are no longer sufficient for the growing number of samples. Therefore, the development of rapid, high-throughput, and low-cost detection methods is essential. Current methods for detecting antibiotic residues include the microbial method, electrochemical method, high-performance liquid chromatography, liquid mass spectrometry, fluorescence method, Raman spectroscopy, etc. [103]. Scientists have detected various antibiotics, such as amoxicillin, clindamycin, and ciprofloxacin, in the inlet and outlet water of wastewater treatment plants. Furthermore, the highest concentrations of triclocarban and triclosan detected in Indian aquatic environments have reached 5860 ng/L [104], indicating the non-negligible potential hazards of residual antibiotics in the environment. It is widely recognized that antibiotic intake can impact the structure, composition, and function of gut microbiota. Moreover, alterations in the gut microbiota have been closely linked with the onset and progression of diabetes [105]. Consequently, there is mounting concern within the scientific community regarding the relationship between antibiotic consumption and diabetes.

A significant body of research has been conducted to investigate the association between antibiotics and diabetes. For instance, several studies have shown that mice on a high-fat diet or those modeled for diabetes, when treated with antibiotics, exhibited reduced levels of endotoxins in plasma and inflammatory factors in adipose tissue. Healthy mice, on the other hand, displayed beneficial effects on glucose metabolism, including reduced fasting blood glucose and decreased area under the glucose tolerance curve [106,107]. Interestingly, another prospective cohort study revealed that patients treated with antibiotics for durations ranging from twenty-five days to two months, or more than two months, saw their risks for type II diabetes increase by 23% and 20%, respectively [108]. In light of this intriguing observation, Fu et al. [109] studied the impact of antibiotic treatment on blood glucose changes in db/db mice. They found that the effects of antibiotics on blood glucose exhibit both immediate and delayed responses: compared to the control group, mice treated with antibiotics for 12 days showed significant declines in body weight and blood glucose levels. However, 24 days post-treatment, these mice experienced weight gains that even surpassed those of the control group, along with elevated levels of plasma and liver total cholesterol and an increase in liver weight. Research on the relationship between antibiotics and diabetes is already quite abundant, with a relatively comprehensive understanding of the relationship between antibiotics and the development of diabetes. It should not be neglected that the timing of antibiotic exposure is also very important and should not be overlooked, such as in pregnant mothers, infants, and adulthood. A comprehensive review of the relationship between antibiotics and diabetes has been presented in an article by Fenneman and will not be further elaborated upon here [110].

7. Role of Gut Microbiota

In recent years, the gut microbiota has garnered unprecedented attention. An increasing corpus of evidence underscores its fundamental role in the digestion of polysaccharides, the biosynthesis of vitamins, and other essential nutrients. This microbial community is inextricably linked with human health [111–113]. Undoubtedly, acting as a novel organ, the gut microbiota functions optimally. However, disruptions in its composition and structure due to external substances can have implications for disease onset and progression [114–116]. Particularly noteworthy are recent studies highlighting how exposure to emergent environmental contaminants can destabilize the gut microbiota, leading to adverse health effects, including disorders in glucose metabolism [99,117–119] (Table 4).

The current scientific discourse is replete with research focusing on the implications of perfluorinated compound exposure on diabetes via its perturbative effects on gut microbiota. Lai et al. [120] embarked on a study exploring the impacts of dietary PFOS exposure on the gut microbiota of adult mice and scrutinized the consequent changes in induced metabolic functions. Their findings delineated a marked increase in the abundance of *Turicibacterales* and *Allobaculum* in the exposed group of mice, juxtaposed with a significant decline in *B. acidifaciens*. Moreover, the researchers discerned that mice exposed to 3 μg/g/day PFOS exhibited a precocious decline in blood glucose levels after oral glucose ingestion. The area under the curve manifested a conspicuous reduction, and after intraperitoneal insulin injection, these mice's blood glucose levels were markedly lower than those in the control group. Delving deeper into another emergent pollutant, microplastics, it has been discerned that, post-exposure, it can influence the onset and trajectory of diabetes via various mechanisms, with the resultant disruption in gut microbiota being non-trivial. Using mice as model organisms, investigations into the aftermath of microplastics exposure on the gut microbiota were conducted. The outcomes indicated that, post-exposure, there was a perturbation in the gut–liver axis of the mice, a pronounced reduction in gut microbiota diversity, diminished richness of *Bacteroidetes* and *Verrucomicrobia*, and an increased abundance of *Firmicute*, *Deferribacteres*, and *Actinobacteria*. Concurrently, scientists observed that microplastic-exposed mice presented elevated fasting blood glucose and insulin levels [99]. These findings echo the results from Huang et al., who, in addition to observing reduced microbial richness, also noted an increased relative abundance of Gram-negative bacteria within the mice [13]. It is common knowledge that antibiotics have had a longstanding history of use, targeting pathogenic strains within the microbiota. Yet, their administration might also inadvertently impact other microbial communities, resulting in a decrease in the host's short-chain fatty acid content. This, in turn, might disrupt metabolic processes and energy assimilation, potentially influencing the onset and progression of diabetes [117,121]. Currently, articles published on the topic of new pollutants and their impact on glucose metabolism through the influence on the structure and composition of the gut microbiota are relatively few, with most studies focusing on perfluorinated compounds and microplastics. Additionally, research in this area remains significantly under-developed. The changes in the composition and structure of the gut microbiota after exposure to new pollutants, alterations in their metabolic pathways and metabolites, and the specific mechanisms of their impact on blood sugar still require more in-depth investigation.

Table 4. Summary of studies related to the effects of emerging contaminants on glucose-metabolism-associated gut microbiota.

Class	Species	Chemical	Changes in Intestinal Microbiota		Reference
Animal Experiments	Female CD-1 mice	PFOS	• Significant changes in the abundance of metabolizing bacteria associated with phyla *Firmicutes*, *Bacteroidetes*, *Proteobacteria*, and *Cyanobacteria* in mouse gut microbiota after PFOS exposure. • A significant increase in the number of *Turicibacterales* and *Allobaculum* and a significant decrease in the number of *B. acidifaciens* were observed in the exposure mice. • Disruption of intestinal metabolism leads to significant changes in the metabolism of amino acids, methane, and short-chain fatty acids in mice, and these metabolites are thought to be associated with inflammation and metabolic functions.	• Disturbances in fat and glucose metabolism. • The OGTT experiment showed that PFOS exposure caused abnormalities in glucose tolerance, which were more pronounced in the high-exposure group. • After insulin treatment, the blood glucose of mice in the high-dose exposure group was significantly lower than that of mice in the control group. • Pyruvate conversion was significantly lower in exposed mice compared to control mice.	[120]
	ICR mice	Polystyrene microplastics (1 μm)	• The gut microbiota of mice in the exposure group was significantly altered and the diversity was significantly reduced. • At the phylum level, the relative abundance of *Bacteroidetes* and *Verrucomicrobia* was significantly reduced in exposure group mice, whereas the relative abundance of *Firmicute*, *Deferribacteres*, and *Actinobacteria* was increased. • At the genus level, the relative abundance of *Lactobacillus* and *Bifidobacterium* increased in exposure groups, while the relative abundance of *Oscillospira*, *Akkermansia*, and *Desulfovibrio* decreased.	• Plasma cortisol levels are elevated and non-hepatic glycogen utilization is inhibited. • Elevated plasma insulin levels and increased levels of hepatic gluconeogenesis.	[99]
	Male ICR mice	Polystyrene microplastics (5, 50, 100, 200 μm)	• Significant decrease in the abundance and diversity of gut microbiota and increase in the relative abundance of Gram-negative bacteria. • At the phylum level, the relative abundance of Bacteroidetes increased and the relative abundance of Firmicutes decreased. • At the family level, the relative abundance of *Muribaculaceae* and *Helicobacteraceae* decreased in exposure groups, while the relative abundance of *Prevotellaceae*, *Enterobacteriaceae*, *Desulfovibrionaceae*, and *Rikenellaceae* increased.	• Inhibition of insulin signaling pathway and reduced expression of IRS1 and PI3K in mouse liver. • Increased fasting blood glucose and insulin levels in exposure group mice compared to the control mice. • Significantly increased area under the curve of OGTT and ITT curve and elevated HOMA-IR in microplastic-exposed group of mice.	[13]
	NOD/Shiltj mice	Antibiotics	• Reduced gut microbiota abundance in exposure group of mice.	• Receiving pulsed therapeutic antibiotics early in life accelerates the development of type I diabetes and islet disease.	[117]

8. Conclusions

Emerging pollutants, characteristically under-monitored and under-regulated in the environmental sphere, harbor potential, both known and speculative, adverse implications for ecological systems and human health. However, the corpus of research delineating the toxicological nexus between emerging contaminants and glucose metabolic processes remains markedly under-developed. This review aims to intricately weave together four distinct classes of emerging pollutants with glucose metabolism. It methodically dissects and analyzes the toxicological profiles and underlying mechanisms of persistent organic pollutants, perfluorinated compounds, microplastics, and antibiotics, drawing upon a synthesis of empirical evidence from animal model studies and epidemiological research (Figure 2). Overall, EDCs, PFAS, microplastics, and antibiotics cause disturbances in glucose metabolism and accelerate diabetes mellitus. Most of the descriptions of the mechanisms in the currently published articles focus on the effects on glucose metabolism through inflammatory responses, oxidative stress, and disturbances in the gut microbiota. The elucidation provided herein seeks not only to augment the current understanding of the deleterious effects of these emerging pollutants on glucose metabolism but also to catalyze a paradigm shift in the toxicological examination of emerging environmental contaminants.

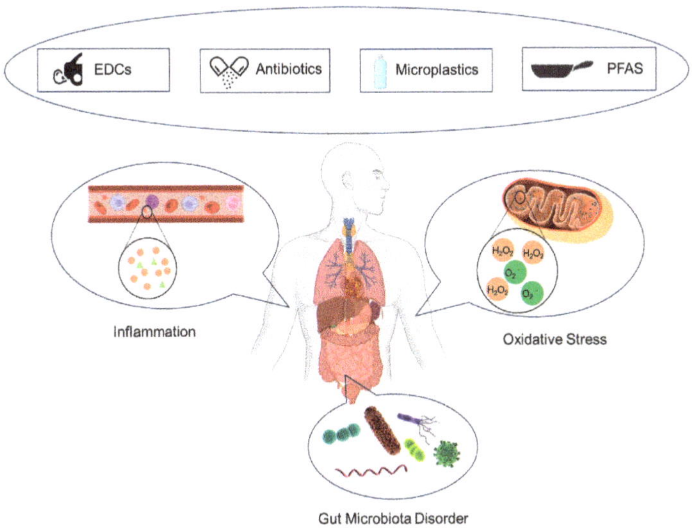

Figure 2. Mechanisms of glucose metabolism disturbance induced by emerging contaminants.

9. Future Research Directions

With the continual evolution of technology and the deepening understanding of contaminants by scientists, an increasing number of novel environmental contaminants are being detected. The persistent accumulation of these contaminants in the environment, coupled with their multifarious exposure pathways, endangers ecosystems and the organisms residing therein, thus progressively capturing the attention of the scientific community. The repercussions of these environmental contaminants, particularly on populations with diabetes or those on high-fat diets, are of heightened concern. Nonetheless, our comprehension in this domain remains somewhat limited:

i. The majority of current epidemiological studies focus on the relationship between the concentration of new pollutants in the serum or urine of the general population and fasting blood sugar, insulin, and glycated hemoglobin concentrations, with some studies involving changes in metabolic pathways related to glucose metabolism and their metabolites. However, studies on the impact of occupational exposure on diabetes, the specific exposure situations of new pollutants in the diet, the correlation with diabetes, and the impact of different geographical locations are relatively scarce. Diet is likely to be a very important exposure, yet rarely assessed in human studies, or when assessed, by questionnaires, often inaccurate. Quantifying the contribution of the human diet by multitargeted metabolomics of food and microbiota-derived metabolites may provide some clues. Therefore, more in-depth and targeted research is still needed to explore the impact of different factors on the development of diabetes.

ii. The current research primarily focuses on the effects of novel environmental contaminants and their exposure on glucose metabolism in human populations, with fewer studies being directed towards animal models. Thus, there remains a pressing need for comprehensive studies to elucidate the specific mechanisms underlying the impact of these contaminants on diabetic or high-fat-diet populations, as well as the potential health outcomes from long-term low exposure.

iii. Additionally, factors influencing the toxic effects of these novel contaminants, such as dose–response relationships, exposure frequency, gender disparities, and attributes like the type and size of the contaminants, have yet to be thoroughly investigated. Hence, there is an urgent need for more in-depth research into the toxicity of these new contaminants, factors modulating their toxicity levels, and their potential hazards. Such insights would furnish policymakers with a robust scientific foundation, aiding in the resolution of environmental challenges and the safeguarding of human health.

iv. It is widely acknowledged that diabetes is influenced not only by genetic, environmental, and lifestyle factors but also by the structure and composition of the gut microbiota. However, current research on the interrelation between novel contaminants, gut microbiota, and diabetes is relatively scant. Consequently, determining whether exposure to these new contaminants might influence glucose metabolism by altering the gut microbiota's structure and composition calls for relentless effort and exploration by researchers.

Author Contributions: Conceptualization: H.N., P.T., L.W. and S.S.; Literature search: H.N., Y.X., M.X. (Manjin Xu) and X.L. (Xueqing Li); Writing—original draft: H.N., M.X. (Mingluan Xing), Z.C., X.W. and X.L. (Xiaoming Lou); Writing—review and editing: All authors; Funding acquisition L.W. All authors have read and agreed to the published version of the manuscript.

Funding: This research was funded by Zhejiang Provincial Project for Medical Research and Health Sciences: 2024KY898 & 2024KY910.

Institutional Review Board Statement: Not applicable.

Informed Consent Statement: Not applicable.

Data Availability Statement: Not applicable.

Conflicts of Interest: The authors declare no conflict of interest.

References

1. Zhao, Y.; Li, Y.; Zhuang, Z.; Song, Z.; Wang, W.; Huang, N.; Dong, X.; Xiao, W.; Jia, J.; Liu, Z.; et al. Associations of polysocial risk score, lifestyle and genetic factors with incident type 2 diabetes: A prospective cohort study. *Diabetologia* **2022**, *65*, 2056–2065. [CrossRef] [PubMed]
2. Papazafiropoulou, A.K.; Papanas, N.; Melidonis, A.; Maltezos, E. Family History of Type 2 Diabetes: Does Having a Diabetic Parent Increase the Risk? *Curr. Diabetes Rev.* **2017**, *13*, 19–25. [CrossRef] [PubMed]
3. Zhou, N.; Qi, H.; Liu, J.; Zhang, G.; Liu, J.; Liu, N.; Zhu, M.; Zhao, X.; Song, C.; Zhou, Z.; et al. Deubiquitinase OTUD3 regulates metabolism homeostasis in response to nutritional stresses. *Cell Metab.* **2022**, *34*, 1023–1041. [CrossRef] [PubMed]
4. Kolb, H.; Martin, S. Environmental/lifestyle factors in the pathogenesis and prevention of type 2 diabetes. *BMC Med.* **2017**, *15*, 131. [CrossRef] [PubMed]
5. Taylor, K.W.; Novak, R.F.; Anderson, H.A.; Birnbaum, L.S.; Blystone, C.; Devito, M.; Jacobs, D.; Köhrle, J.; Lee, D.-H.; Rylander, L.; et al. Evaluation of the association between persistent organic pollutants (POPs) and diabetes in epidemiological studies: A national toxicology program workshop review. *Environ. Health Perspect.* **2013**, *121*, 774–783. [CrossRef] [PubMed]
6. Prata, J.C.; da Costa, J.P.; Lopes, I.; Duarte, A.C.; Rocha-Santos, T. Environmental exposure to microplastics: An overview on possible human health effects. *Sci. Total Environ.* **2020**, *702*, 134455. [CrossRef]

7. Han, D.; Currell, M.J. Persistent organic pollutants in China's surface water systems. *Sci. Total Environ.* **2017**, *580*, 602–625. [CrossRef]
8. Ouda, M.; Kadadou, D.; Swaidan, B.; Al-Othman, A.; Al-Asheh, S.; Banat, F.; Hasan, S.W. Emerging contaminants in the water bodies of the Middle East and North Africa (MENA): A critical review. *Sci. Total Environ.* **2021**, *754*, 142177. [CrossRef]
9. Niu, H.; Liu, S.; Jiang, Y.; Hu, Y.; Li, Y.; He, L.; Xing, M.; Li, X.; Wu, L.; Chen, Z.; et al. Are Microplastics Toxic? A Review from Eco-Toxicity to Effects on the Gut Microbiota. *Metabolites* **2023**, *13*, 739. [CrossRef]
10. Li, X.; Yin Yeung, L.W.; Xu, M.; Taniyasu, S.; Lam, P.K.S.; Yamashita, N.; Dai, J. Perfluorooctane sulfonate (PFOS) and other fluorochemicals in fish blood collected near the outfall of wastewater treatment plant (WWTP) in Beijing. *Environ. Pollut.* **2008**, *156*, 1298–1303. [CrossRef]
11. Yeung, L.W.Y.; So, M.K.; Jiang, G.; Taniyasu, S.; Yamashita, N.; Song, M.; Wu, Y.; Li, J.; Giesy, J.P.; Guruge, K.S.; et al. Perfluorooctanesulfonate and related fluorochemicals in human blood samples from China. *Environ. Sci. Technol.* **2006**, *40*, 715–720. [CrossRef] [PubMed]
12. Zheng, F.; Sheng, N.; Zhang, H.; Yan, S.; Zhang, J.; Wang, J. Perfluorooctanoic acid exposure disturbs glucose metabolism in mouse liver. *Toxicol. Appl. Pharmacol.* **2017**, *335*, 41–48. [CrossRef]
13. Huang, D.; Zhang, Y.; Long, J.; Yang, X.; Bao, L.; Yang, Z.; Wu, B.; Si, R.; Zhao, W.; Peng, C.; et al. Polystyrene microplastic exposure induces insulin resistance in mice via dysbacteriosis and pro-inflammation. *Sci. Total Environ.* **2022**, *838*, 155937. [CrossRef] [PubMed]
14. Jiang, W.; Ding, K.; Huang, W.; Xu, F.; Lei, M.; Yue, R. Potential effects of bisphenol A on diabetes mellitus and its chronic complications: A narrative review. *Heliyon* **2023**, *9*, e16340. [CrossRef] [PubMed]
15. Mariana, M.; Cairrao, E. The Relationship between Phthalates and Diabetes: A Review. *Metabolites* **2023**, *13*, 746. [CrossRef] [PubMed]
16. Bao, L.-J.; Wei, Y.-L.; Yao, Y.; Ruan, Q.-Q.; Zeng, E.Y. Global trends of research on emerging contaminants in the environment and humans: A literature assimilation. *Environ. Sci. Pollut. Res. Int.* **2015**, *22*, 1635–1643. [CrossRef] [PubMed]
17. Naidu, R.; Arias Espana, V.A.; Liu, Y.; Jit, J. Emerging contaminants in the environment: Risk-based analysis for better management. *Chemosphere* **2016**, *154*, 350–357. [CrossRef]
18. Khan, S.; Naushad, M.; Govarthanan, M.; Iqbal, J.; Alfadul, S.M. Emerging contaminants of high concern for the environment: Current trends and future research. *Environ. Res.* **2022**, *207*, 112609. [CrossRef]
19. Mohammadi, A.; Dobaradaran, S.; Schmidt, T.C.; Malakootian, M.; Spitz, J. Emerging contaminants migration from pipes used in drinking water distribution systems: A review of the scientific literature. *Environ. Sci. Pollut. Res. Int.* **2022**, *29*, 75134–75160. [CrossRef]
20. Puri, M.; Gandhi, K.; Kumar, M.S. Emerging environmental contaminants: A global perspective on policies and regulations. *J. Environ. Manag.* **2023**, *332*, 117344. [CrossRef]
21. Chen, Y.; Lin, M.; Zhuang, D. Wastewater treatment and emerging contaminants: Bibliometric analysis. *Chemosphere* **2022**, *297*, 133932. [CrossRef] [PubMed]
22. Bodus, B.; O'Malley, K.; Dieter, G.; Gunawardana, C.; McDonald, W. Review of emerging contaminants in green stormwater infrastructure: Antibiotic resistance genes, microplastics, tire wear particles, PFAS, and temperature. *Sci. Total Environ.* **2023**, *906*, 167195. [CrossRef] [PubMed]
23. Kumar, N.; Shukla, P. Microalgal-based bioremediation of emerging contaminants: Mechanisms and challenges. *Environ. Pollut.* **2023**, *337*, 122591. [CrossRef] [PubMed]
24. Tong, X.; Mohapatra, S.; Zhang, J.; Tran, N.H.; You, L.; He, Y.; Gin, K.Y.-H. Source, fate, transport and modelling of selected emerging contaminants in the aquatic environment: Current status and future perspectives. *Water Res.* **2022**, *217*, 118418. [CrossRef] [PubMed]
25. Dodoo, D.K.; Essumang, D.K.; Jonathan, J.W.A. Accumulation profile and seasonal variations of polychlorinated biphenyls (PCBs) in bivalves Crassostrea tulipa (oysters) and Anadara senilis (mussels) at three different aquatic habitats in two seasons in Ghana. *Ecotoxicol. Environ. Saf.* **2013**, *88*, 26–34. [CrossRef] [PubMed]
26. Bigler, J.; Greene, A. *Guidance for Assessing Chemical Contaminant Data for Use in Fish Advisories*; US EPA Office of Water, Office of Science and Technology: Washington, DC, USA, 1993; Volume 2.
27. Wang, J.; Shi, Y.; Pan, Y.; Cai, Y. Perfluorooctane sulfonate (PFOS) and other fluorochemicals in viscera and muscle of farmed pigs and chickens in Beijing, China. *Chin. Sci. Bull.* **2010**, *55*, 3550–3555. [CrossRef]
28. Yang, D.; Shi, H.; Li, L.; Li, J.; Jabeen, K.; Kolandhasamy, P. Microplastic Pollution in Table Salts from China. *Environ. Sci. Technol.* **2015**, *49*, 13622–13627. [CrossRef]
29. Mason, S.A.; Welch, V.G.; Neratko, J. Synthetic Polymer Contamination in Bottled Water. *Front. Chem.* **2018**, *6*, 407. [CrossRef]
30. Chen, X.; Wang, S.; Mao, X.; Xiang, X.; Ye, S.; Chen, J.; Zhu, A.; Meng, Y.; Yang, X.; Peng, S.; et al. Adverse health effects of emerging contaminants on inflammatory bowel disease. *Front. Public. Health* **2023**, *11*, 1140786. [CrossRef]
31. Snow, D.D.; Cassada, D.A.; Biswas, S.; Malakar, A.; D'Alessio, M.; Marshall, A.H.L.; Sallach, J.B. Detection, occurrence, and fate of emerging contaminants in agricultural environments (2020). *Water Environ. Res.* **2020**, *92*, 1741–1750. [CrossRef]
32. Alonso-Magdalena, P.; Quesada, I.; Nadal, A. Endocrine disruptors in the etiology of type 2 diabetes mellitus. *Nat. Rev. Endocrinol.* **2011**, *7*, 346–353. [CrossRef] [PubMed]

33. Metcalfe, C.D.; Bayen, S.; Desrosiers, M.; Muñoz, G.; Sauvé, S.; Yargeau, V. Methods for the analysis of endocrine disrupting chemicals in selected environmental matrixes. *Environ. Res.* **2022**, *206*, 112616. [CrossRef] [PubMed]
34. Jia, M.; Sha, J.; Li, Z.; Wang, W.; Zhang, H. High affinity truncated aptamers for ultra-sensitive colorimetric detection of bisphenol A with label-free aptasensor. *Food Chem.* **2020**, *317*, 126459. [CrossRef] [PubMed]
35. Jebril, S.; Cubillana-Aguilera, L.; Palacios-Santander, J.M.; Dridi, C. A novel electrochemical sensor modified with green gold sononanoparticles and carbon black nanocomposite for bisphenol A detection. *Mater. Sci. Eng. B* **2021**, *264*, 114951. [CrossRef]
36. Lind, P.M.; Lind, L. Endocrine-disrupting chemicals and risk of diabetes: An evidence-based review. *Diabetologia* **2018**, *61*, 1495–1502. [CrossRef] [PubMed]
37. Porta, M.; Gasull, M.; Puigdomènech, E.; Garí, M.; Bosch de Basea, M.; Guillén, M.; López, T.; Bigas, E.; Pumarega, J.; Llebaria, X.; et al. Distribution of blood concentrations of persistent organic pollutants in a representative sample of the population of Catalonia. *Environ. Int.* **2010**, *36*, 655–664. [CrossRef] [PubMed]
38. Kassotis, C.D.; Vandenberg, L.N.; Demeneix, B.A.; Porta, M.; Slama, R.; Trasande, L. Endocrine-disrupting chemicals: Economic, regulatory, and policy implications. *Lancet Diabetes Endocrinol.* **2020**, *8*, 719–730. [CrossRef]
39. Fisher, B.E. Most unwanted. *Environ. Health Perspect.* **1999**, *107*, A18–A23. [CrossRef]
40. Lee, Y.M.; Kim, K.S.; Jacobs, D.R.; Lee, D.H. Persistent organic pollutants in adipose tissue should be considered in obesity research. *Obes. Rev.* **2017**, *18*, 129–139. [CrossRef]
41. Tian, Y.; Rimal, B.; Gui, W.; Koo, I.; Smith, P.B.; Yokoyama, S.; Patterson, A.D. Early Life Polychlorinated Biphenyl 126 Exposure Disrupts Gut Microbiota and Metabolic Homeostasis in Mice Fed with High-Fat Diet in Adulthood. *Metabolites* **2022**, *12*, 894. [CrossRef]
42. Zhang, Y.; Jiao, Y.; Li, Z.; Tao, Y.; Yang, Y. Hazards of phthalates (PAEs) exposure: A review of aquatic animal toxicology studies. *Sci. Total Environ.* **2021**, *771*, 145418. [CrossRef] [PubMed]
43. Gao, D.; Li, Z.; Wang, H.; Liang, H. An overview of phthalate acid ester pollution in China over the last decade: Environmental occurrence and human exposure. *Sci. Total Environ.* **2018**, *645*, 1400–1409. [CrossRef] [PubMed]
44. Duan, Y.; Sun, H.; Han, L.; Chen, L. Association between phthalate exposure and glycosylated hemoglobin, fasting glucose, and type 2 diabetes mellitus: A case-control study in China. *Sci. Total Environ.* **2019**, *670*, 41–49. [CrossRef] [PubMed]
45. Duan, Y.; Sun, H.; Yao, Y.; Han, L.; Chen, L. Perturbation of serum metabolome in relation to type 2 diabetes mellitus and urinary levels of phthalate metabolites and bisphenols. *Environ. Int.* **2021**, *155*, 106609. [CrossRef] [PubMed]
46. Shaffer, R.M.; Ferguson, K.K.; Sheppard, L.; James-Todd, T.; Butts, S.; Chandrasekaran, S.; Swan, S.H.; Barrett, E.S.; Nguyen, R.; Bush, N.; et al. Maternal urinary phthalate metabolites in relation to gestational diabetes and glucose intolerance during pregnancy. *Environ. Int.* **2019**, *123*, 588–596. [CrossRef] [PubMed]
47. Chen, W.; He, C.; Liu, X.; An, S.; Wang, X.; Tao, L.; Zhang, H.; Tian, Y.; Wu, N.; Xu, P.; et al. Effects of exposure to phthalate during early pregnancy on gestational diabetes mellitus: A nested case-control study with propensity score matching. *Environ. Sci. Pollut. Res. Int.* **2023**, *30*, 33555–33566. [CrossRef] [PubMed]
48. Yang, R.; Zheng, J.; Qin, J.; Liu, S.; Liu, X.; Gu, Y.; Yang, S.; Du, J.; Li, S.; Chen, B.; et al. Dibutyl phthalate affects insulin synthesis and secretion by regulating the mitochondrial apoptotic pathway and oxidative stress in rat insulinoma cells. *Ecotoxicol. Environ. Saf.* **2023**, *249*, 114396. [CrossRef]
49. Liu, X.; Zhang, L.; Li, J.; Meng, G.; Chi, M.; Li, T.; Zhao, Y.; Wu, Y. A nested case-control study of the association between exposure to polybrominated diphenyl ethers and the risk of gestational diabetes mellitus. *Environ. Int.* **2018**, *119*, 232–238. [CrossRef]
50. Ongono, J.S.; Dow, C.; Gambaretti, J.; Severi, G.; Boutron-Ruault, M.-C.; Bonnet, F.; Fagherazzi, G.; Mancini, F.R. Dietary exposure to brominated flame retardants and risk of type 2 diabetes in the French E3N cohort. *Environ. Int.* **2019**, *123*, 54–60. [CrossRef]
51. Liu, Z.-L.; Jiang, S.-R.; Fan, Y.; Wang, J.-S.; Wang, M.-L.; Li, M.-Y. 2,2′,4,4′,5,5′-Hexabromophenyl ether (BDE-153) causes abnormal insulin secretion and disorders of glucose and lipid metabolism in mice. *J. Chin. Med. Assoc.* **2023**, *86*, 388–398. [CrossRef]
52. Ibrahim, M.M.; Fjære, E.; Lock, E.-J.; Naville, D.; Amlund, H.; Meugnier, E.; Le Magueresse Battistoni, B.; Frøyland, L.; Madsen, L.; Jessen, N.; et al. Chronic consumption of farmed salmon containing persistent organic pollutants causes insulin resistance and obesity in mice. *PLoS ONE* **2011**, *6*, e25170. [CrossRef] [PubMed]
53. Gray, S.L.; Shaw, A.C.; Gagne, A.X.; Chan, H.M. Chronic exposure to PCBs (Aroclor 1254) exacerbates obesity-induced insulin resistance and hyperinsulinemia in mice. *J. Toxicol. Environ. Health A* **2013**, *76*, 701–715. [CrossRef] [PubMed]
54. Baker, N.A.; Karounos, M.; English, V.; Fang, J.; Wei, Y.; Stromberg, A.; Sunkara, M.; Morris, A.J.; Swanson, H.I.; Cassis, L.A. Coplanar polychlorinated biphenyls impair glucose homeostasis in lean C57BL/6 mice and mitigate beneficial effects of weight loss on glucose homeostasis in obese mice. *Environ. Health Perspect.* **2013**, *121*, 105–110. [CrossRef] [PubMed]
55. Baker, N.A.; Shoemaker, R.; English, V.; Larian, N.; Sunkara, M.; Morris, A.J.; Walker, M.; Yiannikouris, F.; Cassis, L.A. Effects of Adipocyte Aryl Hydrocarbon Receptor Deficiency on PCB-Induced Disruption of Glucose Homeostasis in Lean and Obese Mice. *Environ. Health Perspect.* **2015**, *123*, 944–950. [CrossRef] [PubMed]
56. Matteo, G.; Hoyeck, M.P.; Blair, H.L.; Zebarth, J.; Rick, K.R.C.; Williams, A.; Gagné, R.; Buick, J.K.; Yauk, C.L.; Bruin, J.E. Prolonged Low-Dose Dioxin Exposure Impairs Metabolic Adaptability to High-Fat Diet Feeding in Female but Not Male Mice. *Endocrinology* **2021**, *162*, bqab050. [CrossRef] [PubMed]
57. Ishida, T.; Kan-o, S.; Mutoh, J.; Takeda, S.; Ishii, Y.; Hashiguchi, I.; Akamine, A.; Yamada, H. 2,3,7,8-Tetrachlorodibenzo-p-dioxin-induced change in intestinal function and pathology: Evidence for the involvement of arylhydrocarbon receptor-mediated alteration of glucose transportation. *Toxicol. Appl. Pharmacol.* **2005**, *205*, 89–97. [CrossRef] [PubMed]

58. Kurita, H.; Yoshioka, W.; Nishimura, N.; Kubota, N.; Kadowaki, T.; Tohyama, C. Aryl hydrocarbon receptor-mediated effects of 2,3,7,8-tetrachlorodibenzo-p-dioxin on glucose-stimulated insulin secretion in mice. *J. Appl. Toxicol.* **2009**, *29*, 689–694. [CrossRef]
59. Hoyeck, M.P.; Merhi, R.C.; Blair, H.L.; Spencer, C.D.; Payant, M.A.; Martin Alfonso, D.I.; Zhang, M.; Matteo, G.; Chee, M.J.; Bruin, J.E. Female mice exposed to low doses of dioxin during pregnancy and lactation have increased susceptibility to diet-induced obesity and diabetes. *Mol. Metab.* **2020**, *42*, 101104. [CrossRef]
60. Vandenberg, L.N.; Hauser, R.; Marcus, M.; Olea, N.; Welshons, W.V. Human exposure to bisphenol A (BPA). *Reprod. Toxicol.* **2007**, *24*, 139–177. [CrossRef]
61. Akash, M.S.H.; Sabir, S.; Rehman, K. Bisphenol A-induced metabolic disorders: From exposure to mechanism of action. *Environ. Toxicol. Pharmacol.* **2020**, *77*, 103373. [CrossRef]
62. Marmugi, A.; Lasserre, F.; Beuzelin, D.; Ducheix, S.; Huc, L.; Polizzi, A.; Chetivaux, M.; Pineau, T.; Martin, P.; Guillou, H.; et al. Adverse effects of long-term exposure to bisphenol A during adulthood leading to hyperglycaemia and hypercholesterolemia in mice. *Toxicology* **2014**, *325*, 133–143. [CrossRef] [PubMed]
63. Ma, Q.; Deng, P.; Lin, M.; Yang, L.; Li, L.; Guo, L.; Zhang, L.; He, M.; Lu, Y.; Pi, H.; et al. Long-term bisphenol A exposure exacerbates diet-induced prediabetes via TLR4-dependent hypothalamic inflammation. *J. Hazard. Mater.* **2021**, *402*, 123926. [CrossRef] [PubMed]
64. Moon, M.K.; Jeong, I.-K.; Jung Oh, T.; Ahn, H.Y.; Kim, H.H.; Park, Y.J.; Jang, H.C.; Park, K.S. Long-term oral exposure to bisphenol A induces glucose intolerance and insulin resistance. *J. Endocrinol.* **2015**, *226*, 35–42. [CrossRef]
65. Lau, C.; Anitole, K.; Hodes, C.; Lai, D.; Pfahles-Hutchens, A.; Seed, J. Perfluoroalkyl acids: A review of monitoring and toxicological findings. *Toxicol. Sci.* **2007**, *99*, 366–394. [CrossRef] [PubMed]
66. Wang, Z.; DeWitt, J.C.; Higgins, C.P.; Cousins, I.T. A Never-Ending Story of Per- and Polyfluoroalkyl Substances (PFASs)? *Environ. Sci. Technol.* **2017**, *51*, 2508–2518. [CrossRef] [PubMed]
67. Domingo, J.L.; Nadal, M. Human exposure to per- and polyfluoroalkyl substances (PFAS) through drinking water: A review of the recent scientific literature. *Environ. Res.* **2019**, *177*, 108648. [CrossRef]
68. Lee, J.-W.; Lee, H.-K.; Lim, J.-E.; Moon, H.-B. Legacy and emerging per- and polyfluoroalkyl substances (PFASs) in the coastal environment of Korea: Occurrence, spatial distribution, and bioaccumulation potential. *Chemosphere* **2020**, *251*, 126633. [CrossRef]
69. Wan, H.T.; Zhao, Y.G.; Wei, X.; Hui, K.Y.; Giesy, J.P.; Wong, C.K.C. PFOS-induced hepatic steatosis, the mechanistic actions on β-oxidation and lipid transport. *Biochim. Biophys. Acta* **2012**, *1820*, 1092–1101. [CrossRef]
70. Sanan, T.; Magnuson, M. Analysis of per- and polyfluorinated alkyl substances in sub-sampled water matrices with online solid phase extraction/isotope dilution tandem mass spectrometry. *J. Chromatogr. A* **2020**, *1626*, 461324. [CrossRef]
71. Olsen, G.W.; Burris, J.M.; Ehresman, D.J.; Froehlich, J.W.; Seacat, A.M.; Butenhoff, J.L.; Zobel, L.R. Half-life of serum elimination of perfluorooctanesulfonate, perfluorohexanesulfonate, and perfluorooctanoate in retired fluorochemical production workers. *Environ. Health Perspect.* **2007**, *115*, 1298–1305. [CrossRef]
72. Lin, C.-Y.; Chen, P.-C.; Lin, Y.-C.; Lin, L.-Y. Association among serum perfluoroalkyl chemicals, glucose homeostasis, and metabolic syndrome in adolescents and adults. *Diabetes Care* **2009**, *32*, 702–707. [CrossRef] [PubMed]
73. Zeeshan, M.; Zhang, Y.-T.; Yu, S.; Huang, W.-Z.; Zhou, Y.; Vinothkumar, R.; Chu, C.; Li, Q.-Q.; Wu, Q.-Z.; Ye, W.-L.; et al. Exposure to isomers of per- and polyfluoroalkyl substances increases the risk of diabetes and impairs glucose-homeostasis in Chinese adults: Isomers of C8 health project. *Chemosphere* **2021**, *278*, 130486. [CrossRef] [PubMed]
74. Xu, C.; Zhang, L.; Zhou, Q.; Ding, J.; Yin, S.; Shang, X.; Tian, Y. Exposure to per- and polyfluoroalkyl substances as a risk factor for gestational diabetes mellitus through interference with glucose homeostasis. *Sci. Total Environ.* **2022**, *838*, 156561. [CrossRef] [PubMed]
75. Sant, K.E.; Jacobs, H.M.; Borofski, K.A.; Moss, J.B.; Timme-Laragy, A.R. Embryonic exposures to perfluorooctanesulfonic acid (PFOS) disrupt pancreatic organogenesis in the zebrafish, Danio rerio. *Environ. Pollut.* **2017**, *220*, 807–817. [CrossRef] [PubMed]
76. Qin, W.-P.; Cao, L.-Y.; Li, C.-H.; Guo, L.-H.; Colbourne, J.; Ren, X.-M. Perfluoroalkyl Substances Stimulate Insulin Secretion by Islet β Cells via G Protein-Coupled Receptor 40. *Environ. Sci. Technol.* **2020**, *54*, 3428–3436. [CrossRef]
77. Zhang, L.; Duan, X.; Sun, W.; Sun, H. Perfluorooctane sulfonate acute exposure stimulates insulin secretion via GPR40 pathway. *Sci. Total Environ.* **2020**, *726*, 138498. [CrossRef]
78. Duan, X.; Sun, W.; Sun, H.; Zhang, L. Perfluorooctane sulfonate continual exposure impairs glucose-stimulated insulin secretion via SIRT1-induced upregulation of UCP2 expression. *Environ. Pollut.* **2021**, *278*, 116840. [CrossRef]
79. Wan, H.T.; Zhao, Y.G.; Leung, P.Y.; Wong, C.K.C. Perinatal exposure to perfluorooctane sulfonate affects glucose metabolism in adult offspring. *PLoS ONE* **2014**, *9*, e87137. [CrossRef]
80. Harada, K.; Nakanishi, S.; Sasaki, K.; Furuyama, K.; Nakayama, S.; Saito, N.; Yamakawa, K.; Koizumi, A. Particle size distribution and respiratory deposition estimates of airborne perfluorooctanoate and perfluorooctanesulfonate in Kyoto area, Japan. *Bull. Environ. Contam. Toxicol.* **2006**, *76*, 306–310. [CrossRef]
81. Barton, C.A.; Butler, L.E.; Zarzecki, C.J.; Flaherty, J.; Kaiser, M. Characterizing perfluorooctanoate in ambient air near the fence line of a manufacturing facility: Comparing modeled and monitored values. *J. Air Waste Manag. Assoc.* **2006**, *56*, 48–55. [CrossRef]
82. Davis, K.L.; Aucoin, M.D.; Larsen, B.S.; Kaiser, M.A.; Hartten, A.S. Transport of ammonium perfluorooctanoate in environmental media near a fluoropolymer manufacturing facility. *Chemosphere* **2007**, *67*, 2011–2019. [CrossRef] [PubMed]

83. Ehresman, D.J.; Froehlich, J.W.; Olsen, G.W.; Chang, S.-C.; Butenhoff, J.L. Comparison of human whole blood, plasma, and serum matrices for the determination of perfluorooctanesulfonate (PFOS), perfluorooctanoate (PFOA), and other fluorochemicals. *Environ. Res.* **2007**, *103*, 176–184. [CrossRef] [PubMed]
84. Lind, L.; Zethelius, B.; Salihovic, S.; van Bavel, B.; Lind, P.M. Circulating levels of perfluoroalkyl substances and prevalent diabetes in the elderly. *Diabetologia* **2014**, *57*, 473–479. [CrossRef] [PubMed]
85. Chung, S.M.; Heo, D.-G.; Kim, J.-H.; Yoon, J.S.; Lee, H.W.; Kim, J.-Y.; Moon, J.S.; Won, K.C. Perfluorinated compounds in adults and their association with fasting glucose and incident diabetes: A prospective cohort study. *Environ. Health* **2022**, *21*, 101. [CrossRef] [PubMed]
86. Yan, S.; Zhang, H.; Zheng, F.; Sheng, N.; Guo, X.; Dai, J. Perfluorooctanoic acid exposure for 28 days affects glucose homeostasis and induces insulin hypersensitivity in mice. *Sci. Rep.* **2015**, *5*, 11029. [CrossRef] [PubMed]
87. He, X.; Wu, D.; Xu, Y.; Zhang, Y.; Sun, Y.; Chang, X.; Zhu, Y.; Tang, W. Perfluorooctanoic acid promotes pancreatic β cell dysfunction and apoptosis through ER stress and the ATF4/CHOP/TRIB3 pathway. *Environ. Sci. Pollut. Res. Int.* **2022**, *29*, 84532–84545. [CrossRef] [PubMed]
88. Liu, T.; Yu, S.; Zhu, X.; Liao, R.; Zhuo, Z.; He, Y.; Ma, H. In-situ Detection Method for Microplastics in Water by Polarized Light Scattering. *Front. Mar. Sci.* **2021**, *8*, 739683. [CrossRef]
89. Dey, T.K.; Uddin, M.E.; Jamal, M. Detection and removal of microplastics in wastewater: Evolution and impact. *Environ. Sci. Pollut. Res. Int.* **2021**, *28*, 16925–16947. [CrossRef]
90. Yong, C.Q.Y.; Valiyaveettil, S.; Tang, B.L. Toxicity of Microplastics and Nanoplastics in Mammalian Systems. *Int. J. Environ. Res. Public. Health* **2020**, *17*, 1509. [CrossRef]
91. Kumar, R.; Manna, C.; Padha, S.; Verma, A.; Sharma, P.; Dhar, A.; Ghosh, A.; Bhattacharya, P. Micro(nano)plastics pollution and human health: How plastics can induce carcinogenesis to humans? *Chemosphere* **2022**, *298*, 134267. [CrossRef]
92. Xia, X.; Sun, M.; Zhou, M.; Chang, Z.; Li, L. Polyvinyl chloride microplastics induce growth inhibition and oxidative stress in Cyprinus carpio var. *larvae. Sci. Total Environ.* **2020**, *716*, 136479. [CrossRef] [PubMed]
93. Lu, Y.; Zhang, Y.; Deng, Y.; Jiang, W.; Zhao, Y.; Geng, J.; Ding, L.; Ren, H. Uptake and Accumulation of Polystyrene Microplastics in Zebrafish (*Danio rerio*) and Toxic Effects in Liver. *Environ. Sci. Technol.* **2016**, *50*, 4054–4060. [CrossRef] [PubMed]
94. Cortés, C.; Domenech, J.; Salazar, M.; Pastor, S.; Marcos, R.; Hernández, A. Nanoplastics as a potential environmental health factor: Effects of polystyrene nanoparticles on human intestinal epithelial Caco-2 cells. *Environ. Sci. Nano* **2020**, *7*, 272–285. [CrossRef]
95. Schirinzi, G.F.; Pérez-Pomeda, I.; Sanchís, J.; Rossini, C.; Farré, M.; Barceló, D. Cytotoxic effects of commonly used nanomaterials and microplastics on cerebral and epithelial human cells. *Environ. Res.* **2017**, *159*, 579–587. [CrossRef] [PubMed]
96. Okamura, T.; Hamaguchi, M.; Hasegawa, Y.; Hashimoto, Y.; Majima, S.; Senmaru, T.; Ushigome, E.; Nakanishi, N.; Asano, M.; Yamazaki, M.; et al. Oral Exposure to Polystyrene Microplastics of Mice on a Normal or High-Fat Diet and Intestinal and Metabolic Outcomes. *Environ. Health Perspect.* **2023**, *131*, 27006. [CrossRef]
97. Fan, X.; Wei, X.; Hu, H.; Zhang, B.; Yang, D.; Du, H.; Zhu, R.; Sun, X.; Oh, Y.; Gu, N. Effects of oral administration of polystyrene nanoplastics on plasma glucose metabolism in mice. *Chemosphere* **2022**, *288*, 132607. [CrossRef]
98. Wang, Y.; Wei, Z.; Xu, K.; Wang, X.; Gao, X.; Han, Q.; Wang, S.; Chen, M. The effect and a mechanistic evaluation of polystyrene nanoplastics on a mouse model of type 2 diabetes. *Food Chem. Toxicol.* **2023**, *173*, 113642. [CrossRef]
99. Shi, C.; Han, X.; Guo, W.; Wu, Q.; Yang, X.; Wang, Y.; Tang, G.; Wang, S.; Wang, Z.; Liu, Y.; et al. Disturbed Gut-Liver axis indicating oral exposure to polystyrene microplastic potentially increases the risk of insulin resistance. *Environ. Int.* **2022**, *164*, 107273. [CrossRef]
100. Klein, E.Y.; Van Boeckel, T.P.; Martinez, E.M.; Pant, S.; Gandra, S.; Levin, S.A.; Goossens, H.; Laxminarayan, R. Global increase and geographic convergence in antibiotic consumption between 2000 and 2015. *Proc. Natl. Acad. Sci. USA* **2018**, *115*, E3463–E3470. [CrossRef]
101. Van Boeckel, T.P.; Gandra, S.; Ashok, A.; Caudron, Q.; Grenfell, B.T.; Levin, S.A.; Laxminarayan, R. Global antibiotic consumption 2000 to 2010: An analysis of national pharmaceutical sales data. *Lancet Infect. Dis.* **2014**, *14*, 742–750. [CrossRef]
102. Tran, N.H.; Chen, H.; Reinhard, M.; Mao, F.; Gin, K.Y.-H. Occurrence and removal of multiple classes of antibiotics and antimicrobial agents in biological wastewater treatment processes. *Water Res.* **2016**, *104*, 461–472. [CrossRef] [PubMed]
103. Lu, N.; Chen, J.; Rao, Z.; Guo, B.; Xu, Y. Recent Advances of Biosensors for Detection of Multiple Antibiotics. *Biosensors* **2023**, *13*, 850. [CrossRef] [PubMed]
104. Gago-Ferrero, P.; Díaz-Cruz, M.S.; Barceló, D. Occurrence of multiclass UV filters in treated sewage sludge from wastewater treatment plants. *Chemosphere* **2011**, *84*, 1158–1165. [CrossRef] [PubMed]
105. Gurung, M.; Li, Z.; You, H.; Rodrigues, R.; Jump, D.B.; Morgun, A.; Shulzhenko, N. Role of gut microbiota in type 2 diabetes pathophysiology. *eBioMedicine* **2020**, *51*, 102590. [CrossRef] [PubMed]
106. Cani, P.D.; Bibiloni, R.; Knauf, C.; Waget, A.; Neyrinck, A.M.; Delzenne, N.M.; Burcelin, R. Changes in gut microbiota control metabolic endotoxemia-induced inflammation in high-fat diet-induced obesity and diabetes in mice. *Diabetes* **2008**, *57*, 1470–1481. [CrossRef] [PubMed]
107. Rodrigues, R.R.; Greer, R.L.; Dong, X.; Dsouza, K.N.; Gurung, M.; Wu, J.Y.; Morgun, A.; Shulzhenko, N. Antibiotic-Induced Alterations in Gut Microbiota Are Associated with Changes in Glucose Metabolism in Healthy Mice. *Front. Microbiol.* **2017**, *8*, 2306. [CrossRef]

108. Yuan, J.; Hu, Y.J.; Zheng, J.; Kim, J.H.; Sumerlin, T.; Chen, Y.; He, Y.; Zhang, C.; Tang, J.; Pan, Y.; et al. Long-term use of antibiotics and risk of type 2 diabetes in women: A prospective cohort study. *Int. J. Epidemiol.* **2020**, *49*, 1572–1581. [CrossRef]
109. Fu, L.; Qiu, Y.; Shen, L.; Cui, C.; Wang, S.; Wang, S.; Xie, Y.; Zhao, X.; Gao, X.; Ning, G.; et al. The delayed effects of antibiotics in type 2 diabetes, friend or foe? *J. Endocrinol.* **2018**, *238*, 137–149. [CrossRef]
110. Fenneman, A.C.; Weidner, M.; Chen, L.A.; Nieuwdorp, M.; Blaser, M.J. Antibiotics in the pathogenesis of diabetes and inflammatory diseases of the gastrointestinal tract. *Nat. Rev. Gastroenterol. Hepatol.* **2023**, *20*, 81–100. [CrossRef]
111. Buffie, C.G.; Bucci, V.; Stein, R.R.; McKenney, P.T.; Ling, L.; Gobourne, A.; No, D.; Liu, H.; Kinnebrew, M.; Viale, A.; et al. Precision microbiome reconstitution restores bile acid mediated resistance to Clostridium difficile. *Nature* **2015**, *517*, 205–208. [CrossRef]
112. Sharon, G.; Garg, N.; Debelius, J.; Knight, R.; Dorrestein, P.C.; Mazmanian, S.K. Specialized metabolites from the microbiome in health and disease. *Cell Metab.* **2014**, *20*, 719–730. [CrossRef] [PubMed]
113. Koh, A.; De Vadder, F.; Kovatcheva-Datchary, P.; Bäckhed, F. From Dietary Fiber to Host Physiology: Short-Chain Fatty Acids as Key Bacterial Metabolites. *Cell* **2016**, *165*, 1332–1345. [CrossRef] [PubMed]
114. Claus, S.P.; Guillou, H.; Ellero-Simatos, S. The gut microbiota: A major player in the toxicity of environmental pollutants? *NPJ Biofilms Microbiomes* **2016**, *2*, 16003. [CrossRef] [PubMed]
115. Lu, K.; Abo, R.P.; Schlieper, K.A.; Graffam, M.E.; Levine, S.; Wishnok, J.S.; Swenberg, J.A.; Tannenbaum, S.R.; Fox, J.G. Arsenic exposure perturbs the gut microbiome and its metabolic profile in mice: An integrated metagenomics and metabolomics analysis. *Environ. Health Perspect.* **2014**, *122*, 284–291. [CrossRef] [PubMed]
116. Bian, X.; Tu, P.; Chi, L.; Gao, B.; Ru, H.; Lu, K. Saccharin induced liver inflammation in mice by altering the gut microbiota and its metabolic functions. *Food Chem. Toxicol.* **2017**, *107*, 530–539. [CrossRef] [PubMed]
117. Livanos, A.E.; Greiner, T.U.; Vangay, P.; Pathmasiri, W.; Stewart, D.; McRitchie, S.; Li, H.; Chung, J.; Sohn, J.; Kim, S.; et al. Antibiotic-mediated gut microbiome perturbation accelerates development of type 1 diabetes in mice. *Nat. Microbiol.* **2016**, *1*, 16140. [CrossRef] [PubMed]
118. Lamichhane, S.; Härkönen, T.; Vatanen, T.; Hyötyläinen, T.; Knip, M.; Orešič, M. Impact of exposure to per- and polyfluoroalkyl substances on fecal microbiota composition in mother-infant dyads. *Environ. Int.* **2023**, *176*, 107965. [CrossRef]
119. Potera, C. POPs and Gut Microbiota: Dietary Exposure Alters Ratio of Bacterial Species. *Environ. Health Perspect.* **2015**, *123*, A187. [CrossRef]
120. Lai, K.P.; Ng, A.H.-M.; Wan, H.T.; Wong, A.Y.-M.; Leung, C.C.-T.; Li, R.; Wong, C.K.-C. Dietary Exposure to the Environmental Chemical, PFOS on the Diversity of Gut Microbiota, Associated With the Development of Metabolic Syndrome. *Front. Microbiol.* **2018**, *9*, 2552. [CrossRef]
121. Crawford, P.A.; Crowley, J.R.; Sambandam, N.; Muegge, B.D.; Costello, E.K.; Hamady, M.; Knight, R.; Gordon, J.I. Regulation of myocardial ketone body metabolism by the gut microbiota during nutrient deprivation. *Proc. Natl. Acad. Sci. USA* **2009**, *106*, 11276–11281. [CrossRef]

toxics

MDPI

Review

Advancements, Challenges, and Future Directions in Aquatic Life Criteria Research in China

Chen Liu [1], Zhaomei Geng [2], Jiayin Xu [1,3], Qingwei Li [1,3], Heng Zhang [1,3] and Jinfen Pan [1,4,*]

1 Key Laboratory of Environment and Ecology (Ministry of Education), Ocean University of China, Qingdao 266100, China; xjy99929@gmail.com (J.X.); liqing-wei@outlook.com (Q.L.); ouczhangheng@outlook.com (H.Z.)
2 School of Mathematics, Sun Yat-Sen University, Guangzhou 510275, China; gengzhm@mail2.sysu.edu.cn
3 Key Laboratory of Marine Eco-Environmental Science and Technology, First Institute of Oceanography, Ministry of Natural Resources, Qingdao 266061, China
4 Laboratory for Marine Ecology and Environmental Science, Laoshan Laboratory, Qingdao 266200, China
* Correspondence: jfpan@ouc.edu.cn

Abstract: Aquatic life criteria (ALC) serve as the scientific foundation for establishing water quality standards, and in China, significant strides have been made in the development of freshwater ALC. This comprehensive review traces the evolution of China's WQC, focusing on the methodological advancements and challenges in priority pollutants selection, test organism screening, and standardized ecotoxicity testing protocols. It also provides a critical evaluation of quality assurance measures, data validation techniques, and minimum data requirements essential for ALC assessments. The paper highlights China's technical guidelines for deriving ALC, and reviews the published values for typical pollutants, assessing their impact on environmental quality standards. Emerging trends and future research avenues are discussed, including the incorporation of molecular toxicology data and the development of predictive models for pollutant toxicity. The review concludes by advocating for a tiered WQC system that accommodates China's diverse ecological regions, thereby offering a robust scientific basis for enhanced water quality management.

Keywords: aquatic life criteria (ALC); water quality criteria (WQC); freshwater; priority pollutants; China

Citation: Liu, C.; Geng, Z.; Xu, J.; Li, Q.; Zhang, H.; Pan, J. Advancements, Challenges, and Future Directions in Aquatic Life Criteria Research in China. *Toxics* **2023**, *11*, 862. https://doi.org/10.3390/toxics11100862

Academic Editor: Gary R. Fones

Received: 10 September 2023
Revised: 9 October 2023
Accepted: 12 October 2023
Published: 16 October 2023

1. Introduction

Water quality criteria (WQC) are essential for protecting aquatic ecosystems and human health. These criteria cover various areas such as aquatic life water quality criteria (ALC), human health water quality criteria, sediment quality criteria, and nutrient criteria [1].

The field of ALC research was first developed in the United States during the 1960s [2]. The U.S. later formalized this research by issuing comprehensive technical guidelines in 1985 [3], which have had a significant influence globally. In parallel, the European Union has made substantial advancements in aquatic risk assessment, with member states like the Netherlands contributing significantly [4]. Beyond the U.S. and EU, countries such as Canada [5], Australia, and New Zealand [6] have also conducted ALC research and developed their own technical guidelines.

When it comes to ALC formulation, the U.S. uniquely uses a dual-value system, incorporating both long-term and short-term ALC for each pollutant [3]. This approach was later adopted by Australia and New Zealand in their 2018 guideline updates [7]. While long-term ALC is used for daily water quality management, short-term ALC is designed to handle sudden water pollution incidents. Most other developed countries focus only on long-term ALC for daily management.

Developing ALC involves a complex process that includes careful screening of ecotoxicity data and choosing the right mathematical models for data analysis. The U.S. guidelines provide a comprehensive framework for this, covering aspects like toxicity endpoints, effect indices, exposure conditions, and data prioritization for both acute and chronic toxicity [3]. Different countries use different mathematical models; for example, the U.S. uses a log-triangle function model [3], the Netherlands uses a log-normal distribution model [4], and Australia and New Zealand use the Burr III model [6].

Quality assurance is crucial in toxicity data for developing reliable ALC. Developed countries have methods for assessing the quality of toxicity data, which can be either qualitative or quantitative. For instance, the U.S. [8] and the E.U. [9] use qualitative methods, while Australia and New Zealand use a quantitative approach [10]. These methods evaluate data quality based on various factors like the properties of the test substance, species characteristics, experimental design, exposure conditions, and statistical methods.

In terms of selecting test species, the U.S. guidelines recommend using native North American aquatic organisms [3]. Guidelines from other countries are less specific, lacking detailed recommendations or requirements about the geographical distribution of test species.

According to U.S. evaluation methods, toxicity data are categorized into quantitative data (used for environmental risk calculations), qualitative data (used to support environmental risk assessments), and invalid data. The E.U. method considers the reliability and relevance of the data and categorizes it into four types: unlimited reliable data, limited reliable data, unreliable data, and uncertain data. In Australia and New Zealand, toxicity data are scored and categorized into unacceptable, acceptable, and high-quality data based on these scores.

In China, ALC research has seen significant progress in recent years. This paper aims to provide a comprehensive overview of China's ALC research, focusing on its historical development, priority pollutants and test species, data collection, technical guidelines, and published ALC values. This review is intended to serve as a valuable reference for the ongoing and future development of ALC. Despite the progress, several challenges continue to persist. These include the need for more expansive toxicity data, the development of reliable and standardized testing protocols, and the creation of a framework that can effectively translate scientific discoveries into actionable policies and standards.

2. The Evolution of Aquatic Life Water Quality Criteria in China

Research on water quality criteria (WQC) in China commenced in the 1980s, initially through the translation of American WQC Red Book and European WQC guidelines focused on fish protection. In the following years, some Chinese researchers conducted studies that utilized toxicity data from resident species in China. However, due to the lack of systematic research, China has largely relied on foreign WQC standards when establishing its own water quality guidelines. A notable example is the "China Surface Water Environmental Quality Standard" (GB 3838-2002), a cornerstone in China's water management policies. This standard comprises 109 water quality criteria, the majority of which are adapted from international guidelines.

The turning point for WQC development in China came in 2005 following a significant water pollution incident involving nitrobenzene leakage in the Songhua River Basin. The emergency response adopted a nitrobenzene standard of 0.017 mg/L, which was based on U.S. criteria at the time. However, its applicability for protecting Chinese bodies of water remains a subject of debate. This incident catalyzed the advancement of WQC in China. The same year, the State Council of China set a national goal for "scientifically determining environmental criteria" in its "Decision on Implementing the Scientific Outlook on Development and Strengthening Environmental Protection." During China's Eleventh Five-Year Plan (2005–2010), several national projects were launched to support systematic WQC research [11].

In 2011, the Ministry of Science and Technology established the State Key Laboratory of Environmental Criteria and Risk Assessment, further boosting WQC research. In 2014, the revised "Environmental Protection Law" explicitly encouraged WQC research, marking the first legal recognition of WQC studies in China.

By 2017, the Ministry of Environmental Protection of China (MEPC) released the country's inaugural batch of technical guidelines for WQC, covering freshwater ALC, human health water quality criteria, and lake nutrient criteria [12]. In 2018, the Ministry of Ecology and Environment of China (MEEC), formerly known as MEPC, included WQC development as part of its regular duties. In 2020, MEEC unveiled the first set of national ALC for substances like cadmium [13], ammonia nitrogen [14], and phenol [15], signifying a landmark achievement in China's ALC research (Table 1).

Table 1. Landmark events in the development of ALC in China.

Year	Events	Related Ministries
2005	National goal for "scientifically determining environmental criteria" set	State Council of China
2011	State Key Laboratory of Environmental Criteria and Risk Assessment established	Ministry of Science and Technology of China
2014	Encouragement of WQC research included in the revised Environmental Protection Law	National People's Congress of China
2017	First batch of technical guidelines for WQC issued	MEPC
2018	WQC development incorporated into MEEC duties	State Commission of Public Sectors Reform
2020	First batch of national ALC was released	MEEC
2022	First technical guidelines for marine organism protection issued	MEEC

3. Methodological Approaches for Priority Pollutants Screening in ALC Studies

Given the labor-intensive and time-consuming nature of environmental criteria research, and considering the multitude of both individual and grouped pollutants in the environment, prioritization is imperative. It is vital to identify not only individual pollutants that pose significant risks, but also to acknowledge and prioritize groups of substances with similar purposes and effects, such as pesticides or PFAS compounds. This nuanced approach ensures comprehensive coverage, addressing both individual pollutants and categories of substances warranting immediate attention, thereby facilitating more effective and encompassing environmental protection strategies. While the topic of priority pollutants screening in water environments is widely discussed, the criteria for selecting priority pollutants for ALC research are diverse (Figure 1). Two key conditions must be met: first, the pollutant should be of concern in water management; second, there should be a significant difference in species sensitivity distribution (SSD) between resident and non-resident species. This ensures that the derived criteria values differ substantially depending on whether resident or non-resident species data are used. If no such SSD difference exists, national water quality standards can be temporarily based on foreign criteria, and the pollutant is not considered a priority for ALC research.

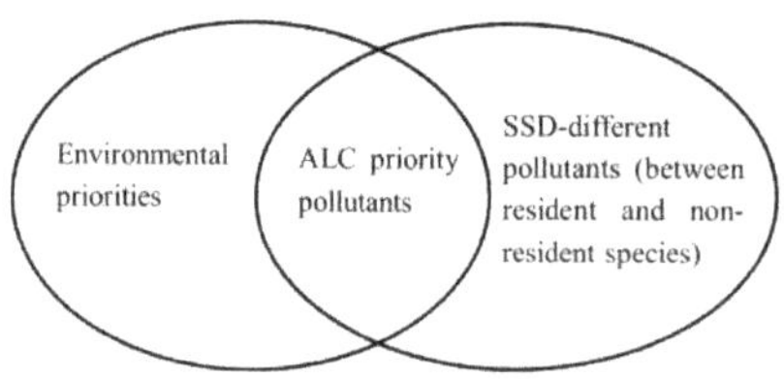

Figure 1. Principles for ALC priority pollutants screening.

Yan et al. [16] conducted a comprehensive study targeting 160 priority pollutants identified by the U.S., the E.U., and China. They collected and analyzed acute toxicity data for these pollutants in freshwater aquatic organisms. Their findings revealed that the HC_5 values (Hazardous Concentration affecting 5% of species, a key metric in ALC derivation) for certain pollutants varied significantly. As a result, 24 pollutants across six categories were identified as priority pollutants for ALC research in China (Table 2). Pesticide compounds were most prevalent, followed by metals and phenols. This distribution is also influenced by the availability of ecotoxicity data; many pollutants could not be adequately assessed due to insufficient data. As more toxicity data become available, it is likely that additional pollutants will be classified as priority pollutants for ALC research.

Table 2. Chinese ALC priority pollutants [16].

No.	CAS Number	Pollutants	Classification
1	7440-41-7	Be(II)	Metal
2	7440-43-9	Cd(II)	Metal
3	7440-47-3	Cr(VI)	Metal
4	7440-02-0	Ni(I)	Metal
5	57-74-9	Chlordane	Pesticide
6	60-57-1	Dieldrin	Pesticide
7	115-29-7	Endosulfan	Pesticide
8	72-20-8	Endrin	Pesticide
9	76-44-8	Heptachlor	Pesticide
10	608-73-1	Hexachlorocyclohexane	Pesticide
11	309-00-2	Aldrin	Pesticide
12	8001-35-2	Toxaphene	Pesticide
13	60-51-5	Dimethoate	Pesticide
14	298-00-0	Parathion-methyl	Pesticide
15	52-68-6	Trichlorfon	Pesticide
16	1912-24-9	Atrazine	Pesticide
17	470-90-6	Chlorfenvinfos	Pesticide
18	1582-09-8	Trifluralin	Pesticide
19	108-92-2	Phenol	Phenol
20	120-83-2	2, 4–Dichlorophenol	Phenol
21	51-28-5	2, 4–Dinitrophenol	Phenol
22	206-44-0	Fluoranthene	PAHs
23	/	Tributyltin compounds	Organotin
24	7664-41-7	Ammonia	Common chemical

Currently, a significant challenge is the scarcity of toxicity data for a broad spectrum of pollutants. This limitation obstructs the process of identifying priority pollutants for ALC research in China. Furthermore, the absence of systematic research and dependence on international WQC standards complicate the development of criteria that are meticulously designed for the distinctive biodiversity and aquatic ecosystems present in China.

4. Criteria for the Selection of Test Organisms in Aquatic Ecotoxicology

The biodiversity of aquatic ecosystems varies significantly across different countries, thereby influencing the target organisms for aquatic life criteria (ALC). Identifying species that are particularly sensitive to pollutants is crucial for the development of accurate ALC. While water quality criteria (WQC) studies have generally lacked a systematic approach to selecting sensitive aquatic organisms, the U.S. ALC guidelines [3] provide a list of recommended North American aquatic species. However, the sensitivity of these listed species has not been rigorously evaluated.

Yan et al. [17] developed a method for screening ALC test organisms based on the distribution characteristics of freshwater species in China. Utilizing species sensitivity analyses, they systematically identified sensitive aquatic organisms across various categories, including amphibians [18], fish [19], crustaceans [20], aquatic insects [21], mollusks [22], annelids [23], and aquatic plants [12]. In total, 46 sensitive freshwater species spanning seven phyla were identified. These include three species of coelenterates, one species of flatworms, three species of rotifers, two species of annelids, three species of mollusks, 13 species of arthropods, 11 species of chordates, three species of green algae, one species of diatoms, one species of ferns, and five species of angiosperms. These species have been recommended as test organisms for China's ALC research and are detailed in the supplementary materials (Table S1) of the Chinese ALC guidelines [12].

5. Standardized Ecotoxicity Testing Protocols

The development of standardized ecotoxicity testing methods is foundational for generating reliable ecotoxicity data. Currently, China has established a range of national standard methods for ecotoxicity testing, encompassing both acute and chronic toxicity tests for fish, chironomids, daphnia, and algae (Table 3). However, for other freshwater organisms like shellfish, annelids, and rotifers, China has yet to establish standard testing protocols. In these cases, researchers rely on international standards or methods published in scientific literature for ALC studies.

Table 3. China national standard toxicity test guidelines for freshwater organisms.

Species Group	Test Guideline	Guideline Number
Fish	Water quality—Determination of the acute toxicity of substances to a freshwater fish (*Brachydanio rerio* Hamilton-Buchanan)	GB/T 13267-1991
Fish	Chemicals—Fish acute toxicity test	GB/T 27861-2011
Fish	Chemicals—Fish (*Oryzias latipes*, d-rR medaka) early life stage toxicity test	GBT 29764-2013
Fish	Chemicals—Fish, juvenile growth test	GB/T 21806-2008
Fish	Testing of chemicals—Fish, short-term toxicity test on embryo and sac-fry stages	GB/T 21807-2008
Fish	Chemicals—Fish, early-life stage toxicity test	GB/T 21854-2008
Fish	Chemicals—Rare minnow (*Gobiocypris rarus*) acute toxicity test	GB/T 29763-2013
Daphnia	Method for acute toxicity test of *Daphnia magna* straus	GB/T 16125-2012
Daphnia	Chemicals—*Daphnia magna* reproduction test	GB/T 21828-2008
Chironomid	Chemicals—Sediment-water chironomid toxicity test—Spiked water method	GB/T 27858-2011
Chironomid	Chemicals—Sediment-water chironomid toxicity test—Spiked sediment method	GB/T 27859-2011
Alga	Chemicals—Algae growth inhibition test	GB/T 21805-2008
Duckweed	Chemicals—*Lemna* sp. growth inhibition test	GB/T 35524-2017

Given that ALC development requires extensive toxicity data, including data from non-standard test organisms, there is an urgent need to develop additional testing methods. Existing Chinese standards do not yet cover the full spectrum of freshwater biological groups. To address this gap, Chinese researchers are in the process of developing standard test methods for rotifers, water worms, mollusks, planaria, and region-specific fish species. In the interim, non-standard test methods continue to be employed for toxicity testing in ALC research. The lack of standardized testing protocols for a variety of freshwater organisms poses a significant challenge. This gap forces researchers to depend on international standards or methods documented in scientific literature. However, these might not always be applicable or reflective of the rich diversity of aquatic life in China.

6. Quality Assurance and Data Validation in Aquatic Ecotoxicological Studies

Ensuring the quality of toxicity data is fundamental for the development of robust water quality standards. As early as the 1990s, Klimisch et al. [24] introduced a method for assessing the quality of toxicity data. Subsequent studies [25–27] have expanded on this, although the reliability of their evaluation outcomes has been questioned [28].

In 2011, the U.S. Environmental Protection Agency released guidelines specifically aimed at quality assessment in ALC-related ecotoxicity studies. These guidelines provide a qualitative framework for toxicity data assessment, covering aspects such as data screening, evaluation, classification, and application [29]. Similarly, the European Union has established the Criteria for Reporting and Evaluating Ecotoxicity Data (CRED), which assesses data quality based on its reliability and relevance [9]. Australia and New Zealand followed suit, issuing their own guidelines for ecotoxicity data assessment in 2018 [30].

Drawing upon methodologies from Western countries, Chinese researchers have proposed a quantitative approach for evaluating the quality of ecotoxicity data. This approach considers five key aspects: data sources, chemical reagents, test organisms, experimental procedures, and experimental outcomes. Based on the evaluation scores, toxicity data are categorized into three levels: high-quality, acceptable, and unacceptable for ALC development in China. These categories are further detailed in the Supplementary Materials (Table S2).

7. Minimum Data Requirements and Data Prioritization Strategies for ALC Development

7.1. Minimum Toxicity Data Requirements (MTDR)

MTDR serve as a cornerstone for deriving ALC values. Developed countries have distinct MTDR frameworks; for example, the U.S. guidelines mandate data from eight families of aquatic animals and one aquatic plant [3], whereas other nations require data from five or six families [4,5]. In China, scholars have tailored MTDR to the nation's nascent ALC development stage. According to China's ALC guideline (HJ 831-2022), the MTDR encompasses data from five aquatic animals—specifically, one Cyprinidae fish, one non-Cyprinidae teleost fish, one zooplankton, one mollusk or benthic crustacean, and one amphibian or another phylum of animals—as well as one aquatic plant. Furthermore, toxicity data for a minimum of 10 species must be collected to derive the water quality criteria (WQC). As the volume of ecotoxicity data for native Chinese species grows, these MTDR are expected to evolve accordingly.

7.2. Data Prioritization Strategies

Both acute and chronic toxicity data are essential for dual-value ALC studies. These data come in various forms, with chronic toxicity indices including no observed effect concentration (NOEC), lowest observed effect concentration (LOEC), maximum acceptable toxic concentration (MATC), lethal concentration of 50% tested species (LC_{50}), and concentration for x% of maximal effect (EC_X), among others. Factors such as the life stage of the test organism, the taxonomic category of the data, the exposure methodology, and the monitoring of pollutant concentrations can all influence toxicity test outcomes. Consequently, establishing data prioritization is crucial in WQC studies. National requirements on this issue vary; for instance, the U.S. prioritizes genus-level toxicity data and favors the use of MATC [3], while most other countries prioritize NOEC for long-term WQC derivation [5,7,31]. Comparative studies have also been conducted to analyze the relationship between EC_{10} and NOEC [32]. In the updated 2022 China Freshwater Biological Water Quality Criteria Guidelines (HJ 831-2022), the prioritization hierarchy for chronic toxicity indices is as follows: MATC > EC_{20} > EC_{10} = NOEC > LOEC > EC_{50} > LC_{50}. Additionally, data from sensitive life stages, monitored pollutant concentrations, and flow toxicity experiments are given precedence in ALC derivation.

8. Technical Guidelines for the Development and Implementation of ALC in China

China's inaugural technical guideline for freshwater ALC was released in 2017, adopting a dual-value system comprising both long-term and short-term ALC [33]. The guideline outlines a structured approach to ALC development, encompassing phases such as target pollutant identification, data collection and screening, ALC derivation, and technical report compilation. It specifies that test species should be those commonly found in various freshwater ecosystems across China. Data for acute and chronic ecotoxicity of target pollutants are sourced from databases like Web of Science, as well as domestic and international toxicity databases like ECOTOX, and are screened based on stringent criteria.

Four statistical models—normal, log-normal, logistic, and log-logistic—are employed to fit the species sensitivity distribution (SSD) curve. The optimal model is selected based on fitting parameter comparisons. The HC_5 value, fundamental for ALC calculation, is then derived using an optimal model and adjusted with a correction factor to reduce uncertainties in real-world conditions. The factor applied depends on the available toxicity data; a factor of two for 15 species, and a factor of three for 10 to 14 species ensures accurate and relevant ALC calculations for China's specific environmental contexts. Acute data inform the short-term ALC, while chronic data are used for the long-term ALC.

In 2020, following these guidelines, the Ministry of Ecology and Environment of China (MEEC) issued national ALC documents for cadmium [13], ammonia nitrogen [14], and phenol [15]. In 2022, the 2017 guidelines underwent a comprehensive revision, culminating in the release of the updated version (HJ 831-2022). This revised edition incorporates various modifications, including changes in criteria derivation methods, the details of which are elaborated on in a published paper [34].

9. A Review of Published ALC Values for Pollutants in Chinese Aquatic Ecosystems

Over a decade of accelerated research has yielded published ALC values for a range of key pollutants in China, including ammonia nitrogen, metals, pesticides, endocrine disruptors, and emerging contaminants (Table 4). These values serve as valuable criteria for updating China's surface water quality standards.

Table 4. Published ALC values in China.

Chemicals		Short-term ALC (μg/L)	Long-term ALC (μg/L)	References
Ammonia nitrogen		12,000 (20 °C, pH 7.0) (National criteria)	1500 (20 °C, pH 7.0) (National criteria)	[14,35]
Cd(II)		4.2 (hardness = 100 mg/L) (National criteria)	0.23 (hardness = 100 mg/L) (National criteria)	[13]
Zn(II)		48.43	20.01	[36]
Zn(II)		230.6	/	[37]
Pb(II)		90.7 (hardness = 100 mg/L)	2.1 (hardness = 100 mg/L)	[38]
Cr(VI)		45.79	14.22	[36]
Cu(II)		1.391	0.495	[39]
		/	0.87–1.49	[40]
Ag(I)		$e^{1.58\ln H - 8.68}$ *	$e^{1.58\ln H - 10.28}$ *	[41]
As	As(III)	167	42	[42]
	As(V)	384	44	
Chloride		/	187,500	[43]
Phenol		2472	316.2	[44]
Benzene		2651	530.2	[36]
Nitrobenzene		18	1	[45]
Phenanthrene		51.4	18.6	[46]
PAEs		/	0.04–41.9	[47]
Pentachlorophenol		13.21 (pH = 7.8)	1.20 (pH = 7.8)	[48]
Atrazine		/	0.044	[49]
2,4-dichlorophenol		1250	212	[50]
		/	9–44	[51]
2,4,6-trichlorophenol		1010	226	[52]
		/	57	[53]

Table 4. *Cont.*

Chemicals	Short-term ALC (μg/L)	Long-term ALC (μg/L)	References
Dichlorvos	1.33	0.132	[54]
Glyphosate	3350	260	[55]
Malathion	0.100	0.008	[54]
DEET	21,530	520	[56]
Triphenyltin	0.396 (Sn)	0.0056 (Sn)	[57]
PFOS	3780	250	[58]
PFOA	45,540	3520	[58]
Triclosan	9	2	[59]
TBBPA	147.5	12.6	[60]
HBCD	2320	128	[61]
PBDEs	49.2–239	10.3–26.7	[62]
TDCPP	877 (HC_5)	0.03333 (HC_5)	[63]

* H: hardness of water.

In 2020, the Ministry of Ecology and Environment of China (MEEC) officially unveiled national ALC values for cadmium and ammonia nitrogen, marking a significant milestone in China's ALC research landscape.

As China contemplates updates to its surface water quality standards, these published ALC values are poised to make a constructive contribution to the revision process.

10. Future Directions and Emerging Trends in Aquatic Life Water Quality Criteria

10.1. A Milestone in Chinese ALC Research

China has made significant strides in establishing its own ALC technical methodology and publishing national criteria. This progress underscores the remarkable advancements in ALC research within the country. Chinese scholars are actively exploring various facets to further refine the WQC methodology, thereby providing a more robust scientific foundation for future developments.

10.2. Innovations in Methodology

Traditionally, international ALC methodologies have relied on individual-level toxicity data. However, Yang et al. [64] have pioneered a new approach that incorporates molecular toxicology and community-level data. Specifically, they developed an ecological threshold for ammonia nitrogen in Lake Tai based on the response of the lake's phytoplankton community to ammonia concentration changes. As molecular toxicological data continue to grow, researchers are investigating how to integrate this information into ALC development [65].

10.3. Predictive Modeling

Chinese scholars have also focused on predictive modeling to estimate pollutant toxicity. Various models have been developed, including those for heavy metal ecotoxicity [66], endocrine-disrupting compound (EDC) reproductive toxicity [67,68], pesticide ecotoxicity [69], and BTEX substances [70]. These efforts contribute to the enrichment of native Chinese ecotoxicity data and the refinement of the country's ALC methodology.

10.4. Bridging the Gap between WQC and Legal Standards

In China, WQC are viewed as scientifically-derived safety thresholds without legal force, while water quality standards are legally binding and consider economic, technical, and management factors. The challenge lies in translating WQC into actionable water quality standards. Currently, emergency standards, which do not factor in economic costs, are easier to establish. However, creating regular standards remains complex, and no universally accepted approach has been proposed yet.

Given China's vast geographical diversity, there is active exploration into establishing a tiered WQC system, such as a "state-basin-region" ALC system. This would support more nuanced and region-specific water management strategies across China's various basins.

Supplementary Materials: The following supporting information can be downloaded at: https://www.mdpi.com/article/10.3390/toxics11100862/s1, Table S1. Recommended Chinese resident freshwater test organisms for the development of ALC; Table S2. Evaluation criteria for toxicity data in China's aquatic life criteria (ALC) development.

Author Contributions: Conceptualization, C.L. and J.P.; methodology, C.L., J.X. and Q.L.; validation, C.L. and J.P.; formal analysis, H.Z. and Z.G.; investigation, Q.L. and H.Z.; writing—original draft preparation, C.L.; writing—review and editing, C.L. and J.P.; supervision, J.P.; project administration, J.P.; funding acquisition, J.P. All authors have read and agreed to the published version of the manuscript.

Funding: This work was financially supported by the National Natural Science Foundation of China (Grant No. 42376146).

Data Availability Statement: Not applicable.

Conflicts of Interest: The authors declare no conflict of interest.

References

1. Feng, C.; Huang, W.; Qiao, Y.; Liu, D.; Li, H. Research Progress and New Ideas on the Theory and Methodology of Water Quality Criteria for the Protection of Aquatic Organisms. *Toxics* **2023**, *11*, 557. [CrossRef] [PubMed]
2. US Department of the Interior. *Report of the Subcommittee of Water Quality Criteria*; US Department of the Interior: Washington DC, USA, 1968.
3. USEPA. *Guidelines for Deriving Numerical National Water Quality Criteria for the Protection of Aquatic Organisms and Their Uses*; PB85-227049; Office of Research and Development: Washington DC, USA, 1985. Available online: https://www.epa.gov/sites/default/files/2016-02/documents/guidelines-water-quality-criteria.pdf (accessed on 30 October 2010).
4. CCME. A protocol for the derivation of water quality guidelines for the protection of aquatic life 2007. In *Canadian Environmental Quality Guidelines*; CCME: Ottawa, ON, Canada, 2007. Available online: https://ccme.ca/en/res/protocol-for-the-derivation-of-water-quality-guidelines-for-the-protection-of-aquatic-life-2007-en.pdf (accessed on 11 August 2008).
5. RIVM. *Guidance Document on Deriving Environmental Risk Limits in The Netherlands*; National Institute of Public Health and the Environment: Bilthoven, The Netherlands, 2001. Available online: https://www.rivm.nl/bibliotheek/rapporten/601501012.html (accessed on 18 July 2001).
6. ANZECC; ARMCANZ. Australian and New Zealand Guidelines for Fresh and Marine Water Quality. Australian and New Zealand Environment and Conservation Council and Agriculture and Resource Management Council of Australia and New Zealand 2000, Canberra:1-103. Available online: https://www.waterquality.gov.au/sites/default/files/documents/anzecc-armcanz-2000-guidelines-vol1.pdf (accessed on 22 April 2001).
7. ANZG; ASTG. Australian and New Zealand Guidelines for Fresh and Marine Water Quality. Australian and New Zealand Governments and Australian State and Territory Governments, Canberra ACT, Australia. 2018. Available online: https://water.dpie.nsw.gov.au/__data/assets/pdf_file/0006/456909/guidelines-for-fresh-and-marine-quality-water-faqs.pdf (accessed on 7 February 2019).
8. USEPA. *Evaluation Guidelines for Ecological Toxicity Data in the Open Literature*; US Environmental Protection Agency: Washington DC, USA, 2011. Available online: https://www.epa.gov/pesticide-science-and-assessing-pesticide-risks/evaluation-guidelines-ecological-toxicity-data-open (accessed on 16 May 2011).
9. Moermond, C.T.A.; Kase, R.; Korkaric, M.; Ågerstrand, M. CRED: Criteria for reporting and evaluating ecotoxicity data. *Environ. Toxicol. Chem.* **2016**, *35*, 1297–1309. [CrossRef]
10. ANZECC; ARMCANZ. Revised Method for Deriving Australian and New Zealand Water Quality Guideline Values for Toxicants, In "Australian and New Zealand Guidelines for Fresh and Marine Water Quality". 2018. Available online: https://www.waterquality.gov.au/sites/default/files/documents/warne-wqg-derivation2018.pdf (accessed on 30 October 2018).
11. Wu, F.; Meng, W.; Zhao, X.; Li, H.; Zhang, R.; Cao, Y.; Liao, H. China Embarking on Development of its Own National Water Quality Criteria System. *Environ. Sci. Technol.* **2010**, *44*, 7992–7993. [CrossRef]
12. MEEC. Technical Guideline for Deriving Water Quality Criteria for the Protection of Freshwater Aquatic Organisms (HJ 831-2017). Ministry of Ecology and Environment of China. 2017. Available online: https://www.mee.gov.cn/ywgz/fgbz/bz/bzwb/shjbh/xgbzh/201706/W020170612540225076024.pdf (accessed on 12 June 2017).
13. MEEC. Freshwater Aquatic Life Criteria for Cadmium. 2020. Ministry of Ecology and Environment of China (Announcement No. 11 of 2020, MEEC). 2020. Available online: https://www.mee.gov.cn/xxgk2018/xxgk/xxgk01/202003/t20200303_766970.html (accessed on 28 February 2020). (In Chinese)

14. MEEC. Freshwater Aquatic Life Criteria for Ammonia. 2020. Ministry of Ecology and Environment of China (Announcement No. 24 of 2020, MEEC). 2020. Available online: https://www.mee.gov.cn/xxgk2018/xxgk/xxgk01/202004/t20200410_773914.html (accessed on 9 April 2020). (In Chinese)
15. MEEC. Freshwater Aquatic Life Criteria for Phenol 2020. Ministry of Ecology and Environment of China (Announcement No. 70 of 2020, MEEC). 2020. Available online: https://www.mee.gov.cn/xxgk2018/xxgk/xxgk01/202012/t20201224_814675.html (accessed on 24 December 2020). (In Chinese)
16. Yan, Z.; Wang, Y. *Evaluation of SSD for Typical Pollutants in Ambient Water Environment*; Chemical Industry Press: Beijing, China, 2015. (In Chinese)
17. Yan, Z.; Zheng, X.; Jiao, C.; Xiong, X. *Chinese Resident Freshwater Test Organisms for Development of Aquatic Life Criteria*; China Environment Press: Beijing, China, 2020. (In Chinese)
18. Cai, J.; Yan, Z.; He, L.; Wang, W.L.; Liu, Z.-T. Screening of native amphibians for deriving aquatic life criteria. *Res. Environ. Sci.* **2014**, *27*, 349–355. (In Chinese) [CrossRef]
19. Wang, X.N.; Zheng, X.; Yan, Z.-G.; Liu, Z.-T. Screening of native fishes for deriving aquatic life criteria. *Res. Environ. Sci.* **2014**, *27*, 341–348. (In Chinese) [CrossRef]
20. Zheng, X.; Yan, Z.-G.; Wang, X.N.; Liu, Z.-T. Screening of native crustaceans for deriving aquatic life criteria. *Res. Environ. Sci.* **2014**, *27*, 356–364. (In Chinese) [CrossRef]
21. Wang, W.L.; Yan, Z.-G.; Liu, Z.-T.; Zheng, X. Screening of native Annelids and aquatic insects for deriving aquatic life criteria. *Res. Environ. Sci.* **2014**, *27*, 365–372. (In Chinese) [CrossRef]
22. Qin, L.-M.; Zhang, Y.-H.; Cao, Y.; Yan, Z.-G.; Zeng, H.-H.; Liu, Z.-T. Screening native freshwater molluscs for establishing aquatic life criteria. *J. Argo-Environ. Sci.* **2014**, *33*, 1791–1801. (In Chinese)
23. Liu, T.; Zheng, X.; Yan, Z.; Liu, Z. Screening of native aquatic macrophytes for deriving aquatic life criteria. *J. Argo-Environ. Sci.* **2014**, *33*, 2204–2212. (In Chinese)
24. Klimisch, H.J.; Andreae, M.; Tillmann, U. A systematic approach for evaluating the quality of experimental toxicological and ecotoxicological data. *Regul. Toxicol. Pharmacol.* **1997**, *25*, 1–5. [CrossRef]
25. Durda, J.L.; Preziosi, D.V. Data quality evaluation of toxicological studies used to derive ecotoxicological benchmarks. *Hum. Ecol. Risk Assess.* **2000**, *6*, 747–765. [CrossRef]
26. Markich, S.; Warne, M.; Westbury, A.-M.; Roberts, C. A compilation of data on the toxicity of chemicals to species in Australasia. Part 3: Metals. *Australas. J. Ecotoxicol.* **2002**, *8*, 1–72.
27. Schneider, K.; Schwarz, M.; Burkholder, I.; Kopp-Schneider, A.; Edler, L.; Kinsner-Ovaskainen, A.; Hartung, T.; Hoffmann, S. "ToxRTool", a new tool to assess the reliability of toxicological data. *Toxicol. Lett.* **2009**, *189*, 138–144. [CrossRef] [PubMed]
28. Ågerstrand, M.; Breitholtz, M.; Ruden, C. Comparison of four different methods for reliability evaluation of ecotoxicity data: A case study of non-standard test data used in environmental risk assessments of pharmaceutical substances. *Environ. Sci. Eur.* **2011**, *23*, 17. [CrossRef]
29. Brady, D. Evaluation Guidelines for Ecological Toxicity Data in the Open Literature. USEPA. 2011. Available online: https://www.epa.gov/pesticide-science-and-assessing-pesticide-risks/evaluation-guidelines-ecological-toxicity-data-open (accessed on 16 May 2011).
30. Warne, M.; Batley, G.E.; Van Dam, R.A.; Chapman, J.C.; Fox, D.R.; Hickey, C.W.; Stauber, J.L. *Revised Method for Deriving Australian and New Zealand Water Quality Guideline Values for Toxicants*; Department of Science, Information Technology and Innovation: Brisbane, Australia, 2018. [CrossRef]
31. Vlaardingen, P.L.A.v.; Verbruggen, E.M.J. Guidance for the Derivation of Environmental Risk Limits within the Framework of 'International and National Environmental Quality Standards for Substances in the Netherlands' (INS). RIVM Report 601782001/2007. 2007. Available online: https://www.rivm.nl/bibliotheek/rapporten/601501012.html (accessed on 13 November 2007).
32. Crane, M.; Newman, M.C. What levels of effects is a no observed effect? *Environ. Toxicol. Chem.* **2000**, *19*, 516–519. [CrossRef]
33. Feng, C.; Li, H.; Yan, Z.; Wang, Y.; Wang, C.; Fu, Z.; Liao, W.; Giesy, J.P.; Bai, Y. Technical study on national mandatory guideline for deriving water quality criteria for the protection of freshwater aquatic organisms in China. *J. Environ. Manag.* **2019**, *250*, 109539. [CrossRef]
34. Yan, Z.-G.; Zheng, X.; Zhang, Y.-Z.; Yang, Z.-H.; Zhou, Q.; Men, S.-H.; Du, J.-Z. Chinese Technical Guideline for Deriving Water Quality Criteria for Protection of Freshwater Organisms. *Toxics* **2023**, *11*, 194. [CrossRef]
35. Yan, Z.; Zheng, X.; Fan, J.; Zhang, Y.; Wang, S.; Zhang, T.; Sun, Q.; Huang, Y. China national water quality criteria for the protection of freshwater life: Ammonia. *Chemosphere* **2020**, *251*, 126379. [CrossRef]
36. Wu, F.; Feng, C.; Zhang, R.; Li, Y.; Du, D. Derivation of water quality criteria for representative water-body pollutants in China. *Sci. China Earth Sci.* **2012**, *55*, 882–891. [CrossRef]
37. Li, X.F.; Wang, P.F.; Feng, C.L.; Liu, D.Q.; Chen, J.K.; Wu, F.C. Acute toxicity and hazardous concentrations of zinc to native freshwater organisms under different pH values in China. *Bull. Environ. Contam. Toxicol.* **2019**, *103*, 120–126. [CrossRef]
38. Liang, W.; Wang, X.; Zhang, X.; Niu, L.; Wang, J.; Wang, X.; Zhao, X. Water quality criteria and ecological risk assessment of lead (Pb) in China considering the total hardness of surface water: A national-scale study. *Sci. Total Environ.* **2023**, *858 Pt 1*, 159554. [CrossRef]

39. Zhang, Y.; Zang, W.; Qin, L.; Zheng, L.; Cao, Y.; Yan, Z.; Yi, X.; Zeng, H.; Liu, Z. Water quality criteria for copper based on the BLM approach in the freshwater in China. *PLoS ONE* **2017**, *12*, e0170105. [CrossRef]
40. Cui, L.; Wang, Y.; Zhang, H.; Lv, X.; Lei, K. Use of non-linear multiple regression models for setting water quality criteria for copper: Consider the effects of salinity and dissolved organic carbon. *J. Hazard. Mater.* **2023**, *450*, 131107. [CrossRef]
41. Jin, Q.; Feng, C.; Xia, P.; Bai, Y. Hardness-dependent water quality criteria for protection of freshwater aquatic organisms for silver in China. *Int. J. Environ. Res. Public Health* **2022**, *19*, 6067. [CrossRef] [PubMed]
42. Zheng, L.; Liu, Z.; Yan, Z.; Yi, X.; Zhang, J.; Zhang, Y.; Zheng, X.; Zhu, Y. Deriving water quality criteria for trivalent and pentavalent arsenic. *Sci. Total Environ.* **2017**, *587–588*, 68–74. [CrossRef] [PubMed]
43. Hong, Y.; Zhu, Z.; Liao, W.; Yan, Z.; Feng, C.; Xu, D. Freshwater water-quality criteria for chloride and guidance for the revision of the water-quality standard in China. *Int. J. Environ. Res. Public Health* **2023**, *20*, 2875. [CrossRef]
44. MEEC. *Freshwater Aquatic Life Criteria for Phenol 2020 (Draft)*; Ministry of Ecology and Environment of China: Beijing, China, 2020. (In Chinese)
45. Yan, Z.-G.; Zhang, Z.-S.; Wang, H.; Liang, F.; Li, J.; Liu, H.-L.; Sun, C.; Liang, L.-J.; Liu, Z.-T. Development of aquatic life criteria for nitrobenzene in China. *Environ. Pollut.* **2012**, *162*, 86–90. [CrossRef] [PubMed]
46. Wu, J.-Y.; Yan, Z.-G.; Liu, Z.-T.; Liu, J.-d.; Liang, F.; Wang, X.-N.; Wang, W.-L. Development of water quality criteria for phenanthrene and comparison of the sensitivity between native and non-native species. *Environ. Pollut.* **2015**, *196*, 141–146. [CrossRef]
47. Zheng, X.; Yan, Z.; Liu, P.; Li, H.; Zhou, J.; Wang, Y.; Fan, J.; Liu, Z. Derivation of aquatic life criteria for four phthalate esters and their ecological risk assessment in Liao River. *Chemosphere* **2019**, *220*, 802–810. [CrossRef] [PubMed]
48. Zheng, L.; Liu, Z.; Yan, Z.; Zhang, Y.; Yi, X.; Zhang, J.; Zheng, X.; Zhou, J.; Zhu, Y. pH-dependent ecological risk assessment of pentachlorophenol in Taihu Lake and Liaohe River. *Ecotoxicol. Environ. Saf.* **2017**, *135*, 216–224. [CrossRef]
49. Zheng, L.; Zhang, Y.; Yan, Z.; Zhang, J.; Li, L.; Zhu, Y.; Zhang, Y.; Zheng, X.; Wu, J.; Liu, Z. Derivation of predicted no-effect concentration and ecological risk for atrazine better based on reproductive fitness. *Ecotoxicol. Environ. Saf.* **2017**, *142*, 464–470. [CrossRef]
50. Yin, D.; Jin, H.; Yu, L.; Hu, S. Deriving freshwater quality criteria for 2,4-dichlorophenol for protection of aquatic life in China. *Environ. Pollut.* **2003**, *122*, 217–222. [CrossRef] [PubMed]
51. Jin, X.; Zha, J.; Xu, Y.; Wang, Z.; Kumaran, S.S. Derivation of aquatic predicted no-effect concentration (PNEC) for 2,4-dichlorophenol: Comparing native species data with non-native species data. *Chemosphere* **2011**, *84*, 1506–1511. [CrossRef] [PubMed]
52. Yin, D.; Hu, S.; Jin, H.; Yu, L. Deriving freshwater quality criteria for 2,4,6-trichlorophenol for protection of aquatic life in China. *Chemosphere* **2003**, *52*, 67–73. [CrossRef] [PubMed]
53. Jin, X.; Zha, J.; Xu, Y.; Giesy, J.P.; Richardson, K.L.; Wang, Z. Derivation of predicted no effect concentrations (PNEC) for 2,4,6-trichlorophenol based on Chinese resident species. *Chemosphere* **2012**, *86*, 17–23. [CrossRef]
54. Ding, T.-t.; Zhang, Y.-h.; Zhu, Y.; Du, S.-L.; Zhang, J.; Cao, Y.; Wang, Y.-Z.; Wang, G.-T.; He, L.-S. Deriving water quality criteria for China for the organophosphorus pesticides dichlorvos and malathion. *Environ. Sci. Pollut. Res. Int.* **2019**, *26*, 34622–34632. [CrossRef]
55. Fan, Y.; Zhang, K.; Zhang, R.; Guo, G.; Li, H.; Bai, Y.; Lin, Y.; Cai, T. Derivation of water quality criteria for glyphosate and its formulations to protect aquatic life in China. *Environ. Sci. Pollut. Res. Int.* **2022**, *29*, 51860–51870. [CrossRef]
56. Gao, X.; Wang, X.; Li, J.; Ai, S.; Fu, X.; Fan, B.; Li, W.; Liu, Z. Aquatic life criteria derivation and ecological risk assessment of DEET in China. *Ecotoxicol. Environ. Saf.* **2020**, *188*, 109881. [CrossRef]
57. Wen, J.; Cui, X.; Gibson, M.; Li, Z. Water quality criteria derivation and ecological risk assessment for triphenyltin in China. *Ecotoxicol. Environ. Saf.* **2018**, *161*, 397–401. [CrossRef]
58. Yang, S.; Xu, F.; Wu, F.; Wang, S.; Zheng, B. Development of PFOS and PFOA criteria for the protection of freshwater aquatic life in China. *Sci. Total Environ.* **2014**, *470–471*, 677–683. [CrossRef]
59. Wang, X.-N.; Liu, Z.-T.; Yan, Z.-G.; Zhang, C.; Wang, W.-L.; Zhou, J.-L.; Pei, S.-W. Development of aquatic life criteria for triclosan and comparison of the sensitivity between native and non-native species. *J. Hazard. Mater.* **2013**, *260*, 1017–1022. [CrossRef] [PubMed]
60. Yang, S.-W.; Yan, Z.-G.; Xu, F.-F.; Wang, S.-R.; Wu, F.-C. Development of freshwater aquatic life criteria for tetrabromobisphenol A in China. *Environ. Pollut.* **2012**, *169*, 59–63. [CrossRef] [PubMed]
61. Dong, L.; Zheng, L.; Yang, S.; Yan, Z.; Jin, W.; Yan, Y. Deriving freshwater safety thresholds for hexabromocyclododecane and comparison of toxicity of brominatedflame retardants. *Ecotoxicol. Environ. Saf.* **2017**, *139*, 43–49. [CrossRef] [PubMed]
62. Lu, C.; Yang, S.; Yan, Z.; Ling, J.; Jiao, L.; He, H.; Zheng, X.; Jin, W.; Fan, J. Deriving aquatic life criteria for PBDEs in China and comparison of species sensitivity distribution with TBBPA and HBCD. *Sci. Total Environ.* **2018**, *640–641*, 1279–1285. [CrossRef]
63. Liu, D.; Yan, Z.; Liao, W.; Bai, Y.; Feng, C. The toxicity effects and mechanisms of tris(1,3-dichloro-2-propyl) phosphate (TDCPP) and its ecological risk assessment for the protection of freshwater organisms. *Environ. Pollut.* **2020**, *264*, 114788. [CrossRef]
64. Yang, J.; Zhang, X.; Xie, Y.; Song, C.; Sun, J.; Zhang, Y.; Giesy, J.P.; Yu, H. Ecogenomics of zooplankton community reveals ecological threshold of ammonia nitrogen. *Environ. Sci. Technol.* **2017**, *51*, 3057–3064. [CrossRef]
65. Yan, Z.; Yang, N.; Wang, X.; Wang, W.; Meng, S.; Liu, Z. Preliminary analysis of species sensitivity distribution based on gene expression effect. *Sci. China Earth Sci.* **2012**, *55*, 907–913. [CrossRef]

66. Wu, F.; Mu, Y.; Chang, H.; Zhao, X.; Giesy, J.P.; Wu, K.B. Predicting water quality criteria for protecting aquatic life from physicochemical properties of metals or metalloids. *Environ. Sci. Technol.* **2013**, *47*, 446–453. [CrossRef]
67. Fan, J.; Yan, Z.; Zheng, X.; Wu, J.; Wang, S.; Wang, P.; Zhang, Q. Development of interspecies correlation estimation (ICE) models to predict the reproduction toxicity of EDCs to aquatic species. *Chemosphere* **2019**, *224*, 833–839. [CrossRef]
68. Feng, C.L.; Wu, F.C.; Dyer, S.D.; Chang, H.; Zhao, X.L. Derivation of freshwater quality criteria for zinc using interspecies correlation estimation models to protect aquatic life in China. *Chemosphere* **2013**, *90*, 1177–1183. [CrossRef]
69. He, J.; He, H.; Yan, Z.; Gao, F.; Zheng, X.; Fan, J.; Wang, Y. Comparative analysis of freshwater species sensitivity distributions and ecotoxicity for priority pesticides: Implications for water quality criteria. *Ecotoxicol. Environ. Saf.* **2019**, *176*, 119–124. [CrossRef] [PubMed]
70. Xu, J.; Zheng, L.; Yan, Z.; Huang, Y.; Feng, C.; Li, L.; Ling, J. Effective extrapolation models for ecotoxicity of benzene, toluene, ethylbenzene, and xylene (BTEX). *Chemosphere* **2020**, *240*, 124906. [CrossRef] [PubMed]

toxics

MDPI

Article

Effects of BPA Exposure and Recovery on the Expression of Genes Involved in the Hepatic Lipid Metabolism in Male Mice

Changqing Li †, Nan Shen †, Shaohua Yang * and Hui-Li Wang *

College of Food and Biological Engineering, Hefei University of Technology, Hefei 230009, China; sn99hfut@163.com (N.S.)
* Correspondence: yangshaohua@hfut.edu.cn (S.Y.); wanghl@hfut.edu.cn (H.-L.W.)
† These authors contributed equally to this work.

Abstract: Exposure to Bisphenol A (BPA) has led to an increased risk of obesity and nonalcoholic fatty liver diseases (NAFLDs). However, it is as yet unclear if the damage caused by BPA is able to be repaired sufficiently after exposure has ceased. Therefore, this project aims to investigate the effects of BPA on the hepatic lipid metabolism function and its potential mechanisms in mice by comparing the BPA exposure model and the BPA exposure + cessation of drug treatment model. Herein, the male C57BL/6 mice were exposed in the dose of 50 μg/kg/day and 500 μg/kg/day BPA for 8 weeks, and then transferred to a standard chow diet for another 8 weeks to recover. Based on our previous RNA-seq study, we examined the expression patterns of some key genes. The results showed that the mice exposed to BPA manifested NAFLD features. Importantly, we also found that there was a significant expression reversion for *SCD1*, *APOD*, *ANGPT4*, *PPARβ*, *LPL* and *G0S2* between the exposure and recovery groups, especially for *SCD1* and *APOD* ($p < 0.01$). Notably, BPA could significantly decrease the level of *APOD* protein ($p < 0.01$) whereas there was an extremely significant increase after the exposure ceased. Meanwhile, *APOD* over-expression suppressed TG accumulation in the AML12 cells. In conclusion, the damage caused by BPA is able to be repaired by the upregulation of *APOD* and exposure to BPA should be carefully examined in chronic liver metabolic disorders or diseases.

Keywords: BPA; lipid accumulation; *APOD*; lipid homeostasis

Citation: Li, C.; Shen, N.; Yang, S.; Wang, H.-L. Effects of BPA Exposure and Recovery on the Expression of Genes Involved in the Hepatic Lipid Metabolism in Male Mice. *Toxics* **2023**, *11*, 775. https://doi.org/10.3390/toxics11090775

Academic Editors: Zhen-Guang Yan, Zhi-Gang Li, Jinzhe Du and Sunmi Kim

Received: 10 July 2023
Revised: 29 August 2023
Accepted: 29 August 2023
Published: 12 September 2023

1. Introduction

Nonalcoholic fatty liver disease (NAFLD) is the most common liver disease and the incidence of NAFLD has sharply increased since the 1990s. It has now become a major public health concern worldwide [1]. A significant proportion of patients with NAFLD may progress to nonalcoholic steatohepatitis (NASH), which is associated with disease progression, such as fibrosis, cirrhosis, or hepatocellular carcinoma [2]. The causes of epidemic NAFLD still remain unclear and it is evident that environmental factors also play an important role in NAFLD [3]. Environmental exposures, including but not limited to insecticides, pesticides, and polychlorinated biphenyls, can increase the risk of developing NAFLD and the reason for this is that these exposures are potential fat metabolic modifiers in the liver [2].

Bisphenol A (BPA) is one of the well-known endocrine-disrupting chemicals (EDCs) [4] and it is now accepted as a factor contributing to the increasing incidence of obesity and metabolic diseases, including NAFLD, insulin resistance, type 2 diabetes, and dyslipidemia [5,6]. Several studies have indicated that metabolic disorders were observed in later life when they were exposed to BPA during the critical period of development [7,8]. Perinatal exposure to BPA leads to disruption of pathways related to adipogenesis and results in increased fat mass and body weight [9,10]. Consistent with these findings, we also observed that an early-life BPA exposure resulted in a higher body weight and fat percentage, a greater fat mass of white adipocytes and displayed a NAFLD-like phenotype in male C57BL/6 mice in our previous study [11]. Therefore, BPA could be considered as an important risk factor related to the progression of obesity and NAFLD.

Until now, the mechanism of BPA-mediated adipogenesis and NAFLD has not been thoroughly understood. Furthermore, it is as yet unclear if the damage caused by BPA is able to be repaired sufficiently for a return to normal levels. Therefore, we further investigated the changes in a series of key genes involved in TG and lipid metabolism detected by our previous RNA-seq analysis [11] during BPA exposure and post-exposure recovery. This provides a favorable basis for the potential mechanism that BPA exposure leads to metabolic diseases and can also be applied to an understanding of more general mechanisms contributing to hepatic steatosis. Therefore, the present study first investigated the effects of drug treatment and drug withdrawal on metabolic phenotypes in mice; Second, the effects of BPA on the expression of genes and proteins related to lipid metabolism in mouse livers under different treatment conditions were determined by Q-PCR and Western blot (WB); Finally, molecular interference is used to validate the function of the relevant molecules in vitro. The aim is to identify potential molecular therapeutic targets for BPA-induced hepatic lipid metabolism disorders.

2. Materials and Methods

2.1. Chemicals and Materials

Bisphenol A (purity $\geq$ 99%), ethanol and dimethyl sulfoxide (DMSO) were purchased from Sigma-Aldrich (Sigma, St. Louis, MO, USA). Triglyceride (TG) quantification, oil red O staining, liver TG and serum TG, total cholesterol (TCHO), high-density lipoprotein cholesterol (HDL-C) and low-density lipoprotein cholesterol (LDL-C) kits were supplied by the Jiancheng Bioengineering Institute (Nanjing, Nanjing, China). Trizol isolation kits were purchased from Invitrogen (Carlsbad, CA, USA). The total proteins lysis buffer, PVDF membrane and a BCA protein detection kit were provided by TIANGEN (Beijing, China). Primary polyclonal antibody APOD (ab236868, Abcam, Cambridge, MA, USA, Dilution ratio: 1/1000) and SCD1 (ab236868, Abcam, Cambridge, MA, USA) secondary antibody were purchased from Abcam (ab6721, Cambridge, MA, USA, Dilution ratio:1/1000). The other analytical reagents were sourced from the Servicebio Technology Co., Ltd. (Wuhan, Hubei, China), if not specified.

2.2. BPA Exposure and Recovery

Animal care and management were approved by the Animal Care and Use Committee of the Hefei University of Technology (approval number HFUT20210413002). Adult (2-month-old) male and female C57BL/6 mice were purchased from the Charles River. Mice were maintained under a constant temperature of 25 $\pm$ 2 °C and relative humidity of 60 $\pm$ 5% under a 12 h light/dark cycle. As perinatal mice cannot be administered by gavage, therefore after childbirth and during breastfeeding the mother mice were exposed to DMSO and BPA with doses of 50 or 500 μg/kg BW/day separately by gavage. Subsequently, all 3-week-old mice were randomly divided intofour groups (eight mice per group), (I) exposure group: vehicle control group (control), treated with DMSO as negative control for 5 weeks; (II) exposure group: BPA exposure with dose of 50(L-BPA) or 500 μg/kg BW/day (H-BPA) for 5 weeks; (III) recovery group: vehicle control group treated with DMSO as negative control for 5 weeks + maintained with a standard chow diet for another 8 weeks to recover (R-control); recovery group: BPA exposure with doses of 50 or 500 μg/kg BW/day (BPA) for 5 weeks + maintained with a standard chow diet for another 8 weeks to recover (R-BPA). The above administration is by gavage. The maximum exposure dose for this topic meets the Food and Drug Administration (FDA) safety limits for BPA doses.

2.3. Cell Culture and Treatment

Normal mouse hepatocytes (alpha mouse liver 12) AML12 were cultured in a high glucose DMEM basic medium containing 10% FBS (Gibco, Waltham, MA, USA) and a solution of 1% penicillin and streptomycin (PS) (Sigma, St. Louis, MO, USA). Cells were treated with DMSO and BPA at a concentration of 30 μM for two days, respectively. Plasmid transfection was then performed using Lip3000 liposome transfection reagent (Sigma, St. Louis, MO, USA). After two days of continued drug treatment, the following experimental manipulations were performed.

2.4. Measurements

Body weight and food intake of the animals were measured weekly. At the end of the study, mice were sacrificed by CO_2 inhalation after overnight fasting. The body weights of the mice were counted after fasting, and the liver tissue and epididymal adipose tissue were collected and counted separately. Fat to body weight is the ratio of the absolute weight of epididymal fat to body weight. Blood was collected from the heart, and serum was obtained by centrifugation at 3000 rpm for 20 min for subsequent analysis. Serum total cholesterol (TCHO), triglyceride (TG), high-density lipoprotein cholesterol (HDL-C), low-density lipoprotein cholesterol (LDL-C), alanine amioTransferase (ALT) and aspartate Transaminase (AST) were measured by using biochemical assay kits provided by the Jiancheng Bioengineering Institute (Nanjing, China). All assays were performed according to the instructions supplied by the manufacturer.

2.5. Histological Analysis

Freshly harvested adipose tissue of the epididymis and liver was collected, then washed with phosphate-buffered saline (PBS), and the tissues were immediately soaked in 4% of paraformaldehyde for 24 h. After dehydration, the specimens were intactly embedded in paraffin and sectioned (5 μm) for H&E staining. The morphological changes in all the slices in the liver and adipose tissue were examined with a high-resolution CS2 slide scanner (Leica Biosystems Inc., Buffalo Grove, IL, USA).

2.6. RT-qPCR Analysis

Total RNA Isolation Reagent was used to extract total RNA from liver tissues and reverse-transcribed into cDNA by a reverse transcription kit (TransGen, Beijing, China). The key genes were amplified using a SYBR-Green kit following the manufacturer's instruction on a LightCycler480 (Roche, Shanghai, China). The amplification procedure was as follows: 95 °C for 8 min; 95 °C for 10 s, 60 °C for 20 s, 72 °C for 20 s (36 cycles), and final extension at 4 °C. All of the primers for the genes are shown in the Supplementary Table S1. The relative expression levels of genes were normalized using *18S* and the fold changes were analyzed using the $2^{-\Delta\Delta ct}$ method.

2.7. Western Blot Analyses

Total proteins of livers in different groups were harvested using RIPA buffer (Sigma-Aldrich, St. Louis, MO, USA) supplemented with a protease/phosphatase inhibitor (Sigma-Aldrich, St. Louis, MO, USA). Protein concentrations were then determined with the BCA protein detection kit (Thermo Fisher Scientific, Beijing, China). A 20 μg sample of protein was separated by 10% SDS-PAGE gels and then electrically transferred onto PVDF membranes (Millipore, Boston, MA, USA). The membranes were then blocked for 2 h with solutions containing 5% of non-fat milk. After washing, the membranes were incubated with a primary antibody (Abcam, Cambridge, MA, USA) overnight at 4 °C and then incubated with a labeled secondary antibody (Boster, Pleasanton, CA, USA) for 1 h at room temperature. Protein bands were visualized via Image J software, version 6.1 (Bio-Rad, Hercules, CA, USA). The *GAPDH* level was used as a loading control.

2.8. Statistical Analysis

Data shown were expressed as mean ± standard deviation of all independent experiments (number of parallel ≥ 3) and one-way analysis of variance (ANOVA) with Duncan's multiple range tests were used to compare the different treatment mice groups through

SPSS 19.0 software (SPSS Inc., Chicago, IL, USA). A p value < 0.05 was considered to be statistically significant.

3. Results

3.1. BPA Exposure Induced Obesity in Mice

The mice from the control and BPA groups were sacrificed at the age of 8 weeks (BPA exposure) and at the age of 16 weeks (BPA exposure for 8 weeks + recovery for 8 weeks), respectively. During the exposure period, compared with the control group, the body weight ($p < 0.05$) and fat-to-body weight ratio ($p < 0.01$) of L-BPA and H-BPA groups were significantly increased. During recovery, we found that there was a significant decrease in the body weight and fat-to-body weight ratio of the male mice between the BPA group and the control group ($p < 0.05$) (Figure 1A,B). Furthermore, as shown in Figure 1C, adipocytes had a clearly larger size and volume both for BPA exposure and recovery groups than those in the vehicle control group. Interestingly, the size and volume of adipocytes were still larger even if at the recovery time (Figure 1D).

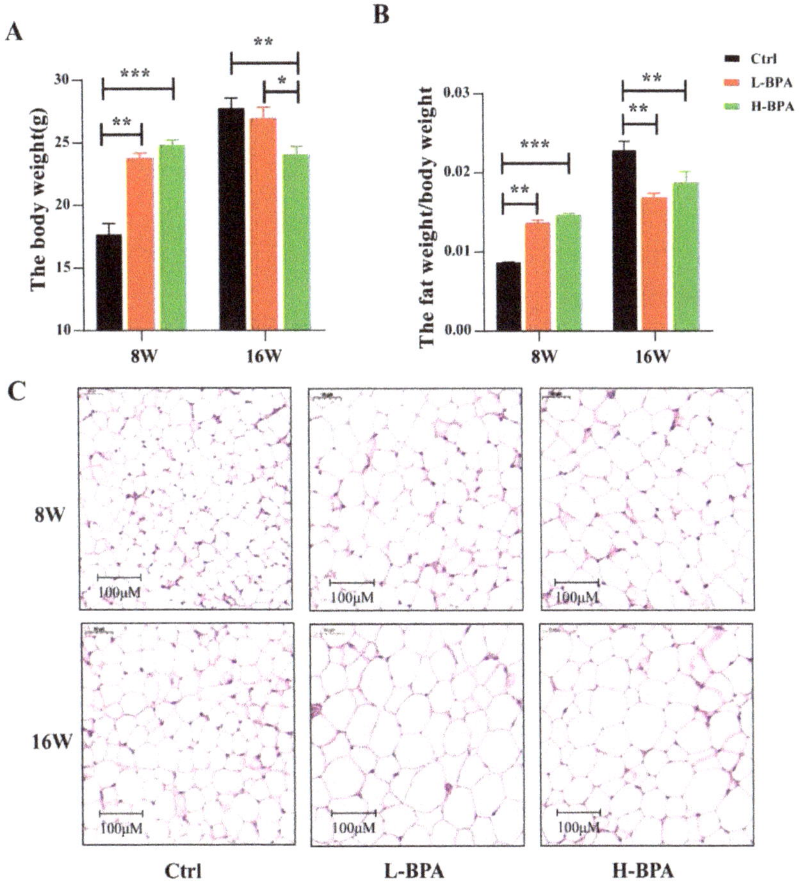

Figure 1. *Cont.*

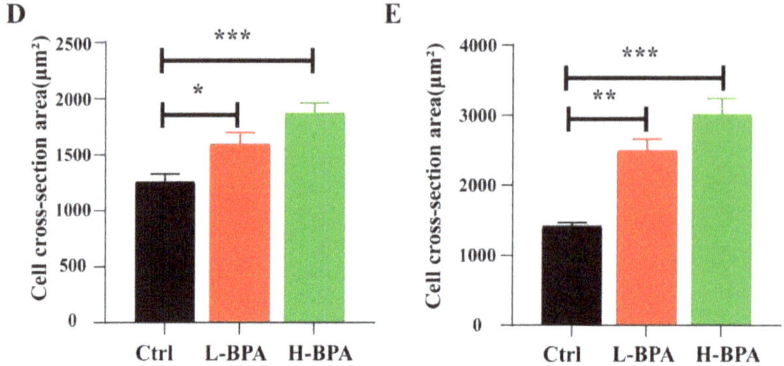

Figure 1. Effects of BPA exposure on the basic changes of parameters in male mice. Note: (**A**) Effects of BPA exposure on the body weight of mice; (**B**) Effects of BPA exposure on fat-to-body weight ratio; (**C**) Representative images of adipose tissue stained with H&E in mice for 8 weeks and 16 weeks (400×); (**D**) Relative area statistics of cells in the adipose tissue of the epididymis for 8 weeks. (**E**) Relative area statistics of cells in the adipose tissue of the epididymis for 16 weeks. Use ImageJ software (Bio-Rad, Hercules, CA, USA). $n = 8$ in each group. "8W" represents the control group treated with 2% of DMSO and the BPA group treated with BPA for 8 weeks, respectively. "16W" represents the control group treated with 2% of DMSO for 16 weeks and the BPA group treated with BPA for 8 weeks followed by an 8-week recovery period (cessation of drug exposure). Statistical significance was determined by one-way ANOVA. * Represents the significance at $p < 0.05$. ** Represents the significance at $p < 0.01$. *** Represents the significance at $p < 0.001$. L-BPA, 50 μg/kg BW/day BPA; H-BPA, 500 μg/kg BW/day BPA.

3.2. BPA Exposure Altered the Homeostasis of Metabolic Outcomes in Mice

After BPA treatment for 8 weeks, compared with the corresponding control groups, serum HDL, LDL, TCHO and TG levels were not obviously reduced in the 50 μg/kg BW/day and 500 μg/kg BW/day-exposed groups, whereas serum LDL and TG levels were only markedly decreased in the 500 μg/kg BW/day exposed group (Figure 2A–D). Importantly, as shown in Figure S2, compared with the corresponding control groups, serum APOD levels for 8 weeks were higher in the BPA exposure, whereas their levels decreased rapidly for another 8 weeks only to recover.

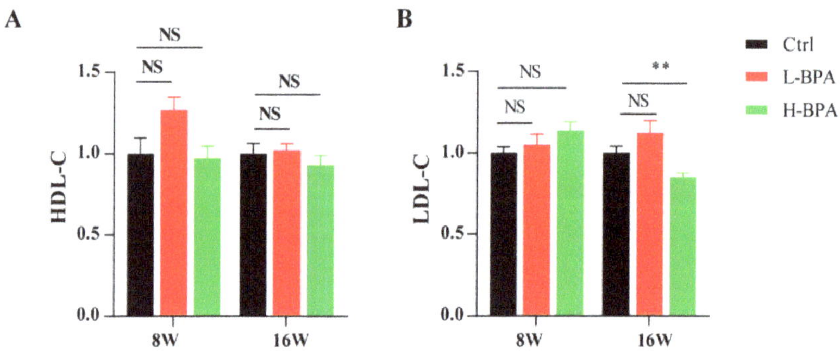

Figure 2. *Cont.*

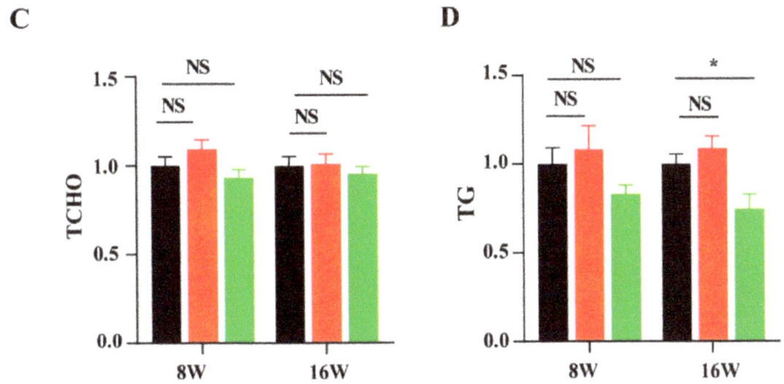

Figure 2. The blood biochemical indicators of mice. Note: (**A**) Relative serum high density lipoprotein cholesterol (HDL-C) levels; (**B**) Relative serum low density lipoprotein cholesterol (LDL-C) levels; (**C**) Relative serum total cholesterol (TCHO) levels; (**D**) Relative serum triglyceride (TG) levels. $n = 6$ in each group. Statistical significance was determined by one-way ANOVA. "8W" represents the control group treated with 2% DMSO and the BPA group treated with BPA for eight weeks, respectively. "16W" represents the control group treated with 2% DMSO for 16 weeks and the BPA group treated with BPA for eight weeks followed by an eight-week recovery period (cessation of drug exposure). * Represents the significance at $p < 0.05$. ** Represents the significance at $p < 0.01$. NS represents the significance at $p > 0.05$.

3.3. BPA Exposure Disrupted Hepatic Lipid Metabolism in Mice

3.3.1. BPA Exposure Induced Liver Injury

The relative liver weights were markedly decreased in male mice after exposure to 500 μg/kg BW/day compared with the control (Figure 3A); during the recovery period, we also found that the high dose group resulted in a significant reduction in the liver weights of the mice (Figure 3B). The data of H&E-stained liver sections revealed normal liver histology in the control group manifesting as normal cell size, with uniform cytoplasm, and a prominent cell nucleus. Meanwhile, liver tissue sections failed to show evidence of lipid droplets or other aberrant changes, such as degeneration or necrosis (Figure 3C). Conversely, the mice exposed to BPA fed on the same diet showed liver pathologies seen in human NAFLDs, including small vacuoles, disordered hepatic cell cords, and increased fat deposition. For the recovery groups, an evident injury was observed in the livers of the mice in both the L-BPA and H-BPA groups. But notably, fat in the liver was mainly observed as macrovesicular droplets in the mice exposed to low-dose BPA (Figure 3D). The quantitative TG assay showed that BPA exposure resulted in a significant increase in TG levels in mouse livers in both the L-BPA and H-BPA groups (Figure 3E). However, this increase in attenuated during recovery (Figure 3F).

3.3.2. BPA Exposure Altered the Genes Expression in Male Mice

Our previous study used RNA-Seq to explore potential mechanisms of BPA-induced adipogenesis in 3T3-L1 preadipocytes and a series of genes associated with de novo lipogenesis and lipid transport were detected to be regulated by BPA [11]. Based on fold change, here we examined the expression patterns of genes *APOD*, *SCD1*, *ANGPT4*, *LPL*, *G0S2*, *FADS2*, *GNAI2*, *PLIN1*, *ELOVL6*, *ACSL3*, *PPARα*, *PPARβ*, *PPARγ*, *FADS1* and *SOD3* in the livers of mice for BPA exposure and recovery groups. Importantly, we also found a significant expression reversion for stearoyl-CoA desaturase 1 (*SCD1*) ($p < 0.01$), apolipoprotein D (*APOD*) ($p < 0.01$), *ANGPT4* ($p < 0.01$), *PPARβ* ($p < 0.05$), *LPL* ($p < 0.01$) and *G0S2* ($p < 0.01$) between the exposure and recovery groups ($p < 0.05$), especially for *SCD1* and *APOD* (Figure 4A). Meanwhile, as shown in Figure S1, BPA can also decrease the expression level of *PPARα*, which was responsible for fatty acid oxidation. Besides, BPA can also increase mRNA expression levels of genes for lipogenesis, such as *FADS1*, *FADS2*, *LPL*, *G0S2*, and *ACSL3*.

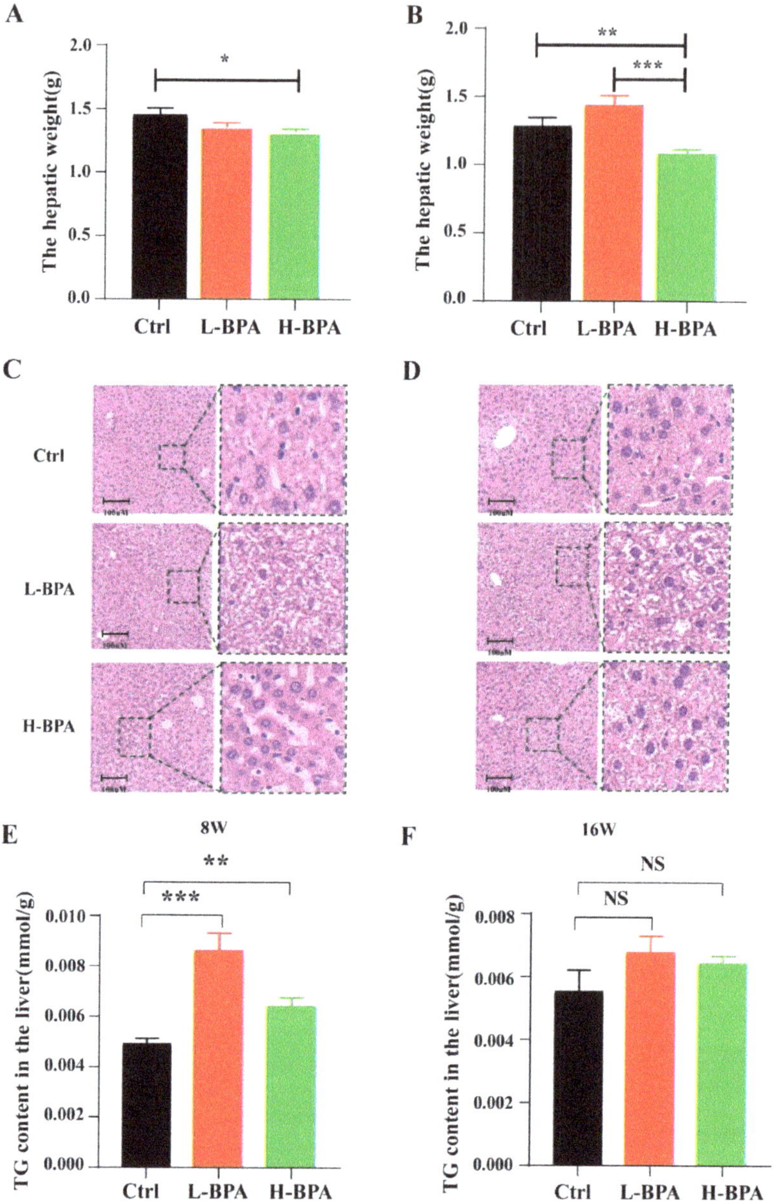

Figure 3. BPA-exposed mice exhibited hepatic lipid accumulation. Note: (**A**) Eight-week mouse liver weights; (**B**) Sixteen-week mouse liver weights; $n = 8$ in each group; (**C**,**D**) Representative images of liver with H&E in mice for 8 weeks and 16 weeks (400×). (**E**,**F**) TG content in mice liver in 8w and 16w groups, respectively. "8W" represents the control group treated with 2% DMSO and the BPA group treated with BPA for eight weeks, respectively. "16W" represents the control group treated with 2% DMSO for 16 weeks and the BPA group treated with BPA for eight weeks followed by an eight-week recovery period (cessation of drug exposure). Statistical significance was determined by one-way ANOVA. * Represents the significance at $p < 0.05$. ** Represents the significance at $p < 0.01$. *** Represents the significance at $p < 0.001$. NS represents the significance at $p > 0.05$.

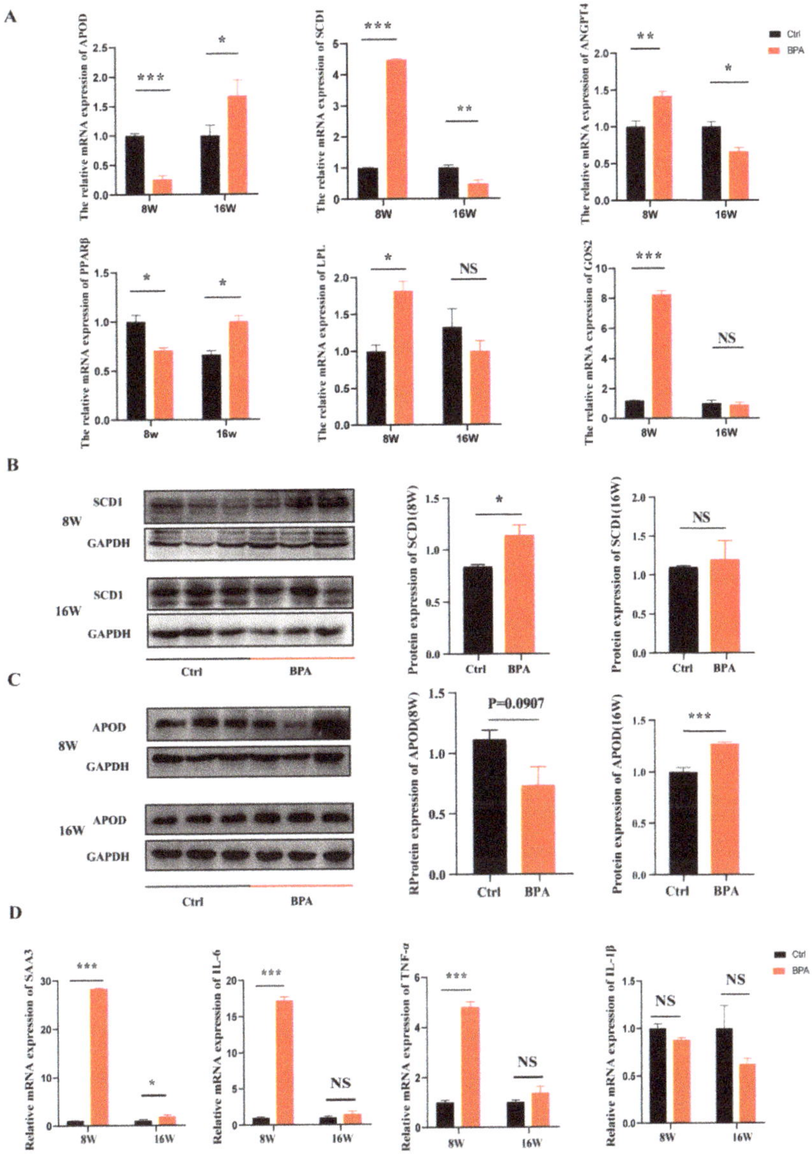

Figure 4. Effects of BPA Exposure on the of Gene Expressions in Liver. Note: (**A**) Effects of BPA on the mRNA expression of genes regulating lipid metabolism. $n = 5$ in each group; (**B**) Expression of SCD1 protein in mouse liver; (**C**) Expression of APOD protein in mouse liver; Quantitative analysis of protein expression was performed using the ImageJ software (Bio-Rad, Hercules, CA, USA); (**D**) Effects of BPA on the mRNA expression of genes regulating inflammatory, $n = 5$ in each group. "8W" represents the control group treated with 2% DMSO and the BPA group treated with BPA for eight weeks, respectively. "16W" represents the control group treated with 2% DMSO for 16 weeks and the BPA group treated with BPA for eight weeks followed by an eight-week recovery period (cessation of drug exposure). * Represents the significance at $p < 0.05$. ** Represents the significance at $p < 0.01$. *** Represents the significance at $p < 0.001$. NS represents the significance at $p > 0.05$.

3.3.3. BPA Exposure Induced Hepatic Inflammation

In comparison to the control group, the expression levels of liver *TNF-α, IL-6* had an obvious increase in the animals exposed to BPA ($p < 0.01$) (Figure 4D). Whereas, the level of *IL-1β* were not increased significantly ($p > 0.05$). After 8 weeks of recovery, the degree of damage gradually decreased. As shown in Figure. 4D, during the recovery period, the expression levels of these genes returned to their initial levels except for *SAA3* ($p < 0.05$).

3.3.4. Effects of BPA Exposure and Recovery on the Expression of SCD1 and APOD

The expression of the proteins (*SCD1*, and *APOD*) were further determined by western blotting for the exposure and recovery groups. The results indicated that the levels of *SCD1* had a significant increase in BPA groups ($p < 0.05$). However, during recovery, there was no significant difference in the *SCD1* protein levels between the BPA group and control group ($p > 0.05$) (Figure 4B). Notably, BPA exposure groups had a lower level of APOD protein level ($p = 0.097$) and accordingly, there was an extremely significant increase after stopping exposure (Figure 4C).

3.4. Effect of APOD Over-Expression on BPA-Induced Dysregulation of Lipid Homeostasis

To assess the potential role of *APOD* in BPA-induced lipid disorders, we investigated whether BPA could increase fat accumulation by regulation of *APOD* in AML12 cells. As shown in Figure 5D,E, TG quantitative assay and oil red O staining results demonstrated that BPA exposure increased TG content, while *APOD* over-expression suppressed TG accumulation caused by BPA in the AML12 cells. These positive effects of *APOD* may be associated with upregulation of *PPARβ*, which plays important roles in angiogenesis, metabolism, and inflammation. These results for the first time demonstrate that *APOD* regulates BPA-induced dysregulation of lipid homeostasis. In summary, as shown in Figure 5, our findings verified that BPA aggravated lipid accumulation in hepatocytes via decreasing *APOD* expression in vitro and in vivo. *APOD* is a potential effective agent for the treatment of obesity and NAFLD caused by BPA exposure.

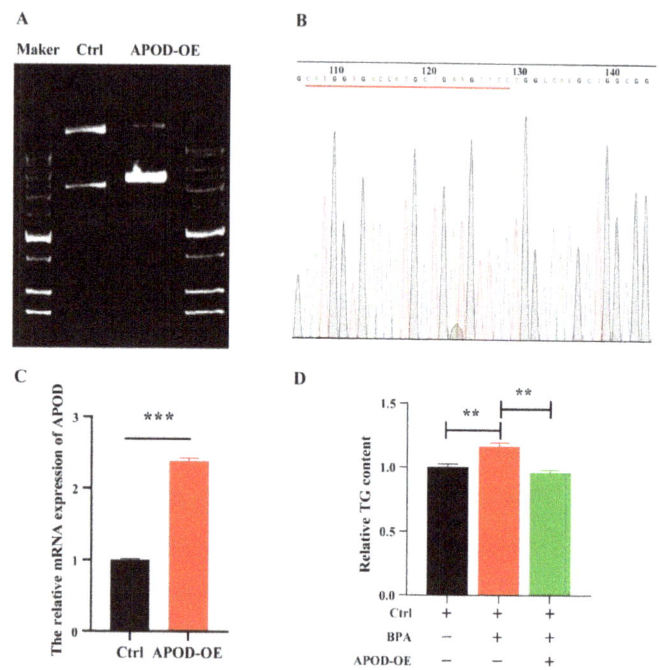

Figure 5. *Cont.*

E

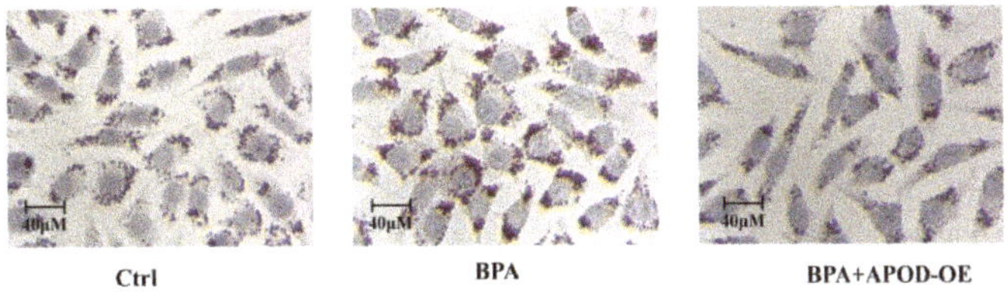

Figure 5. Effect of *APOD* over-expression on the TG accumulation in AML12 cells. Note: (**A**) Electrophoresis of pcDNA3.1 (+) APOD plastimid; (**B**) The sequencing of recombinant plasmid DNAs BPA exposure promoted hepatic inflammatory response in mice; (**C**) The expression of APOD in AML12 cells, $n = 3$ in each group; (**D**) Relative TG content in cells, $n = 4$ in each group; (**E**) Typical image of lipid accumulation in AML12 cells by oil red O staining. ** Represents the significance at $p < 0.01$. *** Represents the significance at $p < 0.001$.

4. Discussion

Epidemiological and experimental studies suggested that the prevalence of NAFLDs may be associated with BPA exposure [12,13], but the mechanism is unclear. Our data demonstrated that the development of NAFLDs induced by BPA exposure was associated with hepatic pro-inflammatory, abnormal lipid metabolism and lipid deposition. Consistent with these results obtained by previous studies [14–16] our study showed that BPA induced hepatic steatosis and fat accumulation [11] revealing that BPA resulted in dose-dependent effects on metabolic parameters [17]. Our study shows that BPA exposure causes weight gain and liver lipoatrophy in mice. Therefore, exposure to BPA in early life should be carefully examined in the etiology of NAFLDs.

As is known, high or prolonged exposure to BPA during early life may exert more long-term adverse outcomes, and increase the risk factors associated with metabolic diseases in adult life [7]. During the exposure period, compared with the control group, the body weight and fat-to-body weight ratio of BPA groups were significantly increased. During the recovery process, we found that there were no significant differences between the BPA exposure group and the control group after the removal of BPA, indicating the recovery of the lipid metabolism. However, the size and volume of adipocytes were still larger even if at the recovery time and the possible cause was irreparable damage to the lipid metabolism. Meanwhile, significant changes in males in serum THCO and HDL-C levels were not found during BPA exposure and post-exposure recovery. But it is worth noting that TG and LDL-C were still higher in low-dose BPA groups, whether during BPA exposure or the recovery period. Meanwhile, it is an interesting finding that fat in the liver was mainly observed as macrovesicular droplets in the mice exposed to low-dose BPA. These data suggested that the live function recovered faster in the high-dose BPA rather than the low-dose BPA. The results in this study are consistent with previous studies that the mice are more sensitive to low-dose BPA exposure as compared to higher doses, thereby contributing to hepatic steatosis [9,14]. We cannot rule out the possibility that other mechanisms, whereas we speculated the dysregulated autophagy by BPA may contribute to the transcriptional impacts of low BPA doses reported here [18]. This raises further questions regarding whether the high-dose BPA-caused impairment to the lipid metabolism mechanism in male mice is identical to that of low-dose BPA.

Our RNA-seq analysis detected some key genes involved in TG and lipid metabolism, including *APOD*, *SCD1*, *ANGPT4*, *LPL*, *G0S2*, *FADS2*, *GNAI2*, *PLIN1*, *ELOVL6*, *ACSL3*, *PPARα*, *PPARβ*, *PPARγ*, *FADS1* and *SOD3* [11]. In this study, we further explored the changes in these genes during BPA exposure and post-exposure recovery. BPA can increase the mRNA expression levels of *FADS1*, *FADS2*, *LPL*, *G0S2*, and *ACSL3*, which were key regulators of lipogenesis [19,20]. Lipid accumulation in liver could be due to the different expression levels of these genes [21,22]. Interestingly, the mRNA expression levels of all these genes nearly reached the control levels during the recovery process, indicating that the recovery speeds of these genes were similar. Meanwhile, these genes were easier to recover in the liver and we believe that the recoveries of different genes follow a specific order after the removal of BPA exposure. This restoration mechanism ensures the recovery of liver function.

Importantly, we found a significant expression reversion for *SCD1*, *APOD*, *ANGPT4*, *PPARβ*, *LPL* and *G0S2* between the exposure and recovery groups, especially for *SCD1* and *APOD*. The mRNA and Western blotting results indicated that the levels of *SCD1* protein levels showed a remarkable increase in the liver of BPA groups while there was no significant difference between the BPA group and control group during recovery. Our results indicated that BPA can affect *SCD1* expression and further impair the process of murine adipogenesis [11]. The decreased expression of *SCD1* showed decreased hepatic triacylglycerol content and reduced body obesity, which was consistent with Zou et al. [23,24]. On the contrary, there was a significantly lower level of APOD protein in the exposure phase. However, in the recovery phase there was an extremely significant increase after exposure had ceased. Additionally, *APOD* over-expression suppressed TG accumulation in the AML12 cells. Therefore, *APOD* played an important role in BPA-induced dysregulation of lipid homeostasis and this increase may be an adaptive mechanism that ensures the recovery of liver function.

Our results support the hypothesis that upregulation of APOD ameliorates dysregulation of lipid homeostasis caused by BPA exposure in male mice. Of relevance, our results showed that the expression of APODs was lower in the BPA group compared with control group mice and suggested an important role for APODs in regulating lipid metabolism caused by BPA exposure. BPA may decrease lipid translocation processes by inhibiting APOD expression in the liver. Interestingly, an important characteristic of BPA-induced hepatic steatosis was that compared to the control group, the trend of APOD concentration in the serum of mice in the BPA group was opposite to the trend of the level of APOD protein expression in the liver. This intriguing phenomenon suggests that the expression patterns are completely different in the liver and blood. Indeed, APOD appears to be a beneficial actor in both lipid metabolisms as it is associated with lipid uptake and inflammation resorption. However, how APOD controls its expression levels remains unknown and needs further attention.

5. Conclusions

In our study, we found that BPA caused severe steatosis in the livers of mice, which was partially alleviated after we stopped exposure. Importantly, BPA could significantly decrease the level of *APOD* protein whereas an extremely significant increase occurred after we stopped exposure. Meanwhile, *APOD* over-expression suppressed TG accumulation in AML12 cells. In conclusion, the damage caused by BPA can be repaired by upregulation of *APOD* and it is a potentially effective biochemical detection indicator for the treatment of obesity or NAFLDs caused by BPA exposure.

Supplementary Materials: The following supporting information can be downloaded at: https://www.mdpi.com/article/10.3390/toxics11090775/s1, Figure S1: Effects of BPA exposure on the expression of key genes in liver; Figure S2: serum APOD levels between comparison groups; Figure S3: Effects of 16 weeks of continuous BPA exposure on mice phenotype; Figure S4: Effects of age factors on the expression of genes in the mice liver; Table S1: PCR primers for qRT-PCR validation.

Author Contributions: C.L.: Conceptualization, Formal analysis, Software, Validation. N.S.: Formal analysis, Methodology, Software, Validation; S.Y.: Methodology, Writing—review and editing, Data curation; H.-L.W.: Conceptualization, Resources, Supervision, Project administration. All authors have read and agreed to the published version of the manuscript.

Funding: This work was supported by the National Science Foundation of China [No. 81773475, 82073592] and the University Synergy Innovation Program of Anhui Province No. GXXT-2020-017.

Institutional Review Board Statement: Ethical approval of all procedures for the animal experiments were approved by the Institutional Animal Use and Care Committee of the Hefei University of Technology (permit number HFUT20210413002).

Informed Consent Statement: Not applicable.

Data Availability Statement: The data are available if requested.

Acknowledgments: The authors thank Ruyue Fang for performing liver samples during WB analysis.

Conflicts of Interest: The authors declare that they have no known competing financial interest or personal relationships that could have appeared to influence the work reported in this paper.

References

1. Farrell, G.C.; Larter, C.Z. Nonalcoholic fatty liver disease: From steatosis to cirrhosis. *Hepatology* **2006**, *43*, S99–S112. [CrossRef] [PubMed]
2. Al-Eryani, L.; Wahlang, B.; Falkner, K.C.; Guardiola, J.J.; Clair, H.B.; Prough, R.A.; Cave, M. Identification of Environmental Chemicals Associated with the Development of Toxicant-associated Fatty Liver Disease in Rodents. *Toxicol. Pathol.* **2015**, *43*, 482–497. [CrossRef]
3. Špačková, J.; Oliveira, D.; Puškár, M.; Ďurovcová, I.; Gaplovská-Kyselá, K.; Oliveira, R.; Ševčovičová, A. Endocrine-Independent Cytotoxicity of Bisphenol A Is Mediated by Increased Levels of Reactive Oxygen Species and Affects Cell Cycle Progression. *J. Agric. Food Chem.* **2020**, *68*, 869–875. [CrossRef] [PubMed]
4. Diamanti-Kandarakis, E.; Bourguignon, J.P.; Giudice, L.C.; Hauser, R.; Prins, G.S.; Soto, A.M.; Zoeller, R.T.; Gore, A.C. Endocrine-disrupting chemicals: An Endocrine Society scientific statement. *Endocr. Rev.* **2009**, *30*, 293–342. [CrossRef]
5. Karsauliya, K.; Bhateria, M.; Sonker, A.; Singh, S.P. Determination of Bisphenol Analogues in Infant Formula Products from India and Evaluating the Health Risk in Infants Asssociated with Their Exposure. *J. Agric. Food Chem.* **2021**, *69*, 3932–3941. [CrossRef] [PubMed]
6. Trasande, L.; Attina, T.M.; Blustein, J. Association between urinary bisphenol A concentration and obesity prevalence in children and adolescents. *JAMA* **2012**, *308*, 1113–1121. [CrossRef]
7. Grohs, M.N.; Reynolds, J.E.; Liu, J.; Martin, J.W.; Pollock, T.; Lebel, C.; Dewey, D. Prenatal maternal and childhood bisphenol a exposure and brain structure and behavior of young children. *Environ. Health* **2019**, *18*, 85. [CrossRef]
8. Hu, C.Y.; Li, F.L.; Hua, X.G.; Jiang, W.; Mao, C.; Zhang, X.J. The association between prenatal bisphenol A exposure and birth weight: A meta-analysis. *Reprod. Toxicol.* **2018**, *79*, 21–31. [CrossRef]
9. Long, Z.; Fan, J.; Wu, G.; Liu, X.; Wu, H.; Liu, J.; Chen, Y.; Su, S.; Cheng, X.; Xu, Z.; et al. Gestational bisphenol A exposure induces fatty liver development in male offspring mice through the inhibition of HNF1b and upregulation of PPARγ. *Cell Biol. Toxicol.* **2021**, *37*, 65–84. [CrossRef]
10. Rubin, B.S.; Soto, A.M. Bisphenol A: Perinatal exposure and body weight. *Mol. Cell Endocrinol.* **2009**, *304*, 55–62. [CrossRef]
11. Fang, R.; Yang, S.; Gu, X.; Li, C.; Bi, N.; Wang, H.L. Early-life exposure to bisphenol A induces dysregulation of lipid homeostasis by the upregulation of SCD1 in male mice. *Environ. Pollut.* **2022**, *304*, 119201. [CrossRef] [PubMed]
12. Shimpi, P.C.; More, V.R.; Paranjpe, M.; Donepudi, A.C.; Goodrich, J.M.; Dolinoy, D.C.; Rubin, B.; Slitt, A.L. Hepatic Lipid Accumulation and Nrf2 Expression following Perinatal and Peripubertal Exposure to Bisphenol A in a Mouse Model of Nonalcoholic Liver Disease. *Environ. Health Perspect.* **2017**, *125*, 087005. [CrossRef] [PubMed]
13. Verstraete, S.G.; Wojcicki, J.M.; Perito, E.R.; Rosenthal, P. Bisphenol a increases risk for presumed non-alcoholic fatty liver disease in Hispanic adolescents in NHANES 2003-2010. *Environ. Health* **2018**, *17*, 12. [CrossRef]
14. Marmugi, A.; Ducheix, S.; Lasserre, F.; Polizzi, A.; Paris, A.; Priymenko, N.; Bertrand-Michel, J.; Pineau, T.; Guillou, H.; Martin, P.G.; et al. Low doses of bisphenol A induce gene expression related to lipid synthesis and trigger triglyceride accumulation in adult mouse liver. *Hepatology* **2012**, *55*, 395–407. [CrossRef] [PubMed]
15. Li, Q.; Zhang, H.; Zou, J.; Mai, H.; Su, D.; Feng, X.; Feng, D. Bisphenol A exposure induces cholesterol synthesis and hepatic steatosis in C57BL/6 mice by down-regulating the DNA methylation levels of SREBP-2. *Food Chem. Toxicol.* **2019**, *133*, 110786. [CrossRef] [PubMed]
16. Kim, K.Y.; Lee, E.; Kim, Y. The Association between Bisphenol A Exposure and Obesity in Children-A Systematic Review with Meta-Analysis. *Int. J. Environ. Res. Public Health* **2019**, *16*, 2521. [CrossRef]
17. Lind, T.; Lejonklou, M.H.; Dunder, L.; Kushnir, M.M.; Öhman-Mägi, C.; Larsson, S.; Melhus, H.; Lind, P.M. Developmental low-dose exposure to bisphenol A induces chronic inflammation, bone marrow fibrosis and reduces bone stiffness in female rat offspring only. *Environ. Res.* **2019**, *177*, 108584. [CrossRef]
18. Yang, S.; Zhang, A.; Li, T.; Gao, R.; Peng, C.; Liu, L.; Cheng, Q.; Mei, M.; Song, Y.; Xiang, X.; et al. Dysregulated Autophagy in Hepatocytes Promotes Bisphenol A-Induced Hepatic Lipid Accumulation in Male Mice. *Endocrinology* **2017**, *158*, 2799–2812. [CrossRef]
19. Fan, J.G.; Zhu, J.; Li, X.J.; Chen, L.; Lu, Y.S.; Li, L.; Dai, F.; Li, F.; Chen, S.Y. Fatty liver and the metabolic syndrome among Shanghai adults. *J. Gastroenterol. Hepatol.* **2005**, *20*, 1825–1832. [CrossRef]

20. Feng, D.; Zhang, H.; Jiang, X.; Zou, J.; Li, Q.; Mai, H.; Su, D.; Ling, W.; Feng, X. Bisphenol A exposure induces gut microbiota dysbiosis and consequent activation of gut-liver axis leading to hepatic steatosis in CD-1 mice. *Environ. Pollut.* **2020**, *265*, 114880. [CrossRef]
21. Vom Saal, F.S.; Nagel, S.C.; Coe, B.L.; Angle, B.M.; Taylor, J.A. The estrogenic endocrine disrupting chemical bisphenol A (BPA) and obesity. *Mol. Cell Endocrinol.* **2012**, *354*, 74–84. [CrossRef] [PubMed]
22. Sun, L.; Ling, Y.; Jiang, J.; Wang, D.; Wang, J.; Li, J.; Wang, X.; Wang, H. Differential mechanisms regarding triclosan vs. bisphenol A and fluorene-9-bisphenol induced zebrafish lipid-metabolism disorders by RNA-Seq. *Chemosphere* **2020**, *251*, 126318. [CrossRef] [PubMed]
23. Zou, Y.; Wang, Y.N.; Ma, H.; He, Z.H.; Tang, Y.; Guo, L.; Liu, Y.; Ding, M.; Qian, S.W.; Tang, Q.Q. SCD1 promotes lipid mobilization in subcutaneous white adipose tissue. *J. Lipid Res.* **2020**, *61*, 1589–1604. [CrossRef] [PubMed]
24. Miyazaki, M.; Flowers, M.T.; Sampath, H.; Chu, K.; Otzelberger, C.; Liu, X.; Ntambi, J.M. Hepatic stearoyl-CoA desaturase-1 deficiency protects mice from carbohydrate-induced adiposity and hepatic steatosis. *Cell Metab.* **2007**, *6*, 484–496. [CrossRef] [PubMed]

toxics

MDPI

Article

Using a Battery of Bioassays to Assess the Toxicity of Wastewater Treatment Plant Effluents in Industrial Parks

Bin Yang [1], Haiyan Cui [1], Jie Gao [1], Jing Cao [1], Göran Klobučar [2] and Mei Li [1,*]

[1] State Key Laboratory of Pollution Control and Resource Reuse, School of Environment, Nanjing University, Nanjing 210023, China
[2] Department of Biology, Faculty of Science, University of Zagreb, Rooseveltov trg 6, 10000 Zagreb, Croatia
* Correspondence: meili@nju.edu.cn

Abstract: Bioassays, as an addition to physico-chemical water quality evaluation, can provide information on the toxic effects of pollutants present in the water. In this study, a broad evaluation of environmental health risks from industrial wastewater along the Yangtze River, China, was conducted using a battery of bioassays. Toxicity tests showed that the wastewater treatment processes were effective at lowering acetylcholinesterase (AChE) inhibition, HepG2 cells' cytotoxicity, the estrogenic effect in T47D-Kbluc cells, DNA damage of Euglena gracilis and the mutagenicity of Salmonella typhimurium in the analyzed wastewater samples. Polycyclic aromatic hydrocarbons (PAHs) were identified as potential major toxic chemicals of concern in the wastewater samples of W, J and T wastewater treatment plants; thus, the potential harm of PAHs to aquatic organisms has been investigated. Based on the health risk assessment model, the risk index of wastewater from the industrial parks along the Yangtze River was below one, indicating that the PAHs were less harmful to human health through skin contact or respiratory exposure. Overall, the biological toxicity tests used in this study provide a good basis for the health risk assessment of industrial wastewater and a scientific reference for the optimization and operation of the treatment process.

Keywords: wastewaters; genotoxicity; bioassays; cytotoxicity

Citation: Yang, B.; Cui, H.; Gao, J.; Cao, J.; Klobučar, G.; Li, M. Using a Battery of Bioassays to Assess the Toxicity of Wastewater Treatment Plant Effluents in Industrial Parks. *Toxics* **2023**, *11*, 702. https://doi.org/10.3390/toxics11080702

Academic Editors: Zhen-Guang Yan, Zhi-Gang Li and Jinzhe Du

Received: 29 June 2023
Revised: 6 August 2023
Accepted: 10 August 2023
Published: 14 August 2023

1. Introduction

The complexity of water pollution is becoming an emerging concern due to the numerous pollutants that enter water bodies in China. The primary sources of water pollution are untreated industrial and agricultural wastewater, domestic sewage and waste [1]. The wastewater discharged from industrial parks mainly manifests in large volumes, complex compositions and high concentrations of pollutants. The persistent pollutants in wastewater can enter the food chain and ultimately endanger human health [2]. In China, the total national wastewater discharge in 2020 was 71.62 billion tons, of which 29% (20.53 billion tons) was contributed from industrial wastewater discharge. Jiangsu Province in China has 58 chemical industry parks of various industrial scales. Therefore, industrial wastewater serves as an essential source of freshwater and marine pollution [3]. This requires an elaborate environmental risk assessment of industrial wastewater pollution using mandatory biological monitoring as an addition to already existing chemical monitoring.

Bioassays are promising methods for studying these sources of pollution, since, on the one hand, all parameters related to the exposed organisms can be controlled as in laboratory experiments. On the other hand, work under environmentally realistic conditions considers the interactions that occur between the chemicals in the effluent and the complexity of the receiving environment [4]. Recent studies have shown that hematological parameters are often used as valuable indicators for assessing fish health, and that the use of pelagic fish data allows for comprehensive monitoring studies of effluents [5,6]. The effects of wastewater treatment plant effluents on biological neurotoxicity, cytotoxicity, genotoxicity and estrogenic effects have been reported in recent decades [7]. These studies of the impact of the effluents on specific physiological functions provide a complete assessment of the overall health of the organism. Additionally, they provide a more comprehensive perspective by accurately predicting the biological effects of wastewater treatment plant effluents [8].

In 2004, the United States Environmental Protection Agency (USEPA) included the Whole Effluent Toxicity Test (WETT) into the implementation guidelines for total wastewater toxicity. The Continental European Organization and Oceania scholars further proposed the Whole Effluent Assessment (WEA) to evaluate the persistence, bioaccumulation and comprehensive toxicity of discharged wastewater using a suite of acute and chronic toxicity tests [9]. Currently, China's monitoring of toxic substance discharge in industrial wastewater is dominated by the physical and chemical monitoring of pollutants. Such methods and monitoring techniques for chemical industry emissions do not fully represent the possible toxicity of industrial wastewater, which jeopardizes environmental safety and industry development [10]. Therefore, performing and standardizing toxicity testing is one of the critical parts of environmental risk assessment and management in China.

This study aims to combine chemical and biological analysis of industrial wastewater before and after WWT (influent and effluent samples) along the Yangtze River in Jiangsu Province. The basic physical and chemical properties of the wastewater samples were measured as a basis for the ecological and health risk assessment of wastewater pollution in these industrialized areas. Biological assays were performed at different biological organizations to evaluate the environmental risk of wastewater from chemical industrial parks. Genotoxicity assays were performed using two organisms, freshwater algal species Euglena gracilis and Salmonella typhimurium. Assays were also performed on human hepatocellular carcinoma cell line HepG2 and human breast cancer cell line T47D-Kbluc to determine cytotoxic and estrogenic cellular responses, respectively, to wastewater pollution. Enzyme activity assays were also performed to elucidate the inhibitory response of acetylcholinesterase.

2. Materials and Methods

2.1. Sample Collection and Preparation

Sampling was performed in accordance with "Water Quality Sampling Technical Guidance" (HJ404–2009). Sampling was conducted using a steel bucket on sunny and rainless days between 10 April and 20 April 2018. Nine inlet and nine outlet samples of the wastewater were collected from the wastewater treatment plants between 9 a.m. and 2 p.m. and samples only contained plant wastewater. Sampling container was sanitized before and after collection to avoid contamination. These WWTPs (W, J and T, as shown in Figure 1) are in the three typical industrial areas along the Yangtze River. A total of nine samples (20 L) were stored under refrigeration at 4 °C for 24 h before carrying out the experiments. The collected samples were processed as follows: filtration (0.7 μm glass fiber filter) was performed to remove particles, followed by acidification at pH 2–3 with concentrated hydrochloric acid. Samples were further processed using solid-phase extraction (Oasis Hydrophile-Lipophile Balance cartridge) and then stored at −20 °C, protected from light, for chemical analysis and bioassays, respectively. Solid-phase extraction was performed in accordance with "Water Quality-Determination of polycyclic aromatic hydrocarbons by liquid-liquid extraction and soild-phase extraction high performance liquid chromatography" (HJ478-2009). The extraction procedure was the following: 10 mL methanol and 10 mL acetone/hexane (*v/v*:1/1) for activating the extraction column (Waters Oasis HLB), 10 mL methanol and 10 mL acetone/hexane (*v/v*:1/1) for eluting, 5 mL/min for solution loading. Before and during the analysis, method blanks, instrumental blanks and solvent blanks were implemented for each batch of samples. Spiked matrices showed 80~110% recovery for compounds.

Figure 1. Sampling location of three WWTPs studied in this study.

2.2. Physical and Chemical Indicators Detection

Water chemistry parameters were measured within 24 h after sampling. The analytical method followed the national surface water environmental quality standards established by the Chinese government (GB3838-2002). The method used for Polycyclic aromatic hydrocarbon (PAH) analysis was from "National Environmental Protection Standards of the People's Republic of China" (HJ478-2009). The high-performance liquid chromatography (HPLC) analysis steps were as follows: the mobile phase was (A) methanol and (B) water with a ratio of 80:20 at a flow rate of 1.0 mL/min, increments of 1.2% methanol/min to 95% methanol + 5% water, hold until the peak was completed and the ultraviolet detection wavelengths were set to 254 nm, 220 nm and 295 nm.

2.3. Bioassays

2.3.1. Acetylcholinesterase Inhibition Assay

The wastewater samples were concentrated or diluted, six concentrations (0.1×, 0.2×, 0.5×, 1×, 2× and 5×) of wastewater samples were used to generate the dose–response curve, and the concentration for 50% of maximal effect (EC_{50}) was calculated via linear regression with Prism 6.0 (GraphPad Software, San Diego, CA, USA).

Methomyl was used as the positive control; the corresponding EC_{50} was obtained. Briefly, water samples with different dilution ratios were sequentially added to 96-well plates in triplicate, each 100 μL, and three wells were set to add an equal amount of phosphate buffer as a blank control. Then, 5, 5′-dithiobis-(2-nitrobenzoic acid), 2-(acetylthio)-N, N, N-trimethylethylammonium iodide and electric eel AChE solution were added to each dilution followed by thorough mixing on a plate shaker. After 15 min, the enzymatic reaction was measured at an OD_{412} nm in a plate reader (iD3, Molecular Devices, San Jose, CA, USA). The inhibition of AChE by the water sample was calculated according to the following formula:

$$E = 1 - \frac{\Delta A_t}{\Delta A_c}$$

where ΔA_t is the change in absorbance of the experimental group compared with the initial and ΔA_c is the change in absorbance of the blank group compared with the initial one.

Toxicity equivalence factor (*TEF*) refers to the index for evaluating the relative toxicity strength or health impact degree of a compound isomer. The toxicity of environmental water samples can be evaluated more intuitively with *TEF*. The calculation method was based on the following formula:

$$TEF = \frac{EC_{50,positive}}{EC_{50,sample}}$$

where $EC_{50,positive}$ and $EC_{50,sample}$ are half the maximum effective concentration of positive control group and sample group and sample groups.

2.3.2. Cytotoxicity Assay

The wastewater samples were concentrated or diluted, and six concentrations (0.25×, 0.5×, 1×, 2×, 5× and 10×) of the wastewater from the three WWTPs were selected for testing. The human hepatocellular carcinoma cell line HepG2 was selected to detect the cytotoxicity of influents and effluents of the chemical industry plants in this study. The HepG2 cells obtained from KeyGen Biotech (Nanjing, China) were maintained in a Dulbecco's modified Eagle medium (DMEM) supplemented with 10% fetal bovine serum (FBS). After being exposed for 24 h, the cell viability was detected using a cell counting kit-8 (CCK8) kit (Dojindo Molecular Technologies, Kumamoto city, Japan). The kit determines the number of living cells by measuring the enzymatic reaction at an OD450 nm in a plate reader (iD3, Molecular Devices). Cell viability was calculated according to the following formula:

$$\text{Cell viability (\%)} = A_{\text{experiment}} / A_{\text{blank}} \times 100\%$$

Graphpad Prism software (GraphPad Software, San Diego, CA, USA) was used to fit the dose–effect relationship to find the EC_{50}.

2.3.3. Estrogenic Effect Assay

The human breast cancer cell line T47D-Kbluc (American Type Culture Collection, Rockville, MD, USA) was chosen as the indicator of estrogenic effects in wastewater samples. Roswell park memorial institute 1640 (RPMI 1640) medium was used for cell culture [11]. Activated carbon was added to reduce the estrogen concentration in the culture medium. The fetal bovine serum in the culture medium was replaced with activated carbon to adsorb the fetal bovine serum to reduce the estrogen residue in the culture medium. The cells were exposed to the cell culture fluid with water samples for 24 h and firefly luciferase was added after lysis. Luminescence was measured using a microplate reader (Synergy H1, BioTek, Santa Clara, CA, USA).

The dose–effect curve was generated according to the measured water luminescence and was used to calculate the corresponding EC_{50} using GraphPad Prism. β-estradiol (E2) was used as a positive control.

2.3.4. Genotoxicity Assays

In this study, the Comet assay was used to detect the genotoxicity of wastewater samples. The Comet assay or single-cell gel electrophoresis assay can detect DNA damage and repair at the single-cell level through qualitative and quantitative measurement of single-strand DNA breaks, and accurately reflects the level of DNA damage and repairability. In this study, the Comet assay was used to detect the genotoxicity of wastewater samples. Algae *Euglena gracilis* was provided by the Freshwater Algae Seed Bank (FACHB) of the Typical Culture Collection Committee of the Institute of Hydrobiology, Chinese Academy of Sciences. *E. gracilis* was cultured on Checcucci medium at 25 °C with a 12 h:12 h light/dark cycle in an incubator.

A total of 1.5 mL of *E. gracilis* culture medium was centrifuged at 4000 rpm for 5 min and the precipitate was exposed to the wastewater for 30 min and centrifuged again to collect the precipitated cells. Cells were embedded in 100 µL of 1% low-melting agarose (LMA) sandwiched between 0.7% normal-melting agarose (NMA) and 1% LMA on microscope slides. The slides were placed in a lysis solution (300 mM NaOH; 30 mM Na_2-EDTA·$2H_2O$; 0.01% SDS; 9% DMSO; and 1% Triton X-100) for 20 min at 4 °C. Then, the slides were placed in a horizontal electrophoresis unit with fresh alkaline electrophoresis buffer (300 mM NaOH and 1 mM N_{a2}-EDTA·$2H_2O$, pH 13.0) added and the liquid level was controlled to be 2 mm above the slide at 4 °C for 20 min to allow for DNA unwinding. Electrophoresis was carried out using the same buffer for 20 min using 20 V (0.8 V/cm) and 300 mA at 4 °C. The slides were neutralized by immersing in 0.4 M Tris buffer at pH 7.5 for 5 min and stained with ethidium bromide. The slides were analyzed under a fluorescence microscope (Nikon Eclipse 50i, Tokyo Metropolitan, Japan). Comet Assay Software Project (CASP, University of Wroclaw, Wroclaw, Poland) image analysis software was used to analyze DNA damage.

Umu/SOS experiments used the method of existing studies [12] and *Salmonella typhimurium* (TA1535/pSK1002, *S. typhimurium*) were used to detect the genotoxicity of wastewater influents and effluents. The specific experimental steps were as follows: (a) Shake the bacteria with TGA medium (tryptone 10 g/L, NaCl 5 g/L, HEPES 11.9 g/L, glucose 2 g/L; final pH adjusted to 7.0 ± 0.2) overnight for 12–16 h, dilute the bacterial solution with fresh medium 10 times the next day and continue to culture for about 1.5 h to OD_{600} = 0.2. (b) Add the diluted samples to 96-well plate A at 180 µL per well, then add 20 µL 10× medium and 70 µL bacterial solution. Add another three wells with 153 µL water, 27 µL 4-nitroquinoline-1-oxide (4-NQO), 20 µL TGA medium and 70 µL bacterial solution as positive controls. Add three wells with 180 µL water, 20 µL TGA medium and 70 µL bacterial solution as negative controls, and 180 µL water, 20 µL 10× medium and 70 µL bacterial solution as blank controls. Incubate at 37 °C for 2 h. (c) Take new plate B, add 270 µL TGA medium, preheat at 37 °C. Take 30 µL of each well in plate A and add it to the corresponding well in plate B. Incubate for 2 h and measure the absorbance of plate B at OD_{600}. (d) Take new plate C, add 120 µL B buffer to each well and preheat at 28 °C. Take 30 µL of each well in plate B and add it to the corresponding well in plate C. Quickly add 30 µL of 2-Nitrophenyl β-D-galactopyranoside (ONPG), mix well and put into the incubator. After shaking at 28 °C for 30 min, add 120 µL of blocking solution to each well of plate C to stop the reaction. Measure the absorbance of plate C at OD_{420}. Calculate bacterial growth factor (*G*) and induction ratio (*IR*) according to the following formula [13,14]:

$$G = \frac{(A_{600,S} - A_{600,B})}{(A_{600,N} - A_{600,B})} (When\ G > 0.5, it\ can\ be\ used\ for\ IR\ value\ calculation)$$
$$IR = \frac{(A_{420,S} - A_{420,B})}{(A_{420,N} - A_{420,B})} \times \frac{1}{G} (When\ IR > 2, the\ test\ result\ is\ judged\ to\ be\ positive)$$
$$\beta\ \text{Galactase activity}(UI) = \frac{(A_{420,S} - A_{420,B})}{(A_{600,S} - A_{600,B})}$$

In the formula, $A_{600,S}$ is the absorbance of water sample at 600 nm, $A_{600,B}$ is the absorbance of blank at 600 nm and $A_{600,N}$ is the absorbance of negative control at 600 nm; $A_{420,S}$ is the absorbance of the water sample at 420 nm, $A_{420,B}$ is the absorbance of the blank at 420 nm and $A_{420,N}$ is the absorbance of the negative control at 420 nm.

Comparing the measured slope of the P-galactosidase curve in the test sample with the slope of the P-galactosidase curve of the 4-NQO sample measured simultaneously, the 4-NQO in the test sample can be obtained in an equivalent concentration ($TEQ_{4\text{-}NQO}$):

$$TEQ_{4\text{-}NQO}(\mu g/L) = \frac{K_{sample}(\text{unit}/L)}{K_{4\text{-}NQO}(\text{unit}/\mu g)}$$

2.4. Data Analysis

2.4.1. Statistical Analysis

All the data were expressed as the mean ± standard deviation. The data analyses were performed using SPSS 13.0 and the graphics were generated and produced using Microsoft Excel and GraphPad Prism 8.0 (San Diego, CA, USA). One-way analysis of variance (one-way ANOVA) with Tukey's post hoc test was used to assess the comparisons between two groups and the correlation between variables. For all analyses, $p < 0.05$ was considered significant, and $p < 0.01$ was considered highly significant. Data analysis was repeated three times to reduce errors.

2.4.2. Health Risk Assessment

The detection adopted 16 PAH congeners listed in the "National Environmental Protection Standard of the People's Republic of China" (HJ478-2009) as evaluation indicators. According to the American National Academy of Sciences (NAS) health risk assessment model, PAHs in industrial wastewater are harmful to human health through oral and skin exposure routes (NAS, 2004). The US Environmental Protection Agency (EPA) has classified 16 PAHs as priority pollutants based on their possible human exposure and toxicity. Most types of PAHs can diffuse through cellular membranes, resulting in toxic effects to organisms [15]. The toxicity parameters of 11 PAHs are listed in Table S1. According to the toxicity of the target PAHs, the human health risk assessment was based on the above exposure routes (Text S1).

3. Results and Discussion

3.1. PAHs in Wastewater Samples

A total of ten different PAHs were detected in the six samples (Table 1), and their monomer concentrations were between not detectable (ND) and 10.58 mg/L. Concentrations of NAP in influents of plants W, J and T were relatively high, reaching 10.58 mg/L, 12.07 mg/L and 9.87 mg/L, respectively. Organic substances such as fluoranthene were not detected in the influents. Still, there were trace concentrations present in the effluents, indicating that in the process of biochemical treatment, both the decomposition of organic substances and the generation of pollutants occurred. This may be a by-product of the WWT process of decomposing organic substances through photochemical and biological transformation. Correspondingly, the basic physical and chemical indicators of the chemical park wastewater along the Yangtze River changed in effluent wastewater (Table S2).

3.2. Toxicity Effects

3.2.1. Neurotoxicity of Wastewater

Wastewater samples from the three wastewater treatment plants showed different inhibitory effects on AChE (Figure 2). Among them, the influent wastewater of the J plant showed the strongest inhibition, reaching 55%. The inhibition of AChE in plants W and T reached 46% and 33%, respectively. Compared with the influent samples, the inhibition rates of AChE in the effluent samples from the three water plants all decreased, indicating that the WWT process reduced the concentrations of substances with AChE inhibition activity in the samples. The inhibition rates of AChE are closely related to neuronal function disorder and death [16–18]. Examples of AChE inhibitors in wastewater include some low-level pollutants, such as heavy metals or detergents that are widely present in urban rivers, and the AChE inhibitors present in wastewater can adversely affect humans and animals. The toxic equivalents of the water samples were used to evaluate the neurotoxicity of wastewater samples more accurately. The *TEF* indicator can be used to analyze the environmental level of pollutants and their potential impact [19,20]. The EC_{50} significantly increased after the WWT process, while the *TEF* decreased (Table 2), indicating reduced neurotoxicity of the wastewater samples. t

Table 1. Concentrations of PAHs in wastewater samples.

(mg/L)	W Plant		J Plant		T Plant	
PAH	**Influent**	**Effluent**	**Influent**	**Effluent**	**Influent**	**Effluent**
Naphthalene (NAP)	10.58	0.04	12.07	0.01	9.87	0.02
Acenaphthylene (ACY)	0.06	N.D.	0.14	N.D.	0.07	N.D.
Fluorene (FLU)	0.02	N.D.	0.08	N.D.	0.05	N.D.
Fluoranthene	N.D.	0.11	0.08	N.D.	0.08	N.D.
Phenanthrene (PHE)	0.05	0.01	0.08	N.D.	N.D.	N.D.
Anthracene (ANT)	1.23	0.02	1.42	N.D.	N.D.	0.02
Benzoanthracene (BaA)	0.08	N.D.	0.19	0.01	0.04	N.D.
Benzofluoranthrene	0.06	0.01	0.25	0.02	0.03	N.D.
Benzo(a)pyrene (BaP)	1.02	0.02	1.26	0.04	1.17	0.06
Benzo(g,h,i)perylene (B(g,h,i) P)	0.05	N.D.	0.08	N.D.	0.06	N.D.
Total	13.28	0.23	15.99	0.1	11.64	0.11

Note: N.D.: Below detection limit.

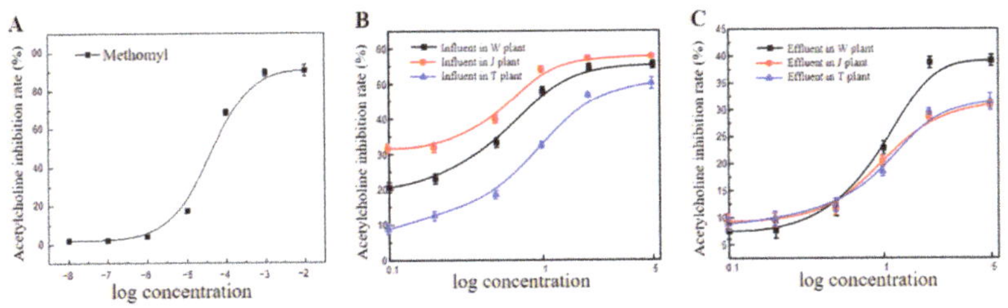

Figure 2. Inhibition rates of AChE in influents and effluents from three wastewater treatment plants: (**A**) methomyl, (**B**) influents, (**C**) effluents.

Table 2. EC_{50} and *TEF* of influents and effluents from three wastewater treatment plants.

		EC_{50}	*TEF*
W plant	Influent	3.06	2.43
	Effluent	80.54	0.10
J plant	Influent	4.04	1.84
	Effluent	65.89	0.11
T plant	Influent	6.88	1.08
	Effluent	85.35	0.10

TEF: Toxicity equivalence factor.

3.2.2. Cytotoxicity of Wastewater

The influents from the WWTPs showed varying concentration-dependent effects on HepG2 cells' viability (Figure 3A). Compared with the J plant, wastewater from the W and T plants showed relatively strong cytotoxicity. According to the Industrial Wastewater Biological Toxicity Classification Standard, all influent samples from the three WWTPs are classified as cytotoxic.

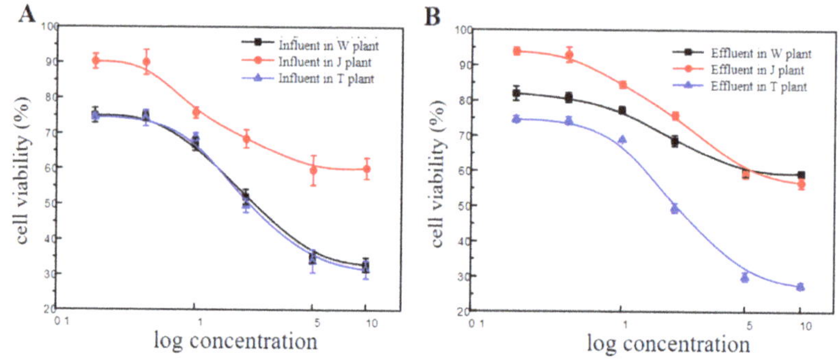

Figure 3. HepG2 cell viability in influents and effluents from three wastewater treatment plants: (**A**) influents, (**B**) effluents.

The cytotoxicity results of the effluents are shown in Figure 3B. The results showed that the WWT decreased the cytotoxicity of the wastewater. The lowered cytotoxicity in the effluent samples may be due to the partial oxidation of organic fractions by the biochemical treatment process [21,22], and the specific causative agent in the wastewater of the cytotoxicity was partially removed [23].

3.2.3. Estrogenic Effect of Wastewater

Environmental endocrine disruptors are exogenous chemicals that can cause abnormalities and disorders in the endocrine system [24–26]. The influent group of wastewater samples showed significant estrogenic effects, and their equivalents reached 0.65 ng/L, 0.74 ng/L and 0.49 ng/L (Figure 4A), respectively. After WWT, the estrogenic effect in plants W, J and T decreased in effluent wastewater by 68%, 59% and 60%, indicating the efficiency of the treatment process. The average removal efficiency of the endocrine-disrupting effect through the WWT was 58–84%. The bacterial action of the wastewater treatment likely facilitates the degradation of endocrine disruptors [27]. Directive 2013/39/EU of the European Parliament and Council proposed a monitoring level for 17β-estradiol (E2) of 0.4 ng/L in the environment [28]. E2 is a compound hormone naturally synthesized in vertebrates which plays an important role in the endocrine and reproductive systems [29]. After the biochemical treatment, the toxic equivalents of each WWTP's effluent were lower than this limit, indicating a low ecological risk to aquatic ecosystems (Table 3).

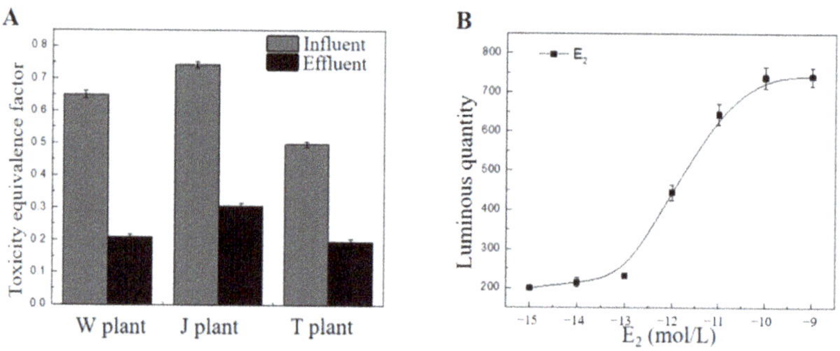

Figure 4. Estrogenic effect in influents and effluents from three wastewater treatment plants: (**A**) toxicity equivalence factor of influents and effluents, (**B**) 17β-estradiol (E2) dose–effect relationship.

Table 3. Estrogenic effect *TEF* of effluent in chemical parks along the Yangtze River.

	W Plant	J Plant	T Plant	E2 Limit
Toxic equivalent factor	0.21	0.31	0.20	0.40

3.2.4. Genotoxicity of Wastewater

It is well-known that the release of genotoxic substances in the environment can damage germinative cells and reduce the abundance and fertility of species in ecosystems [30,31]. Tail length (TL), Tail DNA% (tDNA%), olive tail moment (OTM) and tail moment (TM) were the main parameters of the Comet assay [32]. The four Comet assay parameters showed consistency between the influents and effluents of each sewage treatment plant (Figure 5), and the genotoxicity of the effluent was significantly reduced compared with that of the influents after the WWT. Among the measured parameters, OTM simultaneously reflects the DNA content in the Comet tail and the shape of the Comet tail, and is a commonly used indicator to quantify the degree of DNA damage [19,33]. It can be seen from Figure 5 that the OTMs of the six wastewater samples were significantly higher than those of the control, indicating that each sample caused significant DNA damage to the *E. gracilis* [34–36]. Similarly, the study found, using the Comet assay, that organic extracts from Taihu Lake can induce DNA damage on microalgae cells [37–39]. The genotoxicity of the wastewater samples was J > T > W, and the effluents showed decreases in genotoxicity by 41.06%, 36.12% and 37.15%, respectively.

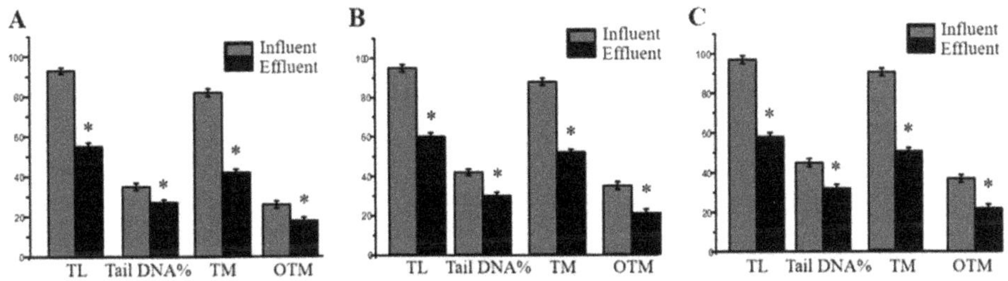

Figure 5. Algae *Euglena gracilis'* DNA damage in three WWTPs: (**A**) influent and effluent in W plant, (**B**) influent and effluent in T plant, (**C**) influent and effluent in J plant. * $p < 0.05$, compared with the influents. TL: Tail length. TM: Tail moment. OTM: Olive tail moment.

Umu/SOS results showed genotoxicity of wastewater influents, and no increase in genotoxicity was observed in wastewater effluents (Table S4). However, both influents and effluents caused growth inhibition and cytotoxicity of *S. typhimurium* (Table S3). The results indicated that the WWT effluents still had potential genotoxicity to aquatic organisms.

3.3. Risk Assessment of PAHs in Wastewater Samples

The PAHs in the ambient air released by the wastewater not only cause strong odor, but also cause a threat to the health of people exposed to these substances [40,41]. According to the risk characterization model, the health risks caused by skin contact and respiratory exposure to PAHs can be calculated. The risk of PAHs from the wastewater samples in the industrial parks along the Yangtze River to human health through the respiratory route was higher than that through skin contact (Tables S5 and S6). According to the findings of the USEPA for the non-carcinogenic risk, PAHs are harmful to human health when the risk index is greater than one. Using the health risk assessment model, the calculated PAH risk index of industrial wastewater along the Yangtze River was below one, indicating that the concentrations of 11 measured PAHs in wastewater were less harmful to human health through skin contact or respiratory exposure.

3.4. Correlation Analysis of Toxicity at Different Endpoints

The toxicity of complex pollutant mixtures in water can be reliably assessed only by applying a suite of bioassays. In this study, five toxicity endpoints involving the use of the HepG2 cell line, T47D-Kbluc cell, *E. gracilis*, *S. typhimurium* and electric eel AChE activity were applied. A comprehensive assessment of the exposure toxicity of the influents and effluents in the chemical parks along the Yangtze River was carried out through heatmaps (Figure 6) using cytotoxicity, AChE inhibition rate, estrogenic effect, mutagenicity, and DNA damage assessments of the influents and effluents. It is important to use different organisms or their cells, as they can produce different reactions to pollutants present in the wastewater [42].

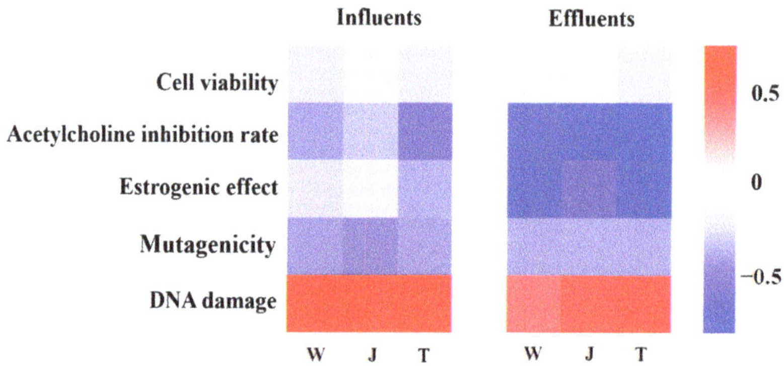

Figure 6. Correlation analysis of toxicity data on different test organisms for influents and effluents from three wastewater treatment plants.

Variations in the results may be caused by organism-specific responses to pollutants in the wastewater, which emphasizes the need to perform a battery of bioassays using different test organisms to perform a comprehensive toxicity and environmental risk assessment of pollutants in wastewater. Traditional physical and chemical monitoring of pollutants is not adequate to provide complete information on the potential toxic effects of pollutants on living organisms, including humans [43]. To assess genotoxicity, this study used *S. typhimurium* and *E. gracilis* as the test organisms to conduct Umu/SOS and Comet assays, respectively. The results indicated that the WWT process reduced the inhibition of AChE, estrogenic effects, mutagenicity and DNA damage. The Comet assay applied to *E. gracilis* was applicable for genotoxicity testing of industrial wastewater. *E. gracilis* can respond rapidly to various pollutants and be a bio-indicator for deteriorating water quality conditions [44].

The specific pollutants in wastewater have the potential to have a range of toxic effects on environmental health. There is a critical need to explore in-depth the potential ecological impacts of specific substances in wastewater and the improvement of wastewater technology for more efficient removal of substances with cytotoxic and genotoxic properties.

4. Conclusions

In this study, risk assessment of the wastewater from WWTPs in chemical parks along the Yangtze River was carried out by detecting specific chemical pollutants and using a battery of bioassays to detect their toxicity. A variety of PAHs were detected in wastewater samples where relatively high concentrations of NAP, ANT and BaP were detected. The bioassays used in this study showed that the WWT process of the W, J and T plants can effectively reduce the cytotoxicity, neurotoxicity and estrogenic and genotoxic effects of the industrial wastewater. Nevertheless, all effluent wastewater from the WWTPs has been characterized by the bioassays as having potentially high ecotoxicity, indicating that their discharge into the environmental water body would potentially cause harm to aquatic organisms. This study provides theoretical support and scientific basis for the environmental risk assessment of industrial wastewater and the progress of wastewater treatment technology. It is envisaged as a guide for the application and development of future industrial wastewater risk assessment standards.

Supplementary Materials: The following supporting information can be downloaded at: https://www.mdpi.com/article/10.3390/toxics11080702/s1, Text S1: Health risk assessment through two exposure routes; Table S1: VF, RfD and CSF values of polycyclic aromatic hydrocarbons; Table S2: Basic physical and chemical indicators of chemical parks; Table S3: Cytotoxicity of water extracts from the chemical parks to *S. typhimurium*; Table S4: Induction rate of the water extracts from the chemical industry parks on *S. typhimurium*; Table S5: Health risk index of non-carcinogenic effects of pollutants via different exposure routes according to the risk characterization model; Table S6: Health risk index of carcinogenic effects of pollutants via different exposure routes according to the risk characterization model [45].

Author Contributions: Conceptualization, M.L. and B.Y.; Data curation, B.Y.; Formal analysis, B.Y.; Funding acquisition, M.L.; Investigation, H.C., J.G. and B.Y.; Supervision, M.L.; Writing—original draft, B.Y.; Writing—review and editing: B.Y., H.C., J.C., G.K. and M.L. All authors have read and agreed to the published version of the manuscript.

Funding: This work was supported by the Science and Technology Support Program of Jiangsu Province (No. BZ2022006, BE2016736).

Institutional Review Board Statement: Not applicable.

Informed Consent Statement: Not applicable.

Data Availability Statement: Not applicable.

Conflicts of Interest: All authors of this article declare that they have no conflict of interest regarding the publication of this study.

References

1. Wang, F.; Wang, B.; Duan, L.; Zhang, Y.; Yu, G. Occurrence and distribution of microplastics in domestic, industrial, agricultural and aquacultural wastewater sources: A case study in Changzhou, China. *Water Res.* **2020**, *182*, 115956.
2. Hakak, S.; Khan, W.-Z.; Gilkar, G.-A. Industrial wastewater management using blockchain technology: Architecture, requirements, and future directions. *IEEE Internet Things* **2020**, *3*, 38–43.
3. Lu, Y.-T.; Gao, J.-F.; Nguyen, H.-T.; Du, P.; Li, X.-Q. Occurrence of per- and polyfluoroalkyl substances (PFASs) in wastewater of major cities across China in 2014 and 2016. *Chemosphere* **2021**, *279*, 130590. [CrossRef]
4. Le, G.-A.; Sanchez, W.; Palluel, O.; Bado, N.-A.; Turies, C.; Chadili, E.; Cavalié, I.; Adam, G.-C.; Porcher, J.-M.; Geffard, A.; et al. In situ experiments to assess effects of constraints linked to caging on ecotoxicity biomarkers of the three-spined stickleback (*Gasterosteus aculeatus* L.). *Sci. Total Environ.* **2016**, *42*, 643–657.
5. Marina, B.; Svetlana, K.; Vincenzo, P. Intra-Decadal (2012–2021) Dynamics of Spatial Ichthyoplankton Distribution in Sevastopol Bay (Black Sea) Affected by Hydrometeorological Factors. *Animals* **2022**, *12*, 3317.
6. Parrino, V.; Cappello, T.; Costa, G.; Cannavà, C.; Sanfilippo, M.; Fazio, F.; Fasulo, S. Comparative study of haematology of two teleost fish (*Mugil cephalus* and *Carassius auratus*) from different environments and feeding habits. *Eur. Zool. J.* **2018**, *85*, 193–199. [CrossRef]
7. Beghin, M.; Paris, P.-S.; Mandiki, S.; Schmitz, M.; Palluel, O.; Gillet, E.; Bonnard, I.; Nott, K.; Robert, C.; Porcher, J.-M.; et al. Integrative multi-biomarker approach on caged rainbow trout: A biomonitoring tool for wastewater treatment plant effluents toxicity assessment. *Sci. Total Environ.* **2022**, *838*, 155912.
8. Whale, G.-F.; Hjort, M.; Di, P.-C.; Redman, A.-D.; Postma, J.-F. Assessment of oil refinery wastewater and effluent integrating bioassays, mechanistic modelling and bioavailability evaluation. *Chemosphere* **2022**, *287*, 132146. [CrossRef]
9. Wolfram, J.; Stehle, S.; Bub, S.; Petschick, L.-L. Water quality and ecological risks in European surface waters-monitoring improves while water quality decreases. *Environ. Int.* **2021**, *152*, 106479. [CrossRef]
10. Liu, H.-P.; Hu, H. The main problems in the development of environmental monitoring and environmental monitoring technology in China. *Environ. Dev.* **2018**, *030*, 129–130.
11. Conley, J.-M.; Iwanowicz, L.-R.; Bradley, P.-M.; Evans, N. Occurrence and in vitro bioactivity of estrogen, androgen, and glucocorticoid compounds in a nationwide screen of United States stream waters. *Environ. Sci. Technol.* **2017**, *51*, 4781–4791. [CrossRef] [PubMed]
12. Dou, J.-X.; Shi, Y.; Song, R.-X. Application of SOS/umu test in the assessment of the genotoxicity and carcinogenic risk of organic extracts from water plants. *J. Environ. Health* **2017**, *3*, 248–251.

13. Ojo, A.-F.; Peng, C.; Ng, J.-C. Combined effects of mixed per- and polyfluoroalkyl substances on the Nrf2-ARE pathway in ARE reporter-HepG2 cells. *J. Haz. Mat.* **2022**, *421*, 126827. [CrossRef]
14. Muthusamy, S.; Peng, C.; Ng, J.-C. Effects of multi-component mixtures of polyaromatic hydrocarbons and heavy metal/loid(s) on Nrf2-antioxidant response element (ARE) pathway in ARE reporter-HepG2 cells. *Toxicol. Res.* **2016**, *4*, 1160–1171. [CrossRef] [PubMed]
15. Rosińska, A. The influence of UV irradiation on PAHs in wastewater. *J. Environ. Manag.* **2021**, *293*, 112760. [CrossRef]
16. Jia, J.-R.; Zhang, Y.-Z.; Yuan, X.; Qin, J.-K.; Yang, G.-K.; Yu, X.-Z. Reactive oxygen species participate in liver function recovery during compensatory growth in zebrafish (*Danio rerio*). *Biochem. Biophys. Res. Commun.* **2018**, *499*, 285–290. [CrossRef]
17. Yang, Y.; Dong, F.-S.; Liu, X.-G.; Xu, J. Crosstalk of oxidative damage, apoptosis, and autophagy under endoplasmic reticulum (ER) stress involved in thifluzamide-induced liver damage in zebrafish (*Danio rerio*). *Environ. Pollut.* **2018**, *243*, 1904–1911. [CrossRef]
18. Ma, J.; Qiao, L.; Ji, L.; Ren, B. The online monitoring and assessment of thallium stress using oxygen consumption rate and carbon dioxide excretion rate of zebrafish (*Danio rerio*). *Chemosphere* **2019**, *216*, 103–109. [CrossRef]
19. Potter, P.-E.; Meek, J.-L.; Neff, N.-H. Acetylcholine and choline in neuronal tissue measured by HPLC with electrochemical detection. *J. Neurochem.* **2015**, *41*, 188–194. [CrossRef]
20. Tom, T.-F.; Robin, J.-L.; Heather, S.-R. Towards a scheme of toxic equivalency factors (TEFs) for the acute toxicity of PAHs in sediment. *Ecotoxicol. Environ. Saf.* **2021**, *74*, 2245–2251.
21. Du, Y.; Yang, Y.; Wang, W.-L.; Zhou, Y.-T.; Wu, Q.-Y. Surrogates for the removal by ozonation of the cytotoxicity and DNA double-strand break effects of wastewater on mammalian cells. *Environ. Int.* **2020**, *135*, 105369. [CrossRef]
22. Reungoat, J.; Escher, B.-I.; Macova, M.; Argaud, F.-X.; Gernjak, W. Ozonation and biological activated carbon filtration of wastewater treatment plant effluents. *Water Res.* **2012**, *46*, 863–872. [CrossRef] [PubMed]
23. Lau, S.-S.; Wei, X.; Bokenkamp, K.; Wagner, E.-D.; Plewa, M.-J. Assessing additivity of cytotoxicity associated with disinfection byproducts in potable reuse and conventional drinking waters. *Environ. Sci. Technol.* **2020**, *54*, 5729–5736. [CrossRef] [PubMed]
24. Gonsioroski, A.; Mourikes, V.-E.; Flaws, J.-A. Endocrine disruptors in water and their effects on the reproductive system. *Int. J. Mol. Sci.* **2020**, *21*, 1929–1995. [CrossRef]
25. Kasonga, T.-K.; Coetzee, M.-A.-A.; Kamika, I.; Ngole, V.-M. Endocrine-disruptive chemicals as contaminants of emerging concern in wastewater and surface water: A review. *J. Environ. Manag.* **2021**, *277*, 111485. [CrossRef] [PubMed]
26. Johnson, A.-C.; Sumpter, J.-P. Removal of endocrine-disrupting chemicals in activated sludge treatment works. *Environ. Sci. Technol.* **2001**, *35*, 4697–4703. [CrossRef] [PubMed]
27. Tang, Z.; Liu, Z.-H.; Wang, H. Occurrence and removal of 17α-ethynylestradiol (EE2) in municipal wastewater treatment plants: Current status and challenges. *Chemosphere* **2021**, *271*, 129551. [CrossRef]
28. Tang, G.; Jie, W.; Yin, J.; Yan, Q.; Gong, Y. Establishment and application of evaluation method for genotoxicity of drinking water. *J. Southeast Univ.* **2018**, *48*, 170–174.
29. Sun, L.; Sun, S.-Q.; Bai, M. Internalization of polystyrene microplastics in *Euglena gracilis* and its effects on the protozoan photosynthesis and motility. *Aquat. Toxicol.* **2021**, *236*, 105840. [CrossRef]
30. Schnell, A.; Steel, P.; Melcer, H.; Hodson, P.-V.; Carey, J.-H. Enhanced biological treatment of bleached kraft mill effluents-II. Reduction of mixed function oxygenase (MFO) induction in fish. *Water Res.* **2000**, *34*, 501–509. [CrossRef]
31. Mišik, M.; Ferk, F.; Schaar, H.; Yamada, M.; Jaeger, W.; Knasmueller, S.; Kreuzinger, N. Genotoxic activities of wastewater after ozonation and activated carbon filtration: Different effects in liver-derived cells and bacterial indicators. *Water Res.* **2021**, *108*, 116328.
32. Kumari, V.; Yadav, A.; Haq, I.; Kumar, S.; Bharagava, R.-N.; Singh, S.-K. Genotoxicity evaluation of tannery effluent treated with newly isolated hexavalent chromium reducing Bacillus cereus. *J. Environ. Manag.* **2016**, *183*, 204–211. [CrossRef] [PubMed]
33. Knopper, L.-D.; Mcnamee, J.-P. Use of the Comet assay in environmental toxicology. *Methods Mol. Biol.* **2008**, *410*, 171. [PubMed]
34. Haq, I.; Raj, A. Biodegradation of Azure-B dye by Serratia liquefaciens and its validation by phytotoxicity, genotoxicity and cytotoxicity studies. *Chemosphere* **2018**, *196*, 58–68. [CrossRef] [PubMed]
35. Luan, H.-Y.; Zhao, J.; Yang, J.; Gao, X.; Song, J.-Y.; Chen, X.-F.; Cai, Q.-Y.; Yang, C. Integrated genotoxicity of secondary and tertiary treatment effluents in North China. *Sci. Total Environ.* **2022**, *865*, 161241. [CrossRef] [PubMed]
36. Kumar, V.; Ameen, F.; Islam, M.-A.; Agrawal, S.; Motghare, A.; Dey, A.; Shah, M.-P. Evaluation of cytotoxicity and genotoxicity effects of refractory pollutants of untreated and biomethanated distillery effluent using *Allium cepa*. *Environ Pollut.* **2022**, *300*, 118975. [CrossRef]
37. Gupta, A.; Kumar, M.; Ghosh, P.-S.; Thakur, I.-S. Risk assessment of a municipal extended aeration activated sludge treatment plant using physico-chemical and in vitro bioassay analyses. *Environ. Technol. Innov.* **2022**, *26*, 102254. [CrossRef]
38. Matejczyk, M.; Ofman, P.; Świsłocka, R.; Parcheta, M.; Lewandowski, W. The study of biological activity of mandelic acid and its alkali metal salts in wastewaters. *Environ. Res.* **2022**, *205*, 112429. [CrossRef]
39. Li, M.; Hu, C.-W.; Gao, X.-Y.; Xu, Y. Genotoxicity of organic pollutants in source of drinking water on microalga *Euglena gracilis*. *Ecotoxicology* **2009**, *18*, 669–676. [CrossRef]
40. Qin, Y.-F.; Liu, Y.-Q.; Wang, J.-B.; Lu, Y.; Xu, Z.-M. Emission of PAHs, PCBs, PBDEs and heavy metals in air, water and soil around a waste plastic recycling factory in an industrial park, Eastern China. *Chemosphere* **2022**, *294*, 133734. [CrossRef]
41. Gao, Y.-H.; Shi, X.; Jin, X.; Wang, X.-C.; Jin, P.-K. A critical review of wastewater quality variation and in-sewer processes during conveyance in sewer systems. *Water Res.* **2023**, *228*, 119398. [CrossRef] [PubMed]

42. Yu, R.-Z.; Mu, Y.-F.; Wang, H.-Y.; Xu, Q.-J.; Gao, J.-F.; Du, L.-N.; Meng, W. Review on the aquatic organisms toxicity test in the whole effluent assessment. *Res. Environ. Sci.* **2014**, *27*, 390–397.
43. Acosta-Lizárraga, L.-G.; Berges-Tiznado, M.-E. Bioaccumulation of mercury and selenium in tissues of the mesopelagic fish Pacific hake (*Merluccius productus*) from the northern Gulf of California and the risk assessment on human health. *Chemosphere* **2020**, *255*, 126941. [CrossRef] [PubMed]
44. Sun, S.-X.; Zhang, Y.-N.; Lu, D.-L.; Li, W.; Samwel, M.-L. Concentration-dependent effects of 17β-estradiol and bisphenol A on lipid deposition, inflammation and antioxidant response in male zebrafish. *Chemosphere* **2019**, *237*, 124422. [CrossRef] [PubMed]
45. Feng, H.-Y.; Fu, X.-Q.; Zhao, Q. Health risk assessment of polycyclic aromatic hydrocarbons in soils of Ningbo area, China. *J. Agro-Environ. Sci* **2011**, *30*, 1998–2004.

toxics

MDPI

Article

Ambient Volatile Organic Compound Characterization, Source Apportionment, and Risk Assessment in Three Megacities of China in 2019

Zhanshan Wang [1,†], Puzhen Zhang [1,†], Libo Pan [1,†], Yan Qian [1], Zhigang Li [1], Xiaoqian Li [1], Chen Guo [1], Xiaojing Zhu [1], Yuanyuan Xie [2] and Yongjie Wei [1,*]

1 State Key Laboratory of Environmental Criteria and Risk Assessment, Chinese Research Academy of Environmental Sciences, Beijing 100012, China; 18701650609@163.com (Z.W.)
2 Foreign Environmental Cooperation Centre, Ministry of Ecology and Environment, Beijing 100035, China
* Correspondence: weiyj@craes.org.cn
† These authors contributed equally to this work.

Abstract: In order to illustrate pollution characterization, source apportionment, and risk assessment of VOCs in Beijing, Baoding, and Shanghai, field observations of CO, NO, NO_2, O_3, and volatile organic compounds (VOCs) were conducted in 2019. Concentrations of VOCs were the highest in Beijing (105.4 ± 52.1 ppb), followed by Baoding (97.1 ± 47.5 ppb) and Shanghai (91.1 ± 41.3 ppb). Concentrations of VOCs were the highest in winter (120.3 ± 61.5 ppb) among the three seasons tested, followed by summer (98.1 + 50.8 ppb) and autumn (75.5 + 33.4 ppb). Alkenes were the most reactive VOC species in all cities, accounting for 56.0%, 53.7%, and 39.4% of ozone formation potential in Beijing, Baoding, and Shanghai, respectively. Alkenes and aromatics were the reactive species, particularly ethene, propene, 1,3,5-trimethylbenzene, and m/p-xylene. Vehicular exhaust was the principal source in all three cities, accounting for 27.0%, 30.4%, and 23.3% of VOCs in Beijing, Baoding, and Shanghai, respectively. Industrial manufacturing was the second largest source in Baoding (23.6%) and Shanghai (21.3%), and solvent utilization was the second largest source in Beijing (25.1%). The empirical kinetic modeling approach showed that O_3 formation was limited by both VOCs and nitric oxides at Fangshan (the suburban site) and by VOCs at Xuhui (the urban site). Acrolein was the only substance with an average hazard quotient greater than 1, indicating significant non-carcinogenic risk. In Beijing, 1,2-dibromoethane had an R-value of 1.1×10^{-4} and posed a definite carcinogenic risk.

Keywords: VOCs; ozone; PSCF; source apportionment; EKMA; risk assessment

Citation: Wang, Z.; Zhang, P.; Pan, L.; Qian, Y.; Li, Z.; Li, X.; Guo, C.; Zhu, X.; Xie, Y.; Wei, Y. Ambient Volatile Organic Compound Characterization, Source Apportionment, and Risk Assessment in Three Megacities of China in 2019. *Toxics* **2023**, *11*, 651. https://doi.org/10.3390/toxics11080651

Academic Editors: David C. Spink and Andrey Y. Khlystov

Received: 5 May 2023
Revised: 24 July 2023
Accepted: 25 July 2023
Published: 27 July 2023

1. Introduction

In September 2013, the Chinese government implemented the Action Plan on Air Pollution Prevention and Control, resulting in significant reductions in ambient concentrations of CO, SO_2, NO_2, and fine particulate levels nationwide [1–3]. However, O_3 pollution has not decreased and appears to be worsening in China. The O_3 concentration showed an increasing trend of 1–3 ppbv/y from 2013 to 2017 in eastern China [4]. Volatile organic compounds (VOCs) and nitrogen oxides (NOx) are the main precursors of O_3. The relationship between O_3 and its precursors is highly nonlinear due to the complex photochemical reactions that occur in the atmosphere [5–7]. Conditions for forming ground-level O_3 can be divided into VOC-limited, NOx-limited, and both VOC- and NOx-limited [8]. In most developed areas of China, including the Yangtze River Delta, Jing-Jin-Ji, and Pearl River Delta regions, O_3 formation is reported to be VOC-limited [9–11]. Thus, controlling VOC emissions is critical for reducing O_3 pollution in China.

High concentrations of VOCs have adverse effects on public health by affecting the respiratory and cardiovascular systems [12–16]. Previous studies showed that cancer risk is greater in high-VOC-exposure areas than in clean areas [17,18]. Hazardous VOCs, including non-carcinogens and carcinogens, account for 20–40% of all non-methane VOCs in China [19]. Many VOC species, including benzene and 1,3-Butadiene, are classified as hazardous air pollutants by the United States Environmental Protection Agency (USEPA) and other international agencies [20–23].

The Beijing–Tianjin–Hebei (BTH) and Yangtze River Delta (YRD) regions are two of the largest urban agglomerations in China. Numerous studies of VOCs have been performed in BTH and YRD. Several studies have examined the general characteristics of VOCs and discussed their spatiotemporal variations [24–28]. Other studies have focused on the relative reactivity and ozone formation potential (OFP) of VOCs [29–32]. Several studies have aimed to reveal the health effects of VOCs [33–38]. Additionally, the emission inventory [39], regional transport [36,40–42], and source apportionment [26,43–46] of VOCs have been discussed.

Beijing, the capital of China, and Baoding, one of the most air-polluted cities, are both located in the BTH region. Shanghai is one of the most economically developed cities in the YRD and has relatively concentrated energy consumption and pollutant emissions. Most previous studies have been limited to a small number of sampling locations or a short sampling period. In this study, field observations of CO, NO, NO_2, O_3, and VOCs were conducted in these three megacities. The main objectives of this study were to: (1) characterize the concentrations and spatiotemporal variations of VOCs; (2) discuss the regional transport and source apportionment of VOCs; (3) determine the roles of VOCs in ground-level O_3 formation; (4) estimate the carcinogenic and non-carcinogenic risks of VOCs; and (5) identify the key hazardous VOCs in the three cities.

2. Methodology

2.1. Sampling Site and Period

Ten sites in the three cities were selected for this study, among which four were in Beijing, three were in Baoding, and three were in Shanghai (Figure 1). The sampling sites in each city included a background site, an urban site, and a suburban site. A continuous 2-week period in each of the four seasons was selected. However, due to the COVID-19 crisis, sampling in the spring was terminated. Thus, the sampling periods were 15–28 August in the summer, 13–26 October in the autumn, and 18–31 December in the winter of 2019.

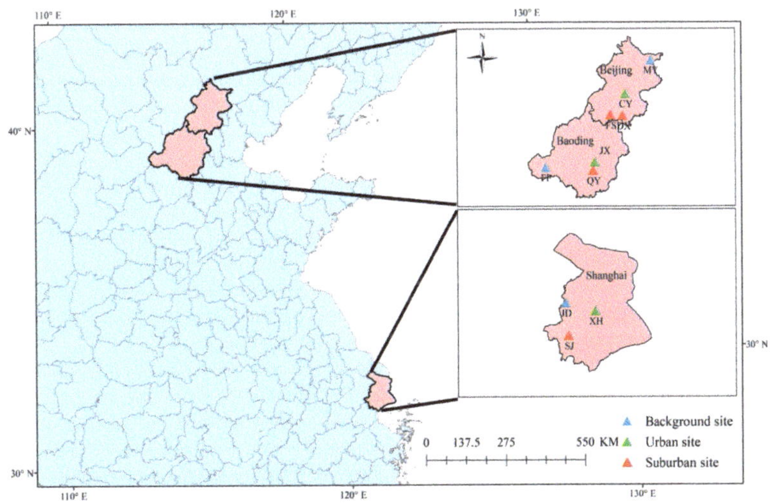

Figure 1. Sites sampled in this study. MY, FP, and JD are background sites. CY, JX, and XH are urban sites. FS, DX, QY, and SJ are suburban sites.

2.2. Sampling and Analysis

CO, NO, NO_2, and O_3 were observed using a sensor-based air monitoring instrument developed by the Chinese Research Academy of Environmental Sciences. This instrument passed an intercomparison assessment with an instrument produced by TSI Corporations. The correlation coefficient was 0.92, and the two instruments' relative error was 8%. VOCs were sampled at 16:00 at each site using a 3.2-L Summa canister (Entech Instruments Inc., Simi Valley, CA, USA). In total, 99 VOCs were observed, including 29 alkanes, 13 alkenes, 1 alkyne (acetylene), 16 aromatics, 32 haloalkanes, and 8 oxygenated VOCs (OVOCs). VOC samples were analyzed using the Agilent 5973N gas chromatography–mass selective detector flame ionization detector (Agilent Technologies, Santa Clara, CA, USA). A liquid nitrogen primary cryogenic trap with glass beads at −160 °C was used to trap VOCs. Then, the trap was heated to 10 °C, and target compounds were transferred to a secondary trap at −50 °C. Next, the VOCs were transferred using helium to a third trap at −170 °C. A DB-1 capillary column (60 m × 0.32 mm × 1.0 μm, Agilent Inc.) was used with helium as the carrier gas. Rigorous quality assurance and quality control procedures were employed. Periodic calibration was performed every 5 d. Calibration curve results for a given target species with less than 10% variation relative to the actual values were considered acceptable. Meteorological data were obtained from the China Meteorological Station Data Sharing Service System (http://cdc.cma.gov.cn/home.do, accessed on 5 January 2021).

2.3. Determination of the Ozone Formation Potential

The OFP can be used to characterize the maximum amount of O_3 production possible from a given VOC species alone under optimal conditions. The key compounds responsible for O_3 formation can be determined from the respective OFPs [38]. The OFPs are calculated based on the maximum incremental reactivity (MIR) of each individual species and given by the following equation:

$$OFP_i = VOCs_i \times MIR_i \tag{1}$$

where OFP_i is the OFP of VOC species i, $VOCs_i$ is the concentration of VOC species i, and MIR_i is the O_3 formation coefficient for VOC species i at the maximum incremental reactivity of O_3 [47].

2.4. Positive Matrix Factorization Receptor Model

The sources of $PM_{2.5}$ were analyzed using the positive matrix factorization (PMF) receptor model. First, the error associated with the chemical component weights of the receptor was determined. Then, the main sources of contamination and their contribution ratios were determined using the least squares method. PMF is a type of multivariate factor analysis in which a mathematical method decomposes matrix X containing sample data for a given species into two matrices: factor contributions (G) and factor spectra (F). This method does not require the input of a source spectrum and ensures that the decomposition factor contribution (G) and factor spectrum (F) are non-negative [48]. The following formula represents the matrix X:

$$x_{ij} = \sum_{k=1}^{p} g_{ik} f_{kj} + e_{ij} \tag{2}$$

where x_{ij} is the concentration of species j in sample i, p is the number of factors, g_{ik} is the contribution of factor k to sample i, f_{kj} is the contribution of factor k to species j, and e_{ij} is the error of species j in sample i.

The uncertainty of a sample was calculated from the error fraction and the method detection limit (MDL). If the concentration was unknown, it was set to 1/2 of the geometric mean value. If the concentration was below the MDL, it was set to 1/2 of the MDL, and the uncertainty was set to 5/6 of the MDL. If the concentration was higher than the MDL, the uncertainty calculation was based on the error fraction as follows:

$$Unc = \sqrt{(Error\ Fraction \times concentration)^2 + (0.5 \times MDL)^2} \tag{3}$$

The PMF analysis depends on the objective function (Q) to minimize the residual and uncertainty, as shown in Equation (4). The calculation of Q_{exp} is shown in Equation (5).

$$Q = \sum_{i=1}^{n} \sum_{j=1}^{m} \left[\frac{x_{ij} - \sum_{k=1}^{p} g_{ik} \times f_{kj}}{u_{ij}} \right]^2 \quad (4)$$

$$Q_{exp} = n \times m - p \times (n + m) \quad (5)$$

where n and m are the numbers of species and samples, respectively, and u_{ij} is the uncertainty of the jth species in the ith sample.

Several aspects were considered to define a reasonable result (Liu et al., 2021) [22]: (1) the value of Q/Q_{exp} from PMF should be close to one; (2) the change rate of Q/Q_{exp} should be stable; and (3) the explored factors should be physically plausible and interpretable. Ultimately, a five-factor solution was determined in this study. The Q/Q_{exp} values were 1.3, 1.4, and 1.4 for Beijing, Baoding, and Shanghai, respectively.

2.5. Potential Source Contribution Function

The potential source contribution function (PSCF) was employed in this study using the software Meteoinfo (3.6.3) to identify the local and long-range transport pathways of VOCs. The PSCF is a backward-trajectory-based method that combines pollutant concentrations, reflecting the potential for this area to become a source of VOC pollution [49]. The PSCF is a position function defined by unit indexes i and j:

$$PSCF_{ij} = \frac{m_{ij}}{n_{ij}} W_{ij} \quad (6)$$

where n_{ij} is the number of trajectory endpoints that fall within the ijth grid cell, m_{ij} is the number of endpoints corresponding to trajectories that exceed the threshold criterion at the receptor site [50], and ij is the grid cell. The arbitrary weighting function W_{ij} was applied to reduce the uncertainty caused by small values of n_{ij}:

$$W_{ij} = \begin{cases} 0.70 & 3n_{ave} > n_{ij} \geq 1.5n_{ave} \\ 0.42 & 1.5n_{ave} > n_{ij} \geq n_{ave} \\ 0.05 & n_{ave} > n_{ij} \end{cases} \quad (7)$$

where n_{ave} is the average value of the endpoints of the trajectory through all the grids. In this study, the n_{ave} was 1.33, 1.31, and 1.19 for Fangshan, Jiading, and Jingxiu sites, respectively. In this study, the 24-h backward trajectory was calculated at 1-h intervals according to Beijing local time (UTC + 8). The arrival height was set to 100 m above the ground. Meteorological data were obtained from the National Oceanic and Atmospheric Administration (ftp://arlftp.arlhq.noaa.gov/pub/archives/gdas0p5/, accessed on 6 January 2021) with a grid resolution of 0.25° × 0.25°. The threshold value was the average VOC concentration during the observation (Beijing 105.4 ppb, Baoding 97.1 ppb, Shanghai 91.1 ppb). The total number of trajectories was 1008 at each site.

2.6. Observation-Based Model

The observation-based model (OBM) was used in this study in combination with the Master Chemical Mechanism (v3.3.1; http://mcm.leeds.ac.uk/MCM/, accessed on 10 January 2021), a near-explicit mechanism describing the oxidation reactions of 146 primary VOCs and the latest inorganic chemistry data from the International Union of Pure and Applied Chemistry evaluation [51]. The OBM has been widely used to identify photochemical reactivity and photochemical products in various environments [52]. Hourly concentrations of the observed VOCs and four trace gases (CO, NO, NO_2, and SO_2) and hourly meteorological parameters (temperature and relative humidity) were used as input data. The instantaneous concentration of VOCs was converted to hourly concentrations according to the linear regressions with CO, following the method by Yang et al. [53]. The OBM assesses the sensitivity of O_3 photochemical production by calculating the relative incremental reactivity and altering the concentrations of its precursors without requiring detailed or accurate knowledge about these emissions [54]:

$$RIR(X) = \frac{\left[P_{O_3}(X) - P_{O_3}(X - \Delta X)\right] / P_{O_3}(X)}{\Delta S(X)/S(X)} \tag{8}$$

where X is a precursor of O_3, ΔX is the change in the concentration of X, $P(O_3)$ represents the net O_3 production rate, S(X) is the measured concentration of precursor X, and $\Delta S(X)/S(X)$ represents the relative change in S(X), which was 20% in this study.

2.7. Human Health Risk Assessment

The USEPA proposed a method that uses the ambient mass concentration of air pollutants as an exposure evaluation parameter. The health risk of VOCs is divided into non-carcinogenic and carcinogenic risks, which are represented by the hazard quotient (HQ) and the lifetime carcinogenic risk (R), respectively [55]. An HQ less than 1 indicates no significant non-carcinogenic risk for adults, and an R-value less than 1×10^{-6} suggests an acceptable carcinogenic risk [38]. The specific calculation is shown in the Supplemental Material File.

3. Results and Discussion

3.1. Chemical Characteristics of Volatile Organic Compounds

The concentrations of VOCs were the highest in winter (120.3 ± 61.5 ppb) among the three seasons assessed, followed by summer (98.1 ± 50.8 ppb) and autumn (75.5 ± 33.4 ppb), as shown in Figure 2. However, total VOCs were similar in the winter. In Baoding, VOC concentrations in the winter were significantly higher than those in the summer and autumn. VOC concentrations in the winter were 1.9 times those in the summer in Baoding. Baoding is a city of heavy industry, and the increase in industrial and heating emissions in the winter has led to an increase in VOC concentrations. VOC concentrations in the winter were close to those in the summer in Beijing and Shanghai. High temperatures and high solar radiation lead to higher solvent volatilization and plant-related VOC emissions in the summer. VOC concentrations in the autumn were the lowest of the three seasons.

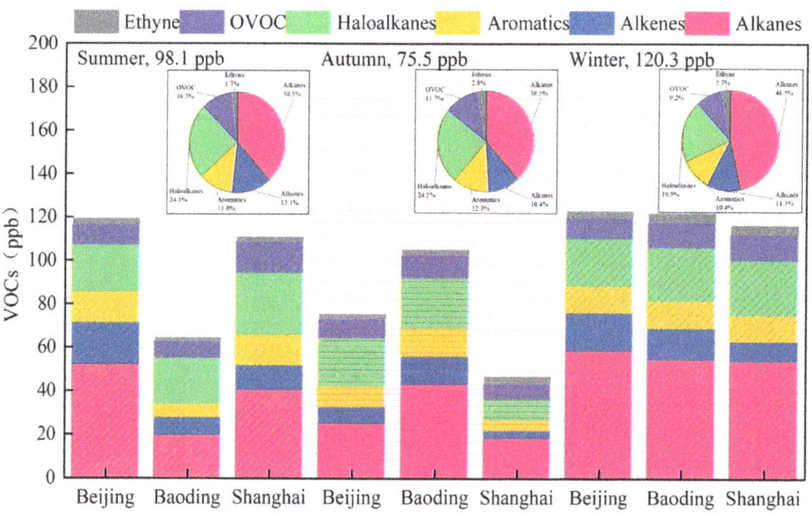

Figure 2. Concentrations and chemical characteristics of VOCs in different seasons during the observation period.

Alkanes were the dominant VOC species in all seasons, especially winter, exceeding 40% of the total (see Figure 3). The concentrations of haloalkanes were the second highest, and their proportion among the other VOCs decreased in the winter. The concentrations of alkenes were highest in the summer, and those of OVOCs and aromatics were highest in the autumn. The concentration of alkenes was higher in Beijing, and that of OVOCs was higher in Shanghai. The aromatic concentrations were at similar proportions in all three cities. VOC concentrations were highest in Beijing (105.4 ± 52.1 ppb), followed by Baoding (97.1 ± 47.5 ppb) and Shanghai (91.1 ± 41.3 ppb). The VOC concentrations in Beijing were 1.09 and 1.16 times those in Baoding and Shanghai, respectively. There were few differences in VOC concentrations throughout the year among the three cities. The differences in VOCs among cities were related to air pollutant emissions, sampling locations, and meteorological conditions.

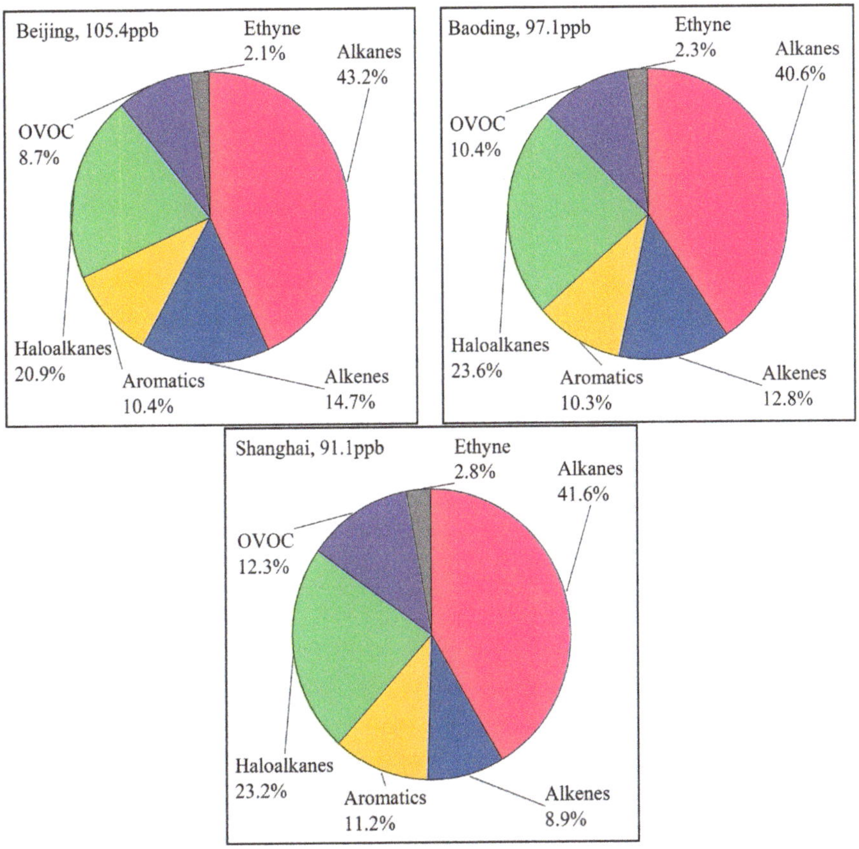

Figure 3. Chemical characteristics of VOCs observed in the three cities.

Table 1 provides a comparison of the monitoring results from this study with previous observations of VOC species. The ethane, ethylene, propane, and acetylene concentrations in Shanghai were higher than those in previous reports. The concentrations of toluene and benzene in Shanghai were lower, and those of ethane and propane in Beijing were higher than those in previous studies. The ethylene, acetylene, and toluene concentrations in Beijing are similar to previously reported levels. The concentrations of VOC species in Baoding were higher than those in previous studies. Although the COVID-19 pandemic lockdown had a certain influence on industrial production in China, the concentration of VOC species did not show an obvious decreasing trend.

Table 1. Comparison of the monitoring results for VOC species between the present study and previous reports.

Reference	Sampling Period	Sampling Site	Site Category	Monitoring Method	Ethane	Ethylene	Propane	Acetylene	Toluene	Benzene
Dai et al. (2010) [33]	2007–2010	Shanghai	Urban	Manual	—	—	4.81	—	4.70	1.81
Zheng et al. (2019) [56]	Autumn 2016	Shanghai	Urban	Online	2.22	1.52	3.59	1.17	5.04	0.70
Zheng et al. (2019) [56]	Autumn 2016	Shanghai	Suburban	Online	3.01	0.99	4.22	0.03	0.96	0.44
Zhang et al. (2020) [7]	7 April to 25 September 2018	Shanghai	Suburban	Online	1.26	1.56	2.93	0.73	1.87	—
This study	Summer to Winter 2019	Shanghai	Urban	Manual	5.98	2.60	6.87	2.89	2.28	0.93
Zhang et al. (2020) [7]	Autumn 2016	Beijing	Urban	Online	3.42	2.13	2.85	0.68	2.00	4.74
Zhang et al. (2020) [7]	Winter 2016	Beijing	Urban	Online	4.60	2.43	6.70	0.26	1.82	6.04
Zhang et al. (2020) [7]	Spring 2017	Beijing	Urban	Online	1.93	0.59	2.33	0.51	1.17	5.41
Zhang et al. (2020) [7]	Summer 2017	Beijing	Urban	Online	2.33	0.57	2.65	0.90	1.34	6.99
Shi et al. (2020) [39]	December 2016 to January 2017	Beijing	Urban	Online	—	12.07	—	8.98	3.63	3.27
This study	Summer to Winter 2019	Beijing	Urban	Manual	7.37	2.59	7.21	2.27	1.81	1.14
Wang et al. (2021) [28]	May to September 2019	Baoding	Urban	Online	3.98	1.51	2.19	0.37	0.58	0.31
This study	Summer to Winter 2019	Baoding	Urban	Manual	5.01	2.16	5.85	2.27	2.61	1.94

3.2. *The Ozone Formation Potential of Volatile Organic Compounds*

The OFP values of VOC species at the sampling sites were calculated (see Figure 4). Both VOCs and VOC OFP were at their maximum in Beijing. The concentrations of VOCs in Baoding and Shanghai were similar, but the OFP values were significantly lower in Shanghai than in Baoding. Alkenes were the most reactive species of VOCs in all cities, accounting for 56.0%, 53.7%, and 39.4% of the OFP in Beijing, Baoding, and Shanghai, respectively. Aromatics were the second most reactive species of VOCs, accounting for 20.7%, 21.0%, and 28.3% of the OFP in Beijing, Baoding, and Shanghai, respectively.

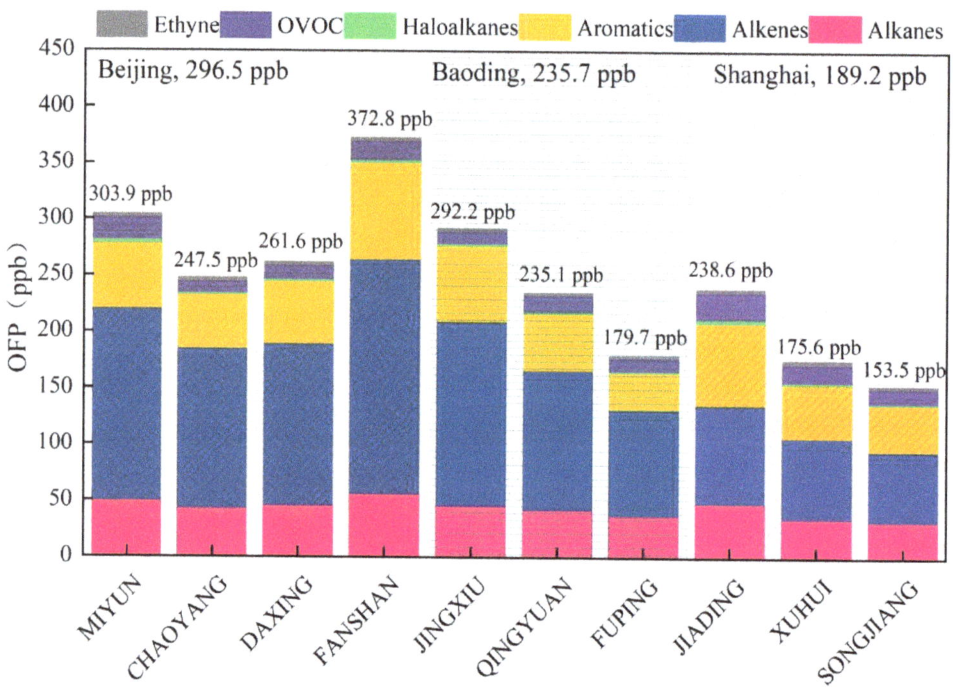

Figure 4. OFPs of VOC species at all sampling sites.

Notably, the OFP values of VOCs at Fangshan, a suburban site near a petrochemical plant, were the highest in Beijing. The OFP of VOCs at the urban Jingxiu site were the highest in Baoding. The OFPs of VOCs at Jiading, a background site, were the highest among sites in Shanghai. The top 10 VOC species with regard to OFP were identified and are shown in Table 2. The most reactive species were alkenes and aromatics, particularly ethene, propene, 1,3,5-trimethylbenzene, and m/p-xylene. The emission sources of these species should be strictly controlled.

Table 2. Top 10 VOC species with the highest OFP values in the three cities.

Cities	VOCs	OFP
Beijing	Ethene	32.6
	Trans-2-butene	29.5
	1-Butene	18.8
	Propene	18.4
	Cis-2-butene	17.4
	1,3,5-Trimethylbenzene	13.3
	1-Pentene	13.1
	Isoprene	12.2
	1-Hexene	10.2
	m/p-xylene	9.7
Baoding	Ethene	26.5
	Cis-2-pentene	16.4
	1,3-Butadiene	16.1
	Propene	15.1
	Trans-2-butene	14.0
	1-Pentene	11.6
	Toluene	11.3
	m/p-xylene	9.1
	1,3,5-Trimethylbenzene	8.6
	1-Butene	7.8
Shanghai	Ethene	24.6
	m/p-xylene	12.7
	Propene	12.4
	Toluene	9.9
	Cis-2-pentene	8.0
	1-Pentene	7.5
	Acrolein	7.1
	1,2,3-Trimethylbenzene	6.9
	1-Hexene	6.4
	O-xylene	5.9

3.3. Potential Source Areas of Volatile Organic Compounds

The Fangshan, Jingxiu, and Jiading sites were selected for source area analysis due to their high VOC concentrations and OFP values. The potential source areas of VOCs for the three cities were simulated, as shown in Figure 5. Three main potential source areas of VOCs for Fangshan were identified: the southeast region along the border of Beijing, Tianjin, and Hebei; the southwest region along the Taihang Mountains; and the western region. Two main potential source areas of VOCs were identified for Jingxiu: the northeast region near Beijing and the southeast region in Hebei. The potential source areas of VOCs for Jiading were located around the site and at sea. VOCs can be transported to and from the sea via airflow and ship emissions.

3.4. Source Apportionment of Volatile Organic Compounds

We did not analyze species with a concentration below the MDL more than 50% of the time or with a significantly low signal-to-noise ratio [2,12]. After screening, 53 compounds in Beijing and Baoding and 47 compounds in Shanghai were selected. Five sources (vehicular exhaust, industrial manufacturing, solvent utilization, fuel combustion, and biogenic VOCs) were identified using the PMF model. Modeled source profiles and the relative contributions of individual sources to each species analyzed are shown in Figure 6.

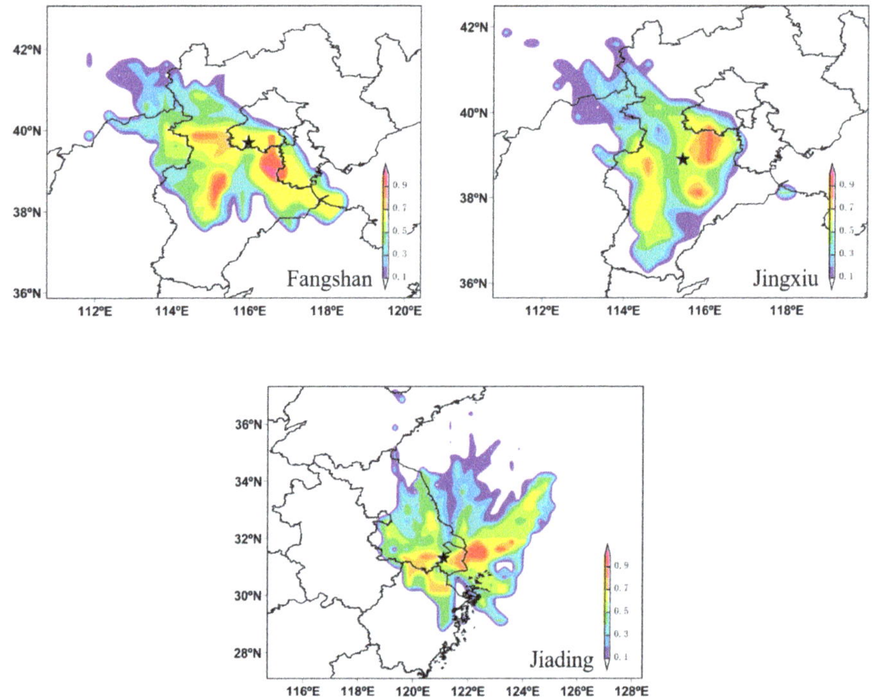

Figure 5. Potential source areas of VOCs for the Fangshan, Jingxiu, and Jiading sites, the five-pointed star referred to the sampling sites.

In the source profiles for Beijing, the first source was characterized by significant amounts of methyl cyclopentane, n-undecane toluene, and 2-butanone, which are representative of industrial manufacturing [57]. The second source was characterized by high concentrations of carbon tetrachloride, tetrachloroethylene, and acetone, which are widely used as solvents [58]. The third source was associated with high concentrations of acetylene and alkane, such as isopentane, n-octane, and n-dodecane, which are major species in vehicular emissions [59]. The fourth source profile was rich in 1-butene, propane, and 2-methylhexane, tracers of fuel combustion [60]. The fifth source represented 97% of the total isoprene, considered the most important biogenic hydrocarbon [61].

In the source profiles for Baoding, the first source was characterized by a high concentration of isoprene(biogenic). The second source was characterized by significant amounts of 3-methylpentane, trans-2-butene, and 1-butene, which are representative of fuel combustion. The third source was associated with high concentrations of 1,2,4-trichlorobenzene and acetone, widely used as solvents. The fourth source profile was rich in benzene, toluene, n-undecane, and n-nonane, major species emitted from industrial manufacturing. The fifth source was characterized by high concentrations of acetylene, propane, and propene, tracers of vehicular exhaust.

In the source profiles for Shanghai, the first source was characterized by significant amounts of acetone, n-propyl benzene, and tetrachloroethylene, which are widely used as solvents. The second source profile was rich in dichloromethane, trichloromethane, toluene, and n-dodecane, major species emitted from industrial manufacturing. The third source represented 92% of the total isoprene, considered the most important biogenic hydrocarbon. The fourth source was characterized by high concentrations of 1-butene and 1-hexene, which are representative of fuel combustion. The fifth source was associated with high concentrations of methyl tertiary butyl ether, ethene, and ethane, major species emitted in vehicular exhaust.

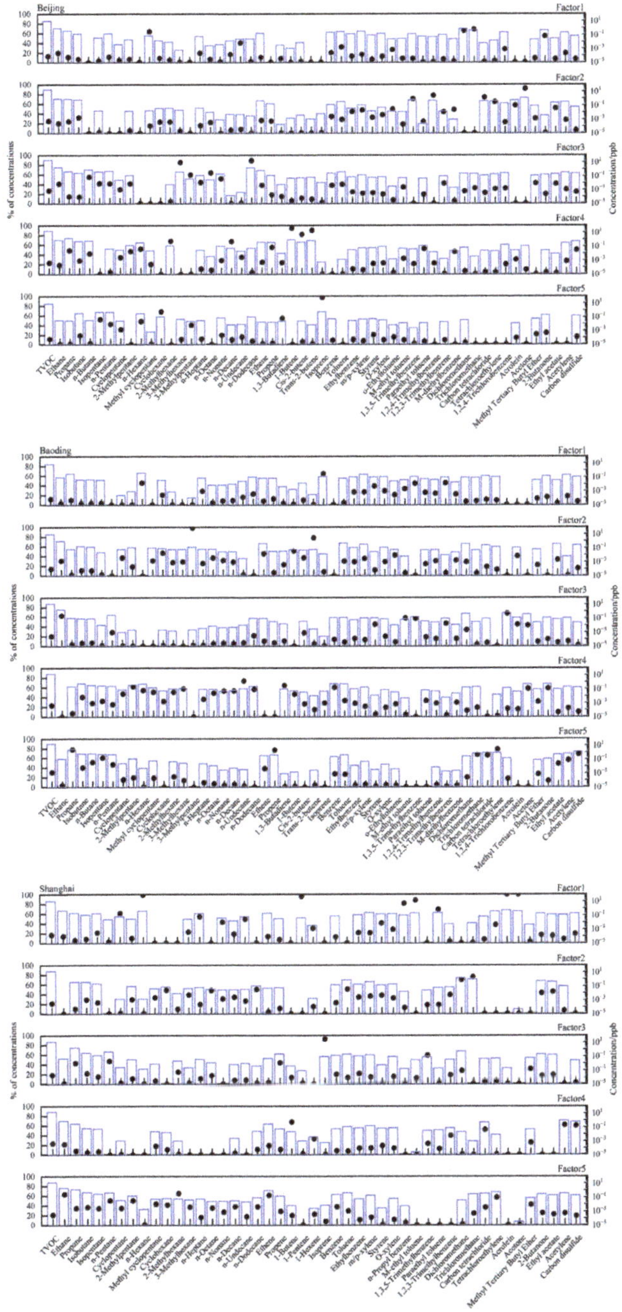

Figure 6. Five source profiles (bars) and contribution percentages (dots) representing each source factor were resolved using the PMF model.

Figure 7 illustrates the percentage source contributions during the sampling period in the three cities. Vehicular exhaust was the largest contributor in all three cities, accounting for 27.0%, 30.4%, and 23.3% of VOCs in Beijing, Baoding, and Shanghai, respectively. Industrial manufacturing was the second largest contributor in Baoding (23.6%) and Shanghai (21.3%), and solvent utilization was the second largest contributor in Beijing (25.1%). Fuel combustion was the third largest contributor in Beijing (23.2%) and Shanghai (20.7%), and solvent utilization was the third largest contributor in Baoding (20.0%). Biogenic sources of VOCs were also important, accounting for 11.5%, 11.9%, and 18.1% of VOCs in Beijing, Baoding, and Shanghai, respectively.

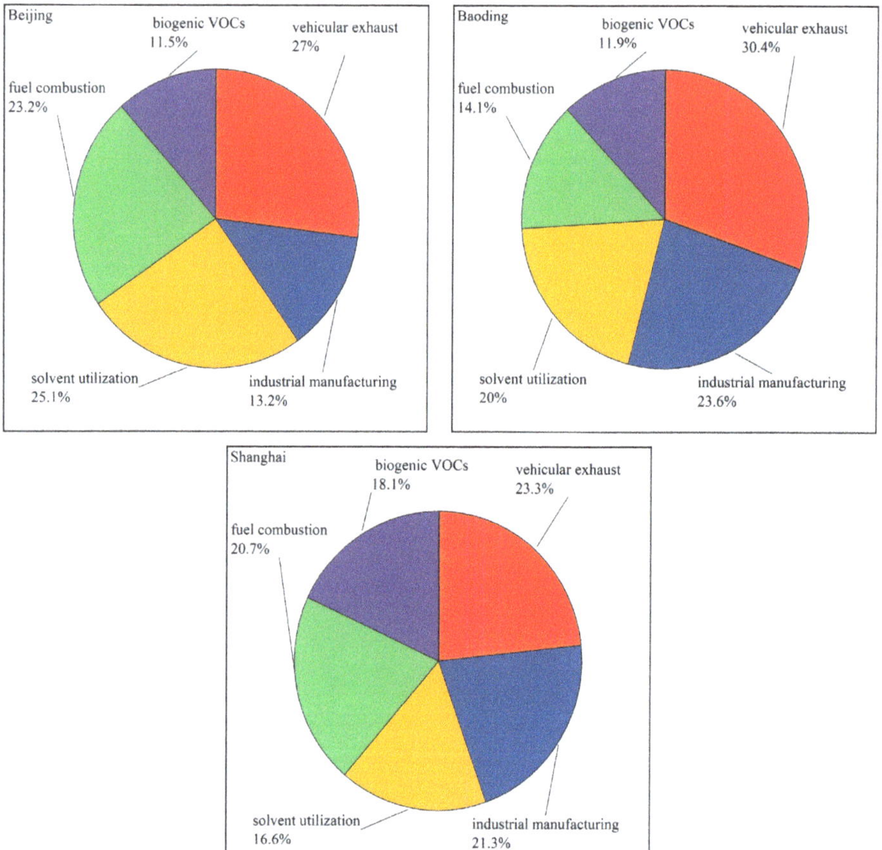

Figure 7. Source apportionment results for VOCs in the three cities.

3.5. Empirical Kinetic Modeling Approach

Meteorological data for the Fangshan and Xuhui sites were obtained from the China Meteorological Station Data Sharing Service System. Thus, the empirical kinetic modeling approach (EKMA) curves for those two sites in the summer period were simulated using the OBM model, as shown in Figure 8. During the sampling period in summer, the average temperature and relative humidity were 30.4 °C and 60% in Fangshan and 31.5 °C and 71% in Xuhui. The EKMA plot was split into two areas by a ridgeline denoting the local maxima in the rate of O_3 formation. The upper–left and lower–right areas represent O_3 formation under VOC-limited and NOx-limited conditions, respectively. The base scenario point for the Fangshan site is located near the ridgeline, indicating a VOCs- and NOx-limited condition. The base scenario point for the Xuhui site is located in the upper-left area, indicating VOCs limitation. Fangshan is a suburban site, and Xuhui is an urban site. Previous studies have reported that urban and suburban areas in China were under VOC-limited and both VOC- and NOx-limited conditions [8,62], respectively.

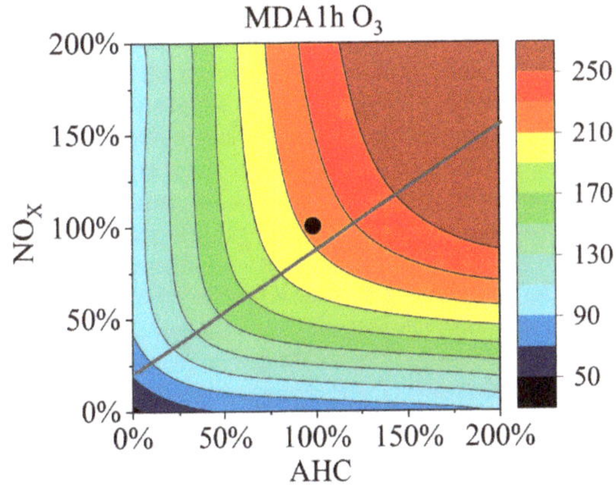

(**a**) Fangshan in Beijing

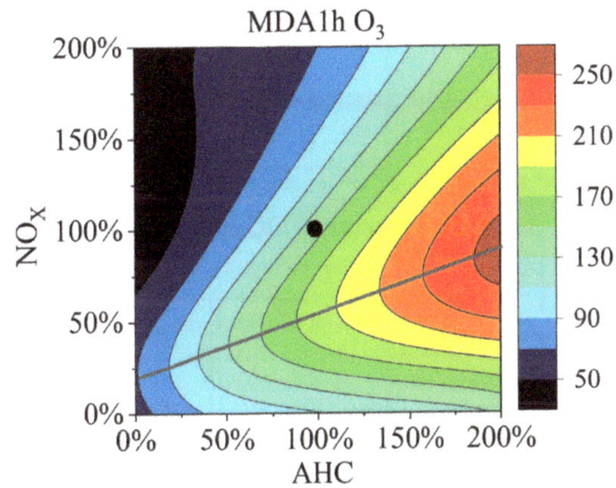

(**b**) Xuhui in Shanghai

Figure 8. EKMA graphs for suburban Beijing and urban Shanghai. The black dot represents the base scenario, and the gray line represents the ridgeline of the EKMA curve. The AHC represents anthropogenic VOCs. The MDA 1 h O_3 represents the daily maximum 1 h average O_3 concentration.

3.6. Health Risk Assessment of Volatile Organic Compounds

Forty-four species were targeted for health risk assessment, and their non-carcinogenic and carcinogenic risks are presented in Figure 9. The USEPA states that pollutants with an HQ of less than 1 pose no significant non-carcinogenic risk to adults. In this study, the average HQ values of the selected VOC species ranged from 5.3×10^{-6} to 16.9×10^{-6} in Beijing, 4.8×10^{-6} to 8.9×10^{-6} in Baoding, and 8.8×10^{-6} to 18.3×10^{-6} in Shanghai. Acrolein was the only substance with an average HQ value greater than 1, indicating significant non-carcinogenic risk. VOC species with carcinogenic risks of $>10^{-4}$, 10^{-5} to 10^{-4}, 10^{-5} to 10^{-6}, and $<10^{-6}$ are classified as definite, probable, possible, and negligible risks, respectively [63]. In this study, the average R-value for the selected VOC species ranged from 4.7×10^{-9} to 1.1×10^{-4} in Beijing, from 6.7×10^{-9} to 6.2×10^{-5} in Baoding,

3. Zheng, H.; Kong, S.; Chen, N.; Niu, Z.; Zhang, Y.; Jiang, S.; Yan, Y.; Qi, S. Source apportionment of volatile organic compounds: Implications to reactivity, ozone formation, and secondary organic aerosol potential. *Atmos. Res.* **2021**, *249*, 105344. [CrossRef]
4. Li, K.; Jacob, D.J.; Liao, H.; Shen, L.; Zhang, Q.; Bates, K.H. Anthropogenic drivers of 2013–2017 trends in summer surface ozone in China. *Proc. Natl. Acad. Sci. USA* **2019**, *116*, 422–427. [CrossRef]
5. Chang, C.-T.; Lee, C.-H.; Wu, Y.-P.; Jeng, F.-T. Assessment of the strategies for reducing volatile organic compound emissions in the automotive industry in Taiwan. *Resour. Conserv. Recycl.* **2002**, *34*, 117–128. [CrossRef]
6. Monti, M.; Perin, E.; Conterosito, E.; Romagnolli, U.; Muscato, B.; Girotto, M.; Scrivani, M.T.; Gianotti, V. Development of an advanced extrusion process for the reduction of volatile and semi-volatile organic compounds of recycled HDPE from fuel tanks. *Resour. Conserv. Recycl.* **2023**, *188*, 106691. [CrossRef]
7. Zhang, K.; Li, L.; Huang, L.; Wang, Y.; Huo, J.; Duan, Y.; Wang, Y.; Fu, Q. The impact of volatile organic compounds on ozone formation in the suburban area of Shanghai. *Atmos. Environ.* **2020**, *232*, 117511. [CrossRef]
8. Wang, T.; Xue, L.; Brimblecombe, P.; Lam, Y.F.; Li, L.; Zhang, L. Ozone pollution in China: A review of concentrations, meteorological influences, chemical precursors, and effects. *Sci. Total Environ.* **2017**, *575*, 1582–1596. [CrossRef] [PubMed]
9. Geng, F.; Tie, X.; Xu, J.; Zhou, G.; Peng, L.; Gao, W.; Tang, X.; Zhao, C. Characterizations of ozone, NOx, and VOCs measured in Shanghai, China. *Atmos. Environ.* **2008**, *42*, 6873–6883. [CrossRef]
10. Shao, M.; Zhang, Y.; Zeng, L.; Tang, X.; Zhang, J.; Zhong, L.; Wang, B. Ground-level ozone in the Pearl River Delta and the roles of VOC and NOx in its production. *J. Environ. Manag.* **2009**, *90*, 512–518. [CrossRef]
11. Xing, J.; Wang, S.X.; Jang, C.; Zhu, Y.; Hao, J.M. Nonlinear response of ozone to precursor emission Changes in China: A modeling study using response surface methodology. *Atmos. Chem. Phys.* **2010**, *11*, 5027–5044. [CrossRef]
12. Li, C.; Liu, Y.; Cheng, B.; Zhang, Y.; Liu, X.; Qu, Y.; An, J.; Kong, L.; Zhang, Y.; Zhang, C.; et al. A comprehensive investigation on volatile organic compounds (VOCs) in 2018 in Beijing, China: Characteristics, sources and behaviours in response to O3 formation. *Sci. Total Environ.* **2022**, *806*, 150247. [CrossRef]
13. Xiu, M.; Wang, X.; Morawska, L.; Pass, D.; Beecroft, A.; Mueller, J.F.; Thai, P. Emissions of particulate matters, volatile organic compounds and polycyclic aromatic hydrocarbons from warm and hot asphalt mixes. *J. Clean. Prod.* **2020**, *275*, 123094. [CrossRef]
14. Yang, Y.; Liu, B.; Hua, J.; Yang, T.; Dai, Q.; Wu, J.; Feng, Y.; Hopke, P.K. Global review of source apportionment of volatile organic compounds based on highly time-resolved data from 2015 to 2021. *Environ. Int.* **2022**, *165*, 107330. [CrossRef] [PubMed]
15. Yin, Y.; He, J.; Zhao, L.; Pei, J.; Yang, X.; Sun, Y.; Cui, X.; Lin, C.-H.; Wei, D.; Chen, Q. Identification of key volatile organic compounds in aircraft cabins and associated inhalation health risks. *Environ. Int.* **2022**, *158*, 106999. [CrossRef] [PubMed]
16. Zhang, K.; Chang, S.; Wang, E.; Zhang, Q.; Fan, Y.; Bai, Y.; Zhang, M.; Fu, Q.; Jia, W. Occurrence, health risk, and removal efficiency assessment of volatile organic compounds in drinking water treatment plants (DWTPs): An investigation of seven major river basins across China. *J. Clean. Prod.* **2022**, *372*, 133762. [CrossRef]
17. Duan, C.; Liao, H.; Wang, K.; Ren, Y. The research hotspots and trends of volatile organic compound emissions from anthropogenic and natural sources: A systematic quantitative review. *Environ. Res.* **2023**, *216*, 114386.
18. Hajizadeh, Y.; Teiri, H.; Nazmara, S.; Parseh, I. Environmental and biological monitoring of exposures to VOCs in a petrochemical complex in Iran. *Environ. Sci. Pollut. Res.* **2018**, *25*, 6656–6667. [CrossRef]
19. Zheng, B.; Tong, D.; Li, M.; Liu, F.; Zhang, Q. Trends in China's anthropogenic emissions since 2010 as the consequence of clean air actions. *Atmos. Chem. Phys.* **2018**, *18*, 14095–14111. [CrossRef]
20. Cheng, C.-A.; Ching, T.-C.; Tsai, S.-W.; Chuang, K.-J.; Chuang, H.-C.; Chang, T.-Y. Exposure and health risk assessment of indoor volatile organic compounds in a medical university. *Environ. Res.* **2022**, *213*, 113644. [CrossRef]
21. Huang, H.; Wang, Z.; Dai, C.; Guo, J.; Zhang, X. Volatile organic compounds emission in the rubber products manufacturing processes. *Environ. Res.* **2022**, *212*, 113485. [CrossRef]
22. Liu, Y.; Kong, L.; Liu, X.; Zhang, Y.; Li, C.; Zhang, Y.; Zhang, C.; Qu, Y.; An, J.; Ma, D.; et al. Characteristics, secondary transformation, and health risk assessment of ambient volatile organic compounds (VOCs) in urban Beijing, China. *Atmos. Pollut. Res.* **2021**, *12*, 33–46. [CrossRef]
23. Singh, B.P.; Sohrab, S.S.; Athar, M.; Alandijany, T.A.; Kumari, S.; Nair, A.; Kumari, S.; Mehra, K.; Chowdhary, K.; Rahman, S.; et al. Substantial Changes in Selected Volatile Organic Compounds (VOCs) and Associations with Health Risk Assessments in Industrial Areas during the COVID-19 Pandemic. *Toxics* **2023**, *11*, 165. [CrossRef]
24. Acton, W.J.F.; Huang, Z.; Davison, B.; Drysdale, W.S.; Fu, P.; Hollaway, M.; Langford, B.; Lee, J.; Liu, Y.; Metzger, S.; et al. Surface–atmosphere fluxes of volatile organic compounds in Beijing. *Atmos. Chem. Phys.* **2020**, *20*, 15101–15125. [CrossRef]
25. Geng, C.; Wang, J.; Yin, B.; Zhao, R.; Li, P.; Yang, W.; Xiao, Z.; Li, S.; Li, K.; Bai, Z. Vertical distribution of volatile organic compounds conducted by tethered balloon in the Beijing-Tianjin-Hebei region of China. *J. Environ. Sci.* **2020**, *95*, 121–129. [CrossRef]
26. Li, J.; Xie, S.D.; Zeng, L.M.; Li, L.Y.; Li, Y.Q.; Wu, R.R. Characterization of ambient volatile organic compounds and their sources in Beijing, before, during, and after Asia-Pacific Economic Cooperation China 2014. Atmos. *Chem. Phys.* **2015**, *15*, 7945–7959. [CrossRef]
27. Ling, Y.; Wang, Y.; Duan, J.; Xie, X.; Liu, Y.; Peng, Y.; Qiao, L.; Cheng, T.; Lou, S.; Wang, H.; et al. Long-term aerosol size distributions and the potential role of volatile organic compounds (VOCs) in new particle formation events in Shanghai. *Atmos. Environ.* **2019**, *202*, 345–356. [CrossRef]

28. Wang, L.; Slowik, J.G.; Tong, Y.; Duan, J.; Gu, Y.; Rai, P.; Qi, L.; Stefenelli, G.; Baltensperger, U.; Huang, R.-J.; et al. Characteristics of wintertime VOCs in urban Beijing: Composition and source apportionment. *Atmos. Environ. X* **2021**, *9*, 100100. [CrossRef]
29. Gao, J.; Zhang, J.; Li, H.; Li, L.; Xu, L.; Zhang, Y.; Wang, Z.; Wang, X.; Zhang, W.; Chen, Y.; et al. Comparative study of volatile organic compounds in ambient air using observed mixing ratios and initial mixing ratios taking chemical loss into account—A case study in a typical urban area in Beijing. *Sci. Total Environ.* **2018**, *628–629*, 791–804. [CrossRef] [PubMed]
30. Han, T.; Ma, Z.; Li, Y.; Pu, W.; Wu, J.; Li, Z.; Shang, J.; He, D.; Zhou, L.; Wang, Y. Chemical characteristics and source apportionments of volatile organic compounds (VOCs) before and during the heating season at a regional background site in the North China Plain. *Atmos. Res.* **2021**, *262*, 105778. [CrossRef]
31. Yang, S.; Li, X.; Song, M.; Liu, Y.; Yu, X.; Chen, S.; Lu, S.; Wang, W.; Yang, Y.; Zeng, L.; et al. Characteristics and sources of volatile organic compounds during pollution episodes and clean periods in the Beijing-Tianjin-Hebei region. *Sci. Total Environ.* **2021**, *799*, 149491. [CrossRef] [PubMed]
32. Zhan, J.; Feng, Z.; Liu, P.; He, X.; He, Z.; Chen, T.; Wang, Y.; He, H.; Mu, Y.; Liu, Y. Ozone and SOA formation potential based on photochemical loss of VOCs during the Beijing summer. *Environ. Pollut.* **2021**, *285*, 117444. [CrossRef] [PubMed]
33. Dai, H.; Jing, S.; Wang, H.; Ma, Y.; Li, L.; Song, W.; Kan, H. VOC characteristics and inhalation health risks in newly renovated residences in Shanghai, China. *Sci. Total Environ.* **2017**, *577*, 73–83. [CrossRef] [PubMed]
34. Gong, Y.; Wei, Y.; Cheng, J.; Jiang, T.; Chen, L.; Xu, B. Health risk assessment and personal exposure to Volatile Organic Compounds (VOCs) in metro carriages—A case study in Shanghai, China. *Sci. Total Environ.* **2017**, *574*, 1432–1438. [CrossRef] [PubMed]
35. Li, C.; Li, Q.; Tong, D.; Wang, Q.; Wu, M.; Sun, B.; Su, G.; Tan, L. Environmental impact and health risk assessment of volatile organic compound emissions during different seasons in Beijing. *J. Environ. Sci.* **2020**, *93*, 1–12. [CrossRef]
36. Liu, Y.; Wang, H.; Jing, S.; Peng, Y.; Gao, Y.; Yan, R.; Wang, Q.; Lou, S.; Cheng, T.; Huang, C. Strong regional transport of volatile organic compounds (VOCs) during wintertime in Shanghai megacity of China. *Atmos. Environ.* **2021**, *244*, 117940. [CrossRef]
37. Xie, F.; Zhou, X.; Wang, H.; Gao, J.; Hao, F.; He, J.; Lü, C. Heating events drive the seasonal patterns of volatile organic compounds in a typical semi-arid city. *Sci. Total Environ.* **2021**, *788*, 147781. [CrossRef]
38. Yao, S.; Wang, Q.; Zhang, J.; Zhang, R.; Gao, Y.; Zhang, H.; Li, J.; Zhou, Z. Ambient volatile organic compounds in a heavy industrial city: Concentration, ozone formation potential, sources, and health risk assessment. *Atmos. Pollut. Res.* **2021**, *12*, 101053. [CrossRef]
39. Shi, Y.; Xi, Z.; Simayi, M.; Li, J.; Xie, S. Scattered coal is the largest source of ambient volatile organic compounds during the heating season in Beijing. *Atmos. Chem. Phys.* **2020**, *20*, 9351–9369. [CrossRef]
40. Wei, W.; Li, Y.; Wang, Y.; Cheng, S.; Wang, L. Characteristics of VOCs during haze and non-haze days in Beijing, China: Concentration, chemical degradation and regional transport impact. *Atmos. Environ.* **2018**, *194*, 134–145. [CrossRef]
41. Xie, G.; Chen, H.; Zhang, F.; Shang, X.; Zhan, B.; Zeng, L.; Mu, Y.; Mellouki, A.; Tang, X.; Chen, J. Compositions, sources, and potential health risks of volatile organic compounds in the heavily polluted rural North China Plain during the heating season. *Sci. Total Environ.* **2021**, *789*, 147956. [CrossRef] [PubMed]
42. Yang, W.; Zhang, Y.; Wang, X.; Li, S.; Zhu, M.; Yu, Q.; Li, G.; Huang, Z.; Zhang, H.; Wu, Z.; et al. Volatile organic compounds at a rural site in Beijing: Influence of temporary emission control and wintertime heating. *Atmos. Chem. Phys.* **2018**, *18*, 12663–12682. [CrossRef]
43. Li, T.; Lu, G.; Lin, J.; Liang, D.; Hong, B.; Luo, S.; Wang, D.; Oeser, M. Volatile organic compounds (VOCs) inhibition and energy consumption reduction mechanisms of using isocyanate additive in bitumen chemical modification. *J. Clean. Prod.* **2022**, *368*, 133070. [CrossRef]
44. Liu, Y.; Wang, H.; Jing, S.; Gao, Y.; Peng, Y.; Lou, S.; Cheng, T.; Tao, S.; Li, L.; Li, Y.; et al. Characteristics and sources of volatile organic compounds (VOCs) in Shanghai during summer: Implications of regional transport. *Atmos. Environ.* **2019**, *215*, 116902. [CrossRef]
45. Norris, C.; Fang, L.; Barkjohn, K.K.; Carlson, D.; Zhang, Y.; Mo, J.; Li, Z.; Zhang, J.; Cui, X.; Schauer, J.J.; et al. Sources of volatile organic compounds in suburban homes in Shanghai, China, and the impact of air filtration on compound concentrations. *Chemosphere* **2019**, *231*, 256–268. [CrossRef]
46. Zhang, Y.; Kong, S.; Sheng, J.; Zhao, D.; Ding, D.; Yao, L.; Zheng, H.; Wu, J.; Cheng, Y.; Yan, Q.; et al. Real-time emission and stage-dependent emission factors/ratios of specific volatile organic compounds from residential biomass combustion in China. *Atmos. Res.* **2021**, *248*, 105189. [CrossRef]
47. Carter, W.P.L. Development of the SAPRC-07 chemical mechanism. *Atmos. Environ.* **2010**, *44*, 5324–5335. [CrossRef]
48. Gu, Y.; Liu, B.; Dai, Q.; Zhang, Y.; Zhou, M.; Feng, Y.; Hopke, P.K. Multiply improved positive matrix factorization for source apportionment of volatile organic compounds during the COVID-19 shutdown in Tianjin, China. *Environ. Int.* **2022**, *158*, 106979. [CrossRef]
49. Li, B.; Ho, S.S.H.; Li, X.; Guo, L.; Chen, A.; Hu, L.; Yang, Y.; Chen, D.; Lin, A.; Fang, X. A comprehensive review on anthropogenic volatile organic compounds (VOCs) emission estimates in China: Comparison and outlook. *Environ. Int.* **2021**, *156*, 106710. [CrossRef]
50. Kong, X.; He, W.; Qin, N.; He, Q.; Yang, B.; Ouyang, H.; Wang, Q.; Xu, F. Comparison of transport pathways and potential sources of PM10 in two cities around a large Chinese lake using the modified trajectory analysis. *Atmos. Res.* **2013**, *122*, 284–297. [CrossRef]

51. Li, Y.; Gao, R.; Xue, L.; Wu, Z.; Yang, X.; Gao, J.; Ren, L.; Li, H.; Ren, Y.; Li, G.; et al. Ambient volatile organic compounds at Wudang Mountain in Central China: Characteristics, sources and implications to ozone formation. *Atmos. Res.* **2021**, *250*, 105359. [CrossRef]
52. An, J.; Zou, J.; Wang, J.; Lin, X.; Zhu, B. Differences in ozone photochemical characteristics between the megacity Nanjing and its suburban surroundings, Yangtze River Delta, China. *Environ. Sci. Pollut. Res.* **2015**, *22*, 19607–19617. [CrossRef] [PubMed]
53. Yang, X.; Xue, L.; Wang, T.; Wang, X.; Gao, J.; Lee, S.; Blake, D.; Chai, F.; Wang, W. Observations and explicit modeling of summertime carbonyl formation in Beijing: Identification of key precursor species and their impact on atmospheric oxidation chemistry. *J. Geophys. Res. Atmos.* **2018**, *123*, 1426–1440. [CrossRef]
54. He, Z.; Wang, X.; Ling, Z.; Zhao, J.; Guo, H.; Shao, M.; Wang, Z. Contributions of different anthropogenic volatile organic compound sources to ozone formation at a receptor site in the Pearl River Delta region and its policy implications. *Atmos. Chem. Phys.* **2019**, *19*, 8801–8816. [CrossRef]
55. Zhang, H.; Ji, Y.; Wu, Z.; Peng, L.; Bao, J.; Peng, Z.; Li, H. Atmospheric volatile halogenated hydrocarbons in air pollution episodes in an urban area of Beijing: Characterization, health risk assessment and sources apportionment. *Sci. Total Environ.* **2022**, *806*, 150283. [CrossRef]
56. Zheng, S.; Xu, X.; Zhang, Y.; Wang, L.; Yang, Y.; Jin, S.; Yang, X. Characteristics and sources of VOCs in urban and suburban environments in Shanghai, China, during the 2016 G20 summit. *Atmos. Pollut. Res.* **2019**, *10*, 1766–1779. [CrossRef]
57. Neal, L.M.; Haribal, V.P.; Li, F. Intensified ethylene production via chemical looping through an exergetically efficient redox scheme. *iScience* **2019**, *19*, 894–904. [CrossRef]
58. Krstic, N.M.; Bjelakovic, M.S.; Pavlovic, V.D.; Robeyns, K.; Juranic, Z.D.; Matic, I.; Novakovic, I.; Sladic, D.M. New androst-4-en-17-spiro-1,3,2-oxathiaphospholanes. Synthesis, assignment of absolute configuration and in vitrocytotoxic and antimicrobial activities. *Steroids* **2012**, *77*, 558–565. [CrossRef]
59. McCarthy, M.C.; Aklilu, Y.-A.; Brown, S.G.; Lyder, D.A. Source apportionment of volatile organic compounds measured in Edmonton, Alberta. *Atmos. Environ.* **2013**, *81*, 504–516. [CrossRef]
60. Yan, Y.; Peng, L.; Li, R.; Li, Y.; Li, L.; Bai, H. Concentration, ozone formation potential and source analysis of volatile organic compounds (VOCs) in a thermal power station centralized area: A study in Shuozhou, China. *Environ. Pollut.* **2017**, *223*, 295–304. [CrossRef]
61. Mo, Z.; Shao, M.; Wang, W.; Liu, Y.; Wang, M.; Lu, S. Evaluation of biogenic isoprene emissions and their contribution to ozone formation by ground-based measurements in Beijing, China. *Sci. Total Environ.* **2018**, *627*, 1485–1494. [CrossRef] [PubMed]
62. Liu, N.; Lin, W.; Ma, J.; Xu, W.; Xu, X. Seasonal variation in surface ozone and its regional characteristics at global atmosphere watch stations in China. *J. Environ. Sci.* **2019**, *77*, 291–302. [CrossRef] [PubMed]
63. Ramírez, N.; Cuadras, A.; Rovira, E.; Borrull, F.; Marcé, R.M. Chronic risk assessment of exposure to volatile organic compounds in the atmosphere near the largest Mediterranean industrial site. *Environ. Int.* **2012**, *39*, 200–209. [CrossRef] [PubMed]
64. Ministry of Environmental Protection of the People's Republic of China (MEP). *Exposure Factors Handbook of Chinese Population*; Chinese Environmental Science Press: Beijing, China, 2013; pp. 798–801.

toxics

Article

The Distribution Characteristics and Ecological Risks of Alkylphenols and the Relationships between Alkylphenols and Different Types of Land Use

Yajun Hong [1,2], Miao Chen [2], Ziwei Zhu [3], Wei Liao [2,3,*], Chenglian Feng [2,*], Zhenfei Yan [2,4], Yu Qiao [2], Yaru Mei [3] and Dayong Xu [1]

[1] School of Chemical and Environmental Engineering, Anhui Polytechnic University, Wuhu 241000, China; hongyajun@mail.ahpu.edu.cn (Y.H.); xdy826@ahpu.edu.cn (D.X.)
[2] State Key Laboratory of Environmental Criteria and Risk Assessment, Chinese Research Academy of Environmental Sciences, Beijing 100012, China; chenmiao@tongji.edu.cn (M.C.); 210205020012@hhu.edu.cn (Z.Y.); qiaoyu202@mails.ucas.ac.cn (Y.Q.)
[3] Wetland Research Center, Jiangxi Academy of Forestry, Nanchang 330032, China; ziweiangel@163.com (Z.Z.); collect21@sina.com (Y.M.)
[4] College of Environment, Hohai University, Nanjing 210098, China
* Correspondence: lovy21@163.com (W.L.); fengcl@craes.org.cn (C.F.); Tel./Fax: +86-010-84931804 (C.F.)

Abstract: In this study, the spatial distribution characteristics of nine alkylphenols (APs) in the Yongding River and Beiyun River were analyzed. The differences in the concentrations and spatial distribution patterns of nine APs were systematically evaluated using principal component analysis (PCA). The relationships between the concentration distribution patterns and the risks associated with nine APs were investigated under various categories of land use conditions in the region. The results demonstrated that the APs were widely present in both rivers, and the pollution risks associated with the APs were more severe in the Yongding River than in the Beiyun River. The results show that the contamination risks associated with 4-NP were the most serious in the two rivers, with detection percentages of 100% and 96.3%, respectively. In the Yongding River, the APs showed a tendency of low concentration levels in the upper reaches and high levels in the middle and lower regions. Meanwhile, the overall concentration levels of the APs in the Beiyun River were relatively high. However, despite the differences between the upper and middle regions of the Yongding River, the distribution pattern of the APs in the Beiyun River was basically stable. The concentration levels and risk quotient of the APs were negatively correlated with the vegetation cover land use type and positively correlated with the cropland and unused land use types within 500 m, 1 km, and 2 km. The purpose of this study was to provide theoretical data support and a basis for AP pollution risk evaluations in the Yongding River and Beiyun River.

Keywords: emerging contaminants; endocrine disrupting chemicals; alkylphenols; distribution characteristics; ecological risks; types of land use

Citation: Hong, Y.; Chen, M.; Zhu, Z.; Liao, W.; Feng, C.; Yan, Z.; Qiao, Y.; Mei, Y.; Xu, D. The Distribution Characteristics and Ecological Risks of Alkylphenols and the Relationships between Alkylphenols and Different Types of Land Use. *Toxics* **2023**, *11*, 579. https://doi.org/10.3390/toxics11070579

Academic Editor: Antonia Concetta Elia

Received: 29 May 2023
Revised: 26 June 2023
Accepted: 1 July 2023
Published: 3 July 2023

1. Introduction

The rivers of the world provide important ecosystems for many forms of life, as well as valuable goods and services to humanity [1]. Regional land use analysis is the main source of information for assessing the extent to which social, economic, and environmental factors influence urbanization processes and spatial structures [2]. Changes in land usage and land coverage will impact the structures and functions of ecosystems and are important driving factors of the changes in ecosystem services. The research regarding such change processes plays a decisive role in maintaining ecosystem services [3,4]. Urbanization levels have been unceasingly increased with the rapid development of social economies, resulting in surges in population. The emissions of industrial wastewater and sewage into nearby water resources have been steadily increasing. Unfortunately, environmental infrastructure construction has lagged, resulting in large amounts of untreated wastewater entering rivers. River water pollution has become a serious threat to the stability of water ecological systems [5,6]. One of the most important rivers in China is the Yongding River, also known as the "mother river". The Beijing-Tianjin-Hebei Region is an important headwater area containing many ecological barriers and corridors [7–9]. However, the healthy development of the economy and society of the Beijing-Tianjin-Hebei Region has been severely restricted by such outstanding problems as damage to the ecosystem, excessive water resource development, and the environmental bearing capacity [10]. The Beiyun River is the largest river system in the Beijing Plain Basin. The Beiyun River is mainly replenished by unconventional water sources, such as effluent from the municipal sewage treatment plant. The sewage discharge volume of the entire river basin is approximately 3 million m^3/d, making it the most seriously polluted river among the five major river systems in Beijing, with high risk of pathogenic microorganisms [11]. Therefore, it is of major significance to strengthen the management and governance of the rivers to improve water quality and fully enhance the utilization rates of the water resources. Such improvements will have beneficial impacts on the ecosystems and the maintenance of ecological stability, as well as protecting the health of the population and increasing life satisfaction.

Alkylphenols (APs) are typical endocrine-disrupting chemicals (EDCs). APs are extensively used as feedstock for alkylphenol polyoxyethylene ethers (APEOs). They are used in the production of phenolic resins, heat stabilizers, antioxidants, and hardeners [12–14]. APs in a water environment are mainly produced by the biodegradation of long-chain ethoxylates. Those ethoxylates can be removed by conventional wastewater treatment technologies [15,16], but APs may be more persistent, lipophilic, and toxic than their precursors [14]. They tend to be present in higher concentrations in the environment compared with other EDCs [17,18]. APs have been continuously detected in river water and sediment, usually at the concentration levels of ng/L or ng/g. The results of the study conducted by Lei et al. [19] confirmed that nonylphenol (NP), octylphenol (OP), and bisphenol A (BPA) could be detected in 100% of the urban rivers in Beijing, Tianjin, and Hebei, with concentration levels between 23 and 255 ng/L. Cheng et al. [20] showed that the concentrations of NP, OP, and BPA in the Yongjiang River Basin were 140–3948, 6–828, and 15–1415 ng/L, respectively. Li et al. [21] determined that the concentrations of three phenolics (NP, OP, and BPA) in the Pearl River sediment varied from 204.4 to 12,604.3, 32.6 to 297.3, and 12.8 to 298.4 ng/g, respectively. In the river sediment of the Duliujian River, the concentration levels of those same substances ranged from 153.5 to 3614.9 ng/g, 90.7 to 990.0 ng/g, and 83.5 to 913.3 ng/g, respectively. Although the above-mentioned concentration levels of APs will not cause serious acute toxicity to aquatic organisms, they will potentially damage their endocrine systems, which has become especially evident in fish sampled from the region [21,22] APs can enter the body in many ways, including through diet and respiration, and they can cause diseases in the human reproductive, cardiovascular, and immune systems [23].

However, the previous research in the Yongding River and Beiyun River has focused on the degrees of heavy metal pollution in the water bodies and the sediment of the rivers [24]. In addition, the water ecological carrying capacities, river restorations, and evaluations [8]; persistent organic pollutants and microplastics [9]; distributions of plankton and microorganisms [11]; and the monitoring of water quality indicators have all been major concerns. To effectively protect the environment and human health, it is also important to determine the pollution levels and characteristics of the distribution patterns of APs in the water environment of the Yongding River and Beiyun River. In this study, 35 water samples were obtained from the Yongding River and Beiyun River (Beijing section). The concentration characteristics of the APs in the river water were analyzed, and the relationships between the concentration distribution pattern and risk potential of the APs and the various land use types in the region were investigated. The aim was to provide a scientific foundation for the study and management of pollution and improve the water quality, as well as protecting aquatic life in the Yongding River and Beiyun River.

2. Materials and Methods

2.1. Chemicals and Reagents

The basic chemical information data are detailed in Table S1. The 4-EP, 2-PPP, 4-t-BP, 2-n-BP, 4-PTP, 4-HXP, 4-HTP, 4-OP, 4-NP, BPA, 4-t-OP, 4-n-NP, and the internal standards (4-EP-d_4, 4-t-OP-$^{13}C_6$, and 4-n-NP-d_4) were obtained from TRC (Toronto, ON, Canada). The HPLC-grade methanol (MeOH), acetonitrile (ACN), and ethyl acetate (ACETATE) were acquired from Fisher Scientific (USA). The hydroxide (NH_4OH, 14 M) used in this study was supplied by Sigma Aldrich (USA). The ultra-pure water (>18.2 MΩ/cm) was obtained using the Milli-Q Advantage Purification System (Millipore, USA). Stock solutions of each compound and the internal standards were taken at a concentration of 1000 mg/L in MeOH and stored in amber brown bottles at −18 °C under dark conditions before use. The working standard solutions (10 mg/L) were obtained by serial dilution prepared from the stock solution and renewed monthly to eliminate destabilizing effects. In addition, calibration standards with gradient concentrations (0, 5, 10, 50, and 100 μg/L) of analytes and 50 μg/L internal standard solutions were also prepared.

2.2. Sample Collection and Preparation

The study targets were the Yongding River and the Beiyun River (Figure 1). Thirty-five samples (27 from the Yongding River and 8 from the Beiyun River) were collected in January of 2022. Pretreatment of the collected samples was performed according to previous methods [21,25,26].

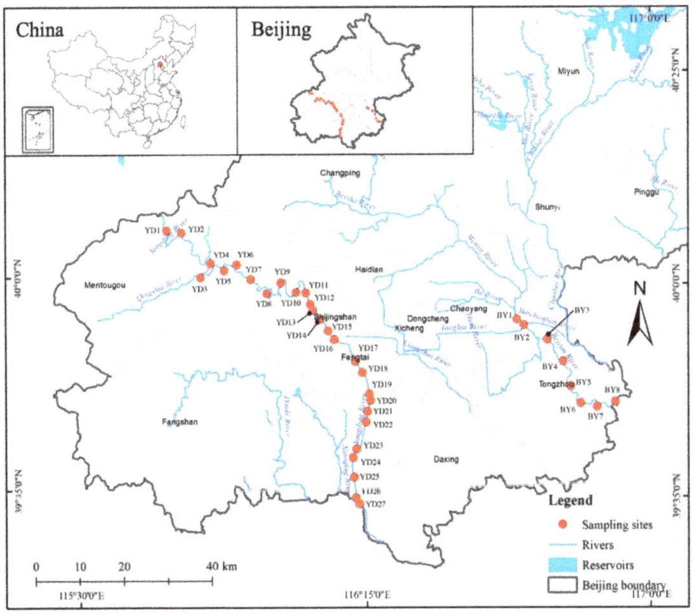

Figure 1. Sampling locations in the Yongding River and the Beiyun River.

2.3. Sample Analysis

Ultra-Performance Liquid Chromatography (UPLC) separation was performed using a Waters ACQUITY UPLC device (Waters, USA). The instrumental conditions were as follows. The column temperature was set at 40 °C, and the sampling volume was 2 μL. MQ water with 0.01% NH_4OH (Solvent A) and MeOH (Solvent B) were used as the flowing phase, with a flow rate of 0.2 mL/min. All of the analytes were identified according to their retention times and targeted ion pairs as per the standards. The optimized parameters of the mass spectrometry for the analytes are listed in Table S1. Quantification of these chemicals was performed using the internal standard method.

2.4. Quality Assurance and Control Measures

It was found that the calibration curves for the selected chemicals showed strong uniformity over a broad range of levels ($R^2 > 0.99$). As shown in Table S1, the method detection limit (MDL) and method quantification limit (MQL) for the surface water samples ranged from 0.05 to 0.15 ng/L and 0.2 to 1.8 ng/L, respectively. The results of the study showed that the sample recoveries of the target analytical analytes in the two rivers were from 74% to 88%, 72% to 78%, and 71% to 87% at the water-spiked levels of 10, 50, and 100 ng/L, respectively. The instrumental quantification limit (IQL) was 10 times that of the signal-to-noise (S/N) ratios, and the MDL and MQL were 3 and 10 times that of the S/N ratios, respectively.

2.5. Principal Component and Difference Analyses

The relationships between the distribution characteristics of the APs in the two rivers were explored in this study. Reductions were made to the data dimensions since correlations may have existed between multiple variables, which increased the difficulty and complexity of the analysis process. This resulted in moderate reductions in the number of indicators to be analyzed. The possibility of loss of information included in the original metrics was minimized to achieve a well-rounded analysis of the gathered data [27]. Principal component analysis (PCA) was used to map the N-dimensional characteristics to k (2–3) dimensions. The obtained feature represented a new orthogonal feature, also called the principal component, which was a reconstruction of the k-dimensional characteristics anchored on the original N-dimensional characteristics [27]. In this study, PCA was used to fit the corresponding functional relationships between the content of nine AP species and the main ranking axis. The results showed the spatial allocation characteristics of the APs in the two examined rivers. PCA dimensionality reduction analysis was completed using Canoco 5 software, and the sample grouping results were verified by ANOVA in R software.

2.6. Spatial Analysis

The geospatial parameters, such as the spatial distances to the outlets and the topological distances between all sites, were obtained using the network analysis tool in ArcGIS software version 10.2 [28]. The digital height model data (250 m resolution) obtained from the Shuttle Radar Topography Mission (SRTM) V4.1 dataset were used to define the watershed basin boundaries and watercourse features with the hydrology tool in ArcGIS software [29]. The category system in CNLUCC (http://www.resdc.cn/data.aspx?DATAID=264, accessed on 11 January 2023) was also referenced in this study. The patterns of the land usage were classified into six first-class types as follows: cropland (paddy fields and dry land areas); forested land; grassland; water areas (rivers, pools, and reservoirs); impervious areas (residential, industrial, and mining cover areas); and unutilized land (desert, marshland, and bare land areas) [28]. Three buffer regions were the focus within a 500 m, 1 km, and 2 km radius, respectively, for the upstream area of each site. The percentages for the six land use types were calculated, and they are detailed in Tables S2 and S3. The range used in this study describes a contiguous continuum of human activity from the local to the regional scale. The buffering tool in ArcGIS software was utilized to extract the land use parameters within the aforementioned buffers [28]

3. Results and Discussion

3.1. Distribution Characteristics of the Concentration Levels of the APs

The detected concentration and frequency values of the APs in the two examined rivers are summarized in Table 1. As can be seen from the table, 4-HXP was not detected at all points in the Yongding River. However, the other eight APs were all detected, among which 4-NP and 4-OP had the highest detection rates at 100% and 88.9%, respectively. The detection rates of 2-PPP and 4-t-BP were over 70%. It was also found that 2-PPP, 4-t-BP, and 4-PTP were detected at all the sampling points, but 4-EP, 4-HXP, and 4-HTP were not. The detection rate of 4-NP was as high as 96.3%, indicating that AP pollution was widely present in the Yongding River and Beiyun River. The concentration levels of the APs were at the ng/L level. AP pollution risks in the Yongding River were more serious than those in the Beiyun River, with the 4-NP pollution found to be the most severe.

Table 1. Concentration and frequency values of the APs detected in the Yongding River and Beiyun River (ng/L).

APs	Yongding River					Beiyun River				
	MAX	MIN	MED	AVE	Detected No.	MAX	MIN	MED	AVE	Detected No.
4-EP	322.9	ND	ND	14.4	6	ND	ND	ND	ND	0
2-PPP	305.4	ND	31.6	57.6	19	125.5	59.0	86.7	86.5	8
4-t-BP	340.6	ND	71.7	88.0	19	271.6	52.4	135.2	151.2	8
4-PTP	5.6	ND	ND	0.8	9	21.1	3.2	11.2	10.8	8
4-HXP	ND	ND	ND	ND	0	ND	ND	ND	ND	0
4-HTP	1.7	ND	ND	0.1	1	ND	ND	ND	ND	0
4-OP	99.9	ND	18.4	22.5	24	24.8	ND	ND	3.1	1
4-NP	580.8	31.4	93.7	133.0	27	306.3	ND	63.1	108.0	7
BPA	5.3	ND	ND	1.2	11	1.3	ND	ND	0.2	1

The detection rates of 4-NP were high in both examined rivers, which was consistent with the results of earlier studies [19]. This was determined to be due to the presence of synthesized nonylphenol polyoxyethylene ether (NPEO), the world's second most abundantly used non-ionic surface-active agent. NPEO is widely used in the pulp and paper making, textile manufacturing, agriculture, metal and plastic manufacturing, and oil refining industries. Products containing NP include detergents, emulsifiers, wetting and dispersing agents, antistatic and emulsifying agents, and solubilizers, which are used in a wide range of industrial, institutional, commercial, and domestic applications [30]. NPEO, as an important component of the product, enters water bodies in various ways. It is easily degraded to NP, with a more stable chemical structure under the combined actions of various environmental factors. Due to recent rapid urbanization, modernization, and industrialization, large amounts of NP have entered rivers, lakes, and reservoirs. It is estimated that approximately 60% of the NP (and its derivatives) produced in the world has been introduced into water resources [22]. Theoretically speaking, there are more than 100 types of NP structural isomers, of which 4-NP accounts for approximately 90% of them. Sewage treatment plants are unable to effectively degrade parts of NP using traditional methods [22]. This study observed that there were many urban residential areas, hospitals, factories, sewage plants, etc. located along the banks of the Yongding River and Beiyun River. The high concentration levels and detection rates of NP in the rivers were related to the large quantities of nondegraded NP in the discharged sewage and wastewater.

3.2. Spatial Distribution of the Concentrations of APs

In this study, the Yongding River was divided into upstream (Points YD1–YD9), middle stream (Points YD10–YD18), and downstream (Points YD19–YD27) sections. The concentration distribution patterns of the APs at each sampling point along the Yongding River and Beiyun River are shown in Figure 2. The concentration distribution of the APs in the Yongding River generally presented a trend of lower concentrations in the upstream region and higher concentrations in the middle and lower regions. The middle section of the Yongding River was found to have high concentration values of APs, with a total concentration above 200 ng/L. This was attributed to the middle section passing through the Mentougou District, Shijingshan District, and Fengtai District, which were characterized by high population densities and a variety of residential communities, hotels, hospitals, sewage plants, and so on. The relatively high concentration levels of APs at the middle section's sampling points were due to the abundant use of APs in the production and living activities of those areas. Point YD20 in the downstream section was determined to be the point with the highest total concentration of APs, possibly due to the location of the tributary of the Yongding River: the Yongding River Main Canal. At that particular sampling point, the river passed through many residential areas and received large amounts of sewage discharge, with very poor water quality and a certain smell detected.

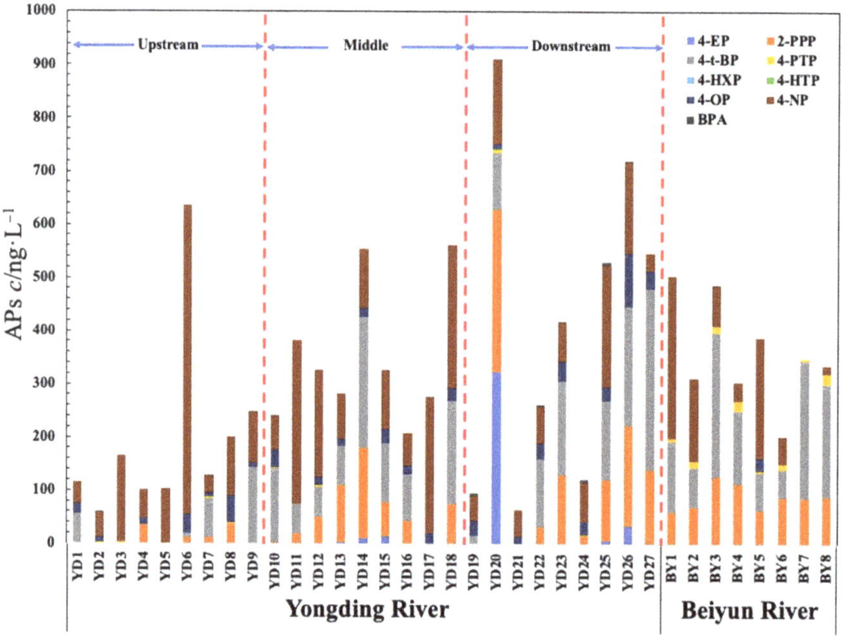

Figure 2. Concentration distribution patterns of the APs at various sampling points in the Yongding River and Beiyun River. Note: In the figure, the concentration levels of each of the nine APs are shown in different colors. Column heights of the stacking plots indicate the total concentration of all APs at those points.

The downstream sampling points revealed that the AP concentrations were lower in those areas. The majority of the points were located in swamps or wetland parks (such as Points YD19 and YD21) or in forested parks and green dam areas upstream (such as Century Park), which were relatively free of pollutants due to water purification and adsorption. In addition, Point YD24 in the upstream section was in an area where the river flowed through various golf clubs and country parks, with lower population densities and scenic environmental conditions, resulting in relatively low concentration levels of APs in the water samples. However, the downstream section also passed through densely populated villages with many agricultural activities, leading to increases in the concentration levels of the APs in those sections. The sampling results also revealed several points along the Beiyun River where the total concentration of APs was relatively high. This was related to the Beiyun River's water distribution in the Beijing Tongzhou District deputy center downstream plains, which accommodates 90% of the central city drainage task [31]. The many tributaries which participate in the Beiyun River's drainage of rain and sewage may have also had influencing effects on the sampling results. Since the drainage water mainly comes from recycled water, the river has both typical and unconventional water sources. The river sediment becomes silted with the discharge of pollutants, resulting in high levels of AP pollution [32]. The pollutants in the river sediment are then rereleased into the water bodies over time.

3.3. Spatial Distribution Differences in the APs

PCA was performed on the distribution of all APs at 27 points along the Yongding River and 8 points along the Beiyun River in order to have a better comprehension of the differences in the spatial distribution patterns. The results are shown in Figure 3. The PCA and regional significance analysis focused on nine sampling sites in the upper reaches of the Yongding River (Points YD1–YD9); the middle section of the Yongding River (Points YD10–YD18); the lower reaches of the Yongding River (Points YD19–YD27); and the Beiyun River, and significant differences were observed (ANOVA, R = 0.186, $p = 0.005 < 0.05$). The sampling sites of the Yongding River were roughly divided into three sections: upper, middle, and lower sections. The upper reaches of the sampling points were basically assessed together, as were the downstream sampling points. It was confirmed that the levels of AP concentrations in the middle and downstream segments were higher than those in the upstream segments. However, there were also some abnormal sampling results observed. With the exception of Point BY5, all the sampling points of the Beiyun River were clustered together, which verified that the water sources were mainly unconventional (wastewater treatment plant return water, agricultural irrigation return water, etc.), and the pollutant concentration levels were basically stable throughout the Beiyun River [32].

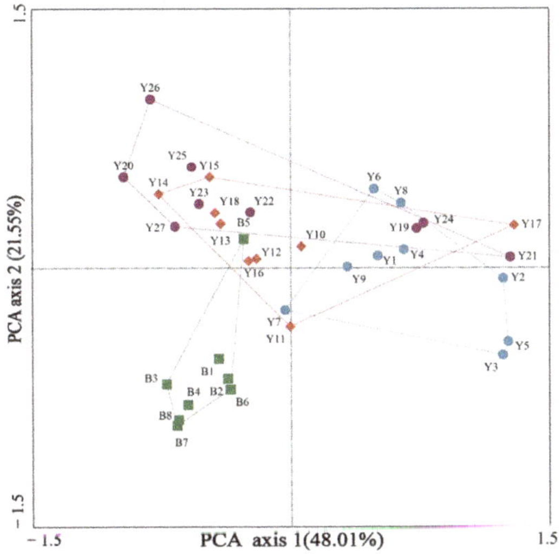

Figure 3. Principal component analysis diagram of the sampling points in the Yongding River and the Beiyun River.

The differences in the distribution patterns of the AP concentrations in the Yongding River and Beiyun River sampling sites were further analyzed to examine the distribution differences in the AP concentration levels at the different sampling sites, as detailed in Figure 4. Sampling Points YD20 and YD26 were in the lower reaches of the Yongding River, with many residential areas, hospitals, and sewage plants nearby. The generated sewage discharge was considered to be one of the important sources of the high concentrations of APs [33]. However, Sampling Point YD6 was located in the upper reaches of the Yongding River, with no serious sources of pollution nearby. As can be seen in Figures 2 and 4b, the sampling results revealed that 4-OP, 4-NP, and BPA displayed high values at Sampling Points BY5, BY1, and BY3 in the Beiyun River. Sampling Point BY1 was located in the upper reaches of the Beiyun River and Wenyu River, with many schools, residential communities, and hospitals nearby. Sewage containing 4-NP may have been discharged into the river at that location, resulting in the observed higher AP concentration values in that section.

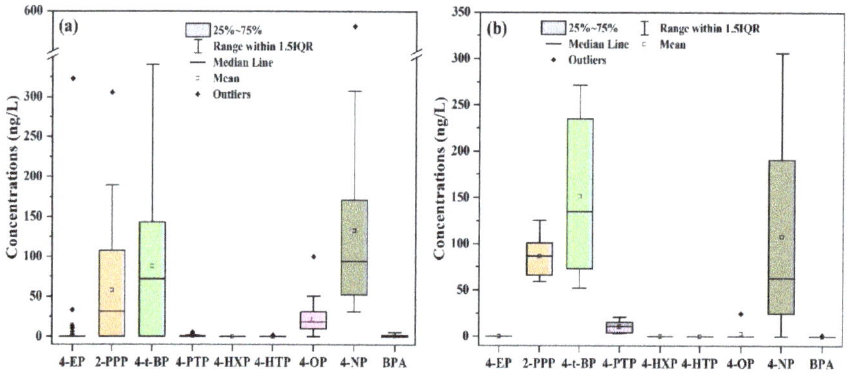

Figure 4. Analysis of the distribution differences of the AP concentrations at different sampling points in the (**a**) Yongding River and (**b**) Beiyun River.

As can be seen in Figure 3, the PCA and sampling site difference analysis results indicated that detection differences existed between the upper and lower reaches of the Yongding River. The high urbanization rates and high proportions of industrial and domestic land in the middle and lower reaches of the Yongding River resulted in higher AP values, while the upper reaches were in mountainous areas with less pollution. However, the distribution patterns of the APs in the Beiyun River remained generally stable. The research results obtained by Yu et al. [34] showed that there are certain correlations between the water quality and types of land use. The effects of new pollutants with different spatial distributions were observed in other research conducted in the area, and the types of land use were determined to be the main influencing factors [35,36]. However, it was considered that in addition to the types of land use, the distribution patterns of the APs and other new pollutants could also be related to the plant cover density, population density, proportion of farmland, and the economic living standards in different regions [37]. Therefore, this study further analyzed the influencing effects and weight correlations of each factor based on relevant data.

3.4. Relationships between the Types of Land Use and the Concentrations and Risk Quotient of the APs

The extent to which land use affects the ecological risks of pollutants may vary from region to region. Therefore, the relationships between the potential risks of new pollutants, as well as the management of buffer landscapes and land use scales in different watersheds, require further consideration. This process can provide a scientific basis for maintaining or improving living standards and formulating land management policies. In our previous studies [26], the ecological risks of AP pollution in the Yongding River and Beiyun River were reported in detail (Figure 5). In this investigation, three buffer zones at each site (500 m, 1 km, and 2 km) were the focus in the upstream areas, and the percentages of the six main land use types (cropland, forested land, grassland, water areas, impervious areas, and unutilized land) were calculated, as detailed in Tables S2 and S3. The relationships between the land use types and the concentration levels and risk quotient of the APs were examined, as shown in Figure 6.

Figure 6 shows that the concentration levels and risk quotient of the APs were negatively correlated with the forested land use type within 500 m, 1 km, and 2 km, which indicates that forested land use may reduce the concentrations and risk quotient of APs. Moreover, the concentration levels and risk quotient of the APs were found to be positively correlated with the cropland and unutilized land use types within 500 m, 1 km, and 2 km, indicating that the cropland areas were important sources of APs in the water. Other sources of APs were not represented, such as industrial, medical, and other land use types, and may have been unutilized land.

Land use types are major factors influencing the exposure and distribution of new contaminants. Land use types associated with human activities, such as croplands and impervious surfaces, may increase pollutant concentrations. On a global scale, the intensification and expansion of anthropogenic land use types are the most important drivers of water quality degradation [38,39]. The higher the proportion of urban areas among the land use types, the higher the potential for pollution by APs [40]. It has been determined that agricultural land and urban sewage are the largest sources of diffuse pollution in freshwater systems [41]. In addition, pollutant concentrations are significantly and negatively correlated with vegetation cover, such as forests and grasslands [40]. On one hand, forests and grasslands promote the uptake of pollutants and play important roles in improving water quality and reducing pollution by APs [41,42]. On the other hand, vegetation cover provides a buffer zone which slows down the diffusion of pollutants to some extent.

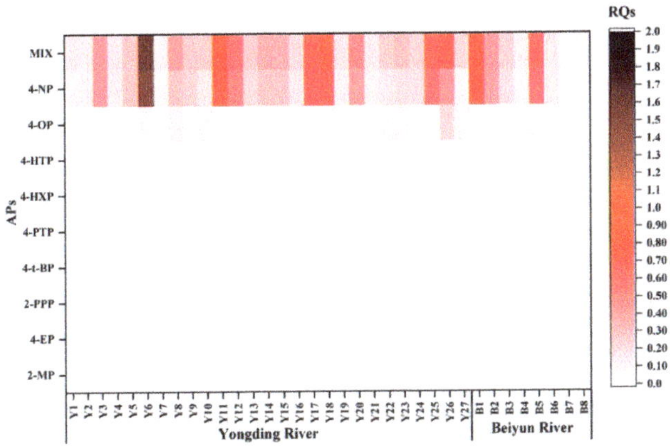

Figure 5. Heat plot of the ecotoxicological risks (represented by chronic RQs) of each AP (y axis) for aquatic organisms in the Yongding River and Beiyun River sampling sites (x axis).

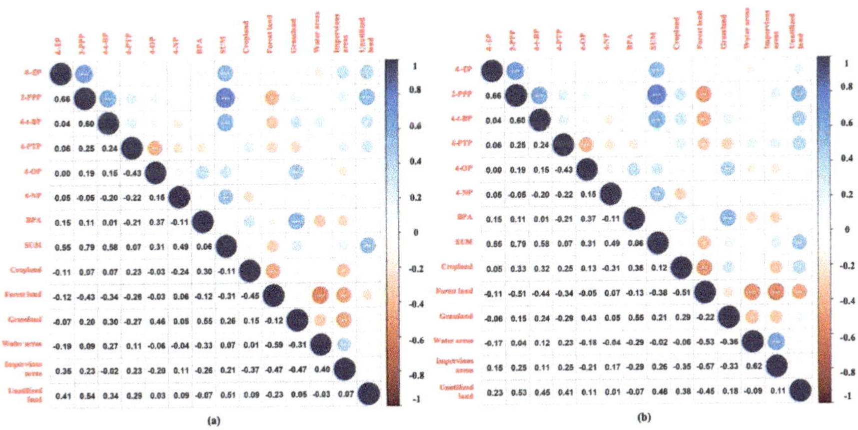

Figure 6. *Cont.*

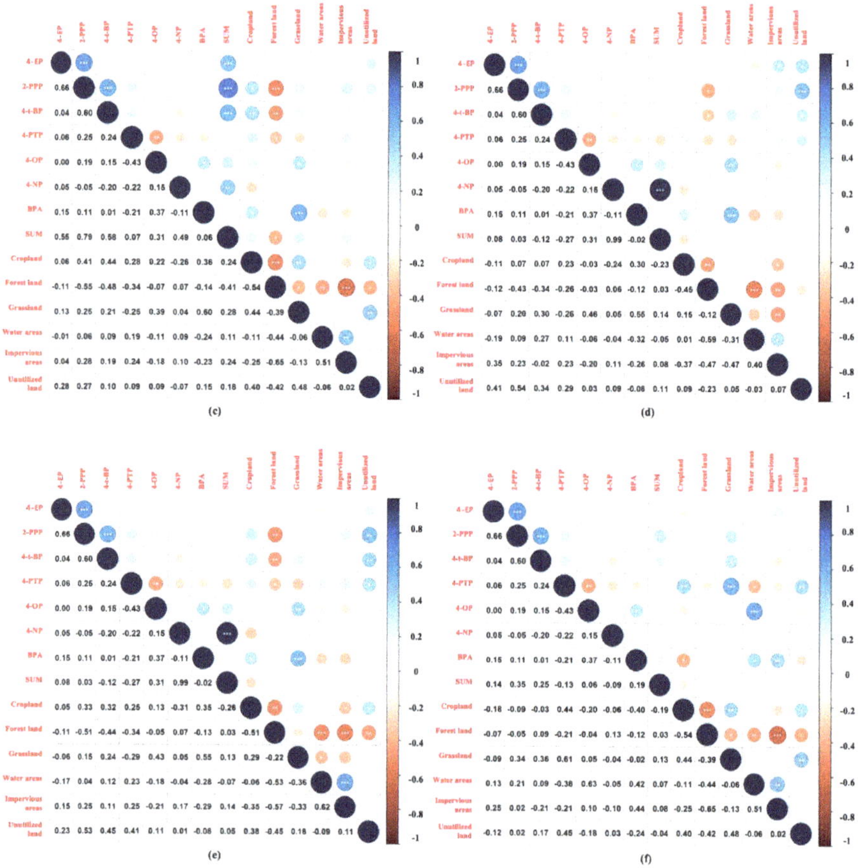

Figure 6. Relationships between the land use patterns and the concentration levels and risk quotient of the APs. (**a–c**) Relationships between the land use modes and concentration distributions of the APs within 500 m, 1 km, and 2 km, respectively. (**d–f**) Relationships between the land use modes and the risk quotient of the APs within 500 m, 1 km, and 2 km, respectively. Note: In the figure, the blue circles indicate positive correlations, and red circles indicate negative correlations. The larger the circle, the greater the correlation. * $p < 0.05$. ** $p < 0.01$. *** $p < 0.001$.

In general, agricultural land and built-up areas are major sources of nutrients and pollutants in surface water, which increases their chemical risks to various species [43]. As the proportion of cropland increases, biodiversity decreases, and habitats change [44]. In addition, runoff increases as urban or agricultural land cover expands and decreases as vegetated land increases, and the loss of forested areas increases erosion and sedimentation, alters water flow and thermal conditions, and affects carbon and nutrient cycling, which may impact the chemical risks in aquatic ecosystems [45].

4. Conclusions

This study's research results confirmed that alkylphenols (APs) are widely present in the Yongding River and Beiyun River. The concentration levels of AP pollution in the Yongding River were determined to be more serious than those in the Beiyun River. Among the nine examined APs, 4-NP pollution was the most severe in the two rivers, with detection rates of 100% and 96.3%, respectively. Based on this study's findings, it will be necessary to examine effective measures for the reduction of AP pollution, especially 4-NP pollution, in order to promote the normal growth and reproduction of aquatic organisms, avoid damaging the normal structures and functions of aquatic ecosystems, and ensure the sustainability of the future use of river water resources.

Supplementary Materials: The following supporting information can be downloaded at https://www.mdpi.com/article/10.3390/toxics11070579/s1: Table S1: Basic information of target analytes and parameters for mass spectrometry optimization. Table S2: The relationship between six land-use types and concentration at three buffer regions interpreted from Landsat 8 remote-sensing images. Table S3: The relationship between six land-use types and risk quotient at three buffer regions interpreted from Landsat 8 remote-sensing images.

Author Contributions: Conceptualization, data curation, formal analysis, investigation, methodology, funding acquisition, and writing—original draft, Y.H.; conceptualization, methodology, and resources, M.C.; methodology, software, visualization, and writing—review and editing, Z.Z.; validation, funding acquisition, and writing—review and editing, W.L.; conceptualization, methodology, supervision, and writing—review and editing, C.F.; methodology and software, Z.Y.; data curation and investigation, Y.Q.; data curation and investigation, Y.M.; conceptualization, formal analysis, and supervision, D.X. All authors have read and agreed to the published version of the manuscript.

Funding: This research was supported by the Scientific Research Start-up Fund for Introduced Talents of Anhui Polytechnic University (Yajun Hong, No. 2022YQQ082); the Open Research Fund of the State Environmental Protection Key Laboratory of Ecological Effects and Risk Assessments of Chemicals (Yajun Hong, No. 2022KFYB02); the National Key Research and Development Program (Chenglian Feng, No. 2021YFC3200104); the Jiangxi Provincial Natural Science Foundation (Wei Liao, No. 20224BAB213050); and the Scientific Research Foundation of the Jiangxi Academy of Forestry (Wei Liao, No. 2022521602).

Institutional Review Board Statement: Not applicable.

Informed Consent Statement: Not applicable.

Data Availability Statement: The data presented in this study are available on request from the corresponding author upon reasonable request.

Conflicts of Interest: The authors declare no conflict of interest.

References

1. Vörösmarty, C.J.; McIntyre, P.B.; Gessner, M.O.; Dudgeon, D.; Prusevich, A.; Green, P.; Glidden, S.; Bunn, S.E.; Sullivan, C.A.; Reidy Liermann, C.; et al. Global threats to human water security and river biodiversity. *Nature* **2010**, *467*, 555. [CrossRef] [PubMed]
2. Li, X.; Yeh, A.G.-O. Analyzing spatial restructuring of land use patterns in a fast growing region using remote sensing and GIS. *Landsc. Urban Plan.* **2004**, *69*, 335–354. [CrossRef]
3. Costanza, R.; D'arge, R.; De Groot, R.; Farber, S. The value of the world's ecosystem services and natural capital. *Ecol. Econ.* **1998**, *25*, 3–15. [CrossRef]
4. Xu, G.Y.; Wng, Z.J.; Hu, Z.D.; Habets, F. Comprehensive index evaluation of land use and cover in Qaidam Basin. *J. Hydroelectr.* **2019**, *38*, 44–55.
5. Lei, K.; Zhu, Y.; Chen, W.; Pan, H.Y.; Cao, Y.X.; Zhang, X.; Guo, B.B.; Sweetman, A.; Lin, C.Y.; Ouyang, W.; et al. Spatial and seasonal variations of antibiotics in river waters in the Haihe River Catchment in China and ecotoxicological risk assessment. *Environ. Int.* **2019**, *130*, 104919. [CrossRef]
6. Zhang, S.; Zheng, Y.T.; Zhan, A.B.; Dong, C.X.; Zhao, J.D.; Yao, M. Environmental DNA captures native and non-native fish community variations across the lentic and lotic systems of a megacity. *Sci. Adv.* **2022**, *8*, eabk0097. [CrossRef]
7. Dai, D.; Sun, M.D.; Lv, X.B.; Lei, K. Evaluating water resource sustainability from the perspective of water resource carrying capacity, a case study of the Yongding River watershed in Beijing-Tianjin-Hebei region, China. *Environ. Sci. Pollut. Res.* **2020**, *27*, 21590–21603. [CrossRef]
8. Jiang, B.; Wong, C.P.; Lu, F.; Ouyang, Z.Y.; Wang, Y.J. Drivers of drying on the Yongding River in Beijing. *J. Hydrol.* **2014**, *519*, 69–79. [CrossRef]
9. Meng, J.; Zhou, Y.Q.; Liu, S.F.; Chen, S.Q.; Wang, T.Y. Increasing perfluoroalkyl substances and ecological process from the Yongding Watershed to the Guanting Reservoir in the Olympic host cities, China. *Environ. Int.* **2019**, *133*, 105224. [CrossRef]
10. Guo, W.; Fu, Y.; Ruan, B.; Ge, H.; Zhao, N. Agricultural non-point source pollution in the Yongding River Basin. *Ecol. Indic.* **2014**, *36*, 254–261. [CrossRef]

11. Guo, Z.; Su, Z.; Di, Y.; Guo, X. Analysis on Diversity of Plankton Microbial Community in the Beijing-Tianjin-Hebei Section of the North Canal River. *Environ. Sci.* **2022**, *43*, 803–812.
12. Esteban, S.; Gorga, M.; Petrovic, M.; Gonzalez-Alonso, S.; Barcelo, D.; Valcarcel, Y. Analysis and occurrence of endocrine-disrupting compounds and estrogenic activity in the surface waters of Central Spain. *Sci. Total Environ.* **2014**, *466–467*, 939–951. [CrossRef]
13. Klosterhaus, S.L.; Grace, R.; Hamilton, M.C.; Yee, D. Method validation and reconnaissance of pharmaceuticals, personal care products, and alkylphenols in surface waters, sediments, and mussels in an urban estuary. *Environ. Int.* **2013**, *54*, 92–99. [CrossRef]
14. Liao, C.Y.; Kannan, K. A survey of alkylphenols, bisphenols, and triclosan in personal care products from China and the United States. *Arch. Environ. Contam. Toxicol.* **2014**, *67*, 50–59. [CrossRef]
15. Bergé, A.; Cladière, M.; Gasperi, J.; Coursimault, A.; Tassin, B.; Moilleron, R. Meta-analysis of environmental contamination by alkylphenols. *Environ. Sci. Pollut. Res.* **2012**, *19*, 3798–3819. [CrossRef]
16. Bergé, A.; Gasperi, J.; Rocher, V.; Gras, L.; Coursimault, A.; Moilleron, R. Phthalates and alkylphenols in industrial and domestic effluents: Case of Paris conurbation (France). *Sci. Total Environ.* **2014**, *488–489*, 26–35. [CrossRef]
17. Urbatzka, R.; Rocha, E.; Reis, B.; Cruzeiro, C.; Monteiro, R.A.F.; Rocha, M.J. Effects of ethinylestradiol and of an environmentally relevant mixture of xenoestrogens on steroidogenic gene expression and specific transcription factors in zebrafish. *Environ. Pollut.* **2012**, *164*, 28–35. [CrossRef]
18. Wang, B.; Dong, F.Q.; Chen, S.; Chen, M.J.; Bai, Y.C.; Tan, J.Y.; Li, F.C.; Wang, Q. Phenolic endocrine disrupting chemicals in an urban receiving river (Panlong river) of Yunnan-Guizhou plateau: Occurrence, bioaccumulation and sources. *Ecotoxicol. Environ. Saf.* **2016**, *128*, 133–142. [CrossRef]
19. Lei, K.; Pan, H.Y.; Zhu, Y.; Chen, W.; Lin, C.Y. Pollution characteristics and mixture risk prediction of phenolic environmental estrogens in rivers of the Beijing-Tianjin-Hebei urban agglomeration, China. *Sci. Total Environ.* **2021**, *787*, 147646. [CrossRef]
20. Cheng, J.R.; Wang, K.; Yu, J.; Yu, Z.X.; Yu, X.B.; Zhang, Z.Z. Distribution and fate modeling of 4-nonylphenol, 4-t-octylphenol, and bisphenol A in the Yong River of China. *Chemosphere* **2018**, *195*, 594–605. [CrossRef]
21. Li, Z.H.; Zhang, W.Q.; Shan, B.Q. The effects of urbanization and rainfall on the distribution of, and risks from, phenolic environmental estrogens in river sediment. *Environ. Pollut.* **2019**, *250*, 1010–1018. [CrossRef] [PubMed]
22. Hong, Y.J.; Feng, C.L.; Yan, Z.F.; Wang, Y.; Liu, D.Q.; Liao, W.; Bai, Y.C. Nonylphenol occurrence, distribution, toxicity and analytical methods in freshwater. *Environ. Chem. Lett.* **2020**, *18*, 2095–2106. [CrossRef]
23. Qiu, W.H.; Liu, S.; Chen, H.H.; Luo, S.S.; Xiong, Y.; Wang, X.J.; Xu, B.T.; Zheng, C.M.; Wang, K.J. The comparative toxicities of BPA, BPB, BPS, BPF, and BPAF on the reproductive neuroendocrine system of zebrafish embryos and its mechanisms. *J. Hazard. Mater.* **2021**, *406*, 124303. [CrossRef] [PubMed]
24. Jiang, J.; Zhao, G.; Xu, Y.; Zhao, J.; Liu, L.; Liu, C.; Wang, D.; Li, Y. Occurrence and distribution characteristics of heavy metals in the surface water of Yongding River Basin, China. *Environ. Sci. Pollut. Res. Int.* **2021**, *29*, 17821–17831. [CrossRef]
25. Gu, Y.Y.; Yu, J.; Hu, X.L.; Yin, D.Q. Characteristics of the alkylphenol and bisphenol A distributions in marine organisms and implications for human health: A case study of the East China Sea. *Sci. Total Environ.* **2016**, *539*, 460–469. [CrossRef] [PubMed]
26. Hong, Y.J.; Feng, C.L.; Jin, X.W.; Xie, H.Y.; Liu, N.; Bai, Y.C.; Wu, F.C.; Raimondo, S. A QSAR-ICE-SSD model prediction of the PNECs for alkylphenol substances and application in ecological risk assessment for rivers of a megacity. *Environ. Int.* **2022**, *167*, 107367. [CrossRef]
27. Li, P.; Zhang, W.; Lu, C.; Zhang, R.; Li, X. Robust kernel principal component analysis with optimal mean. *Neural Netw.* **2022**, *152*, 347–352. [CrossRef]
28. Li, F.L.; Altermatt, F.; Yang, J.H.; An, S.Q.; Li, A.M.; Zhang, X.W. Human activities' fingerprint on multitrophic biodiversity and ecosystem functions across a major river catchment in China. *Glob. Change Biol.* **2020**, *26*, 6867–6879. [CrossRef]
29. Seymour, M.; Deiner, K.; Altermatt, F. Scale and scope matter when explaining varying patterns of community diversity in riverine metacommunities. *Basic Appl. Ecol.* **2016**, *17*, 134–144. [CrossRef]
30. Noorimotlagh, Z.; Kazeminezhad, I.; Jaafarzadeh, N.; Ahmadi, M.; Ramezani, Z.; Silva Martinez, S. The visible-light photodegradation of nonylphenol in the presence of carbon-doped TiO_2 with rutile/anatase ratio coated on GAC: Effect of parameters and degradation mechanism. *J. Hazard. Mater.* **2018**, *350*, 108–120. [CrossRef]
31. Zhi, X.; Chen, L.; Shen, Z. Impacts of urbanization on regional nonpoint source pollution: Case study for Beijing, China. *Environ. Sci. Pollut. Res.* **2018**, *25*, 9849–9860. [CrossRef]
32. Jing, H.; Zhang, Z.; Guo, J. Water pollution characteristics and pollution sources of Bei Canal river system in Beijing. *China Environ. Sci.* **2013**, *33*, 319–327.
33. Soares, A.; Guieysse, B.; Jefferson, B.; Cartmell, E.; Lester, J.N. Nonylphenol in the environment: A critical review on occurrence, fate, toxicity and treatment in wastewaters. *Environ. Int.* **2008**, *34*, 1033–1049. [CrossRef]
34. Yu, M.; Cai, Y.; Liu, H.; Gong, L.; Leng, X.; An, S. Impact of Land Use on Water Quality Along Inflow Rivers in Taihu Basin. *J. Agro-Environ. Sci.* **2014**, *33*, 1024–1032.
35. Liu, S.; Wang, C.; Wang, P.F.; Chen, J.; Wang, X.; Yuan, Q.S. Anthropogenic disturbances on distribution and sources of pharmaceuticals and personal care products throughout the Jinsha River Basin, China. *Environ. Res.* **2021**, *198*, 110449. [CrossRef]
36. Xie, H.J.; Chen, J.W.; Huang, Y.; Zhang, R.H.; Chen, C.E.; Li, X.H.; Kadokami, K. Screening of 484 trace organic contaminants in coastal waters around the Liaodong Peninsula, China: Occurrence, distribution, and ecological risk. *Environ. Pollut.* **2020**, *267*, 115436. [CrossRef]

37. Xiong, T.; Qian, Y.; Wei, S.; Geng, J.; Yu, Q.; Ren, H. High-throughput screening and source analysis of emerging contaminants in surface water of Wujin River Basin. *Acta Sci. Circumstantiae* **2022**, *42*, 260–270.
38. Giri, S.; Qiu, Z. Understanding the relationship of land uses and water quality in Twenty First Century: A review. *J. Environ. Manag.* **2016**, *173*, 41–48. [CrossRef]
39. Zhang, M.; Shi, Y.J.; Lu, Y.L.; Johnson, A.C.; Sarvajayakesavalu, S.; Liu, Z.Y.; Su, C.; Zhang, Y.Q.; Juergens, M.D.; Jin, X.W. The relative risk and its distribution of endocrine disrupting chemicals, pharmaceuticals and personal care products to freshwater organisms in the Bohai Rim, China. *Sci. Total Environ.* **2017**, *590–591*, 633–642. [CrossRef]
40. Hong, B.; Lin, Q.Y.; Yu, S.; Chen, Y.S.; Chen, Y.M.; Chiang, P.C. Urbanization gradient of selected pharmaceuticals in surface water at a watershed scale. *Sci. Total Environ.* **2018**, *634*, 448–458. [CrossRef]
41. Ferreira, P.; van Soesbergen, A.; Mulligan, M.; Freitas, M.; Vale, M.M. Can forests buffer negative impacts of land-use and climate changes on water ecosystem services? The case of a Brazilian megalopolis. *Sci. Total Environ.* **2019**, *685*, 248–258. [CrossRef] [PubMed]
42. Sliva, L.; Williams, D.D. Buffer zone versus whole catchment approaches to studying land use impact on river water quality. *Water Res.* **2001**, *35*, 3462–3472. [CrossRef] [PubMed]
43. Paerl, H.W.; Paul, V.J. Climate change: Links to global expansion of harmful cyanobacteria. *Water Res.* **2012**, *46*, 1349–1363. [CrossRef] [PubMed]
44. Allan, J.D.; Erickson, D.L.; Fay, J. The influence of catchment land use on stream integrity across multiple spatial scales. *Freshw. Biol.* **1997**, *37*, 149–161. [CrossRef]
45. Thomson, J.R.; Bond, N.R.; Cunningham, S.C.; Metzeling, L.; Reich, P.; Thompson, R.M.; Nally, R.M. The influences of climatic variation and vegetation on stream biota: Lessons from the Big Dry in southeastern Australia. *Glob. Change Biol.* **2012**, *18*, 1582–1596. [CrossRef]

toxics

MDPI

Review

Research Progress and New Ideas on the Theory and Methodology of Water Quality Criteria for the Protection of Aquatic Organisms

Chenglian Feng [1], Wenjie Huang [1], Yu Qiao [1], Daqing Liu [1,2] and Huixian Li [1,*]

1 State Key Laboratory of Environmental Criteria and Risk Assessment, Chinese Research Academy of Environmental Sciences, Beijing 100012, China; fengcl@craes.org.cn (C.F.)
2 College of Water Science, Beijing Normal University, Beijing 100875, China
* Correspondence: lihuix111@126.com

Abstract: Water quality criteria (WQC) for the protection of aquatic organisms mainly focus on the maximum threshold values of the pollutants that do not have harmful effects on aquatic organisms. The WQC value is the result obtained based on scientific experiments in the laboratory and data fitting extrapolation and is the limit of the threshold value of pollutants or other harmful factors in the water environment. Until now, many studies have been carried out on WQC for the protection of aquatic organisms internationally, and several countries have also issued their own relevant technical guidelines. Thus, the WQC method for the protection of aquatic organisms has been basically formed, with species sensitivity distribution (SSD) as the main method and the assessment factor (AF) as the auxiliary method. In addition, in terms of the case studies on WQC, many scholars have conducted relevant case studies on various pollutants. At the national level, several countries have also released WQC values for typical pollutants. This study systematically discusses the general situation, theoretical methodology and research progress of WQC for the protection of aquatic organisms, and deeply analyzes the key scientific issues that need to be considered in the research of WQC. Furthermore, combined with the specific characteristics of the emerging pollutants, some new ideas and directions for future WQC research for the protection of aquatic organisms are also proposed.

Keywords: water quality criteria (WQC); species sensitivity distribution (SSD); assessment factor; aquatic organisms; model prediction; emerging pollutants

Citation: Feng, C.; Huang, W.; Qiao, Y.; Liu, D.; Li, H. Research Progress and New Ideas on the Theory and Methodology of Water Quality Criteria for the Protection of Aquatic Organisms. *Toxics* **2023**, *11*, 557. https://doi.org/10.3390/toxics11070557

Academic Editor: Daniel Drage

Received: 8 June 2023
Revised: 21 June 2023
Accepted: 22 June 2023
Published: 25 June 2023

1. Introduction

Water quality criteria (WQC) refers to the maximum concentration or level at which pollutants or environmental factors in water environments do not have harmful effects on human health or water ecosystems [1–3]. WQC is the scientific basis for water quality standards and plays an important role in environmental protection and management. According to the different protective receptors, WQC can be roughly divided into the protection of aquatic organisms and the protection of human health. The most important factors affecting the WQC for aquatic organisms are the toxic effects of pollutants, biota differences, water quality parameter and the extrapolation methods of the WQC. There are differences in biota in different countries/regions, which leads to different WQC values even for the same pollutant [4–6]. WQC studies in different countries are carried out on the basis of their own regional environment. The environmental behavior and bioavailability of pollutants in different regions may be different; thus, WQC values are also obviously regional [7,8]. Some countries have issued corresponding guidelines for the protection of aquatic life WQC and technical documents for several pollutants.

Currently, the two mainstream WQC research systems internationally are based on those of the United States and the European Union (EU). The United States puts forward the toxicity percentage ranking method, which is a two-value criteria system [9]. It pointed out that the toxicity data used to derive WQC should cover at least three phyla and eight families of organisms and provide adequate protection for most biological species (more than 95%). In recent years, WQC methods have been improved, and the species sensitivity distribution (SSD) method has been gradually adopted to extrapolate the WQC. Some countries or organizations represented by the EU adopted the SSD method as the extrapolation method of WQC [10]. With the deepening of research, data screening and model optimization have been developed and improved [11]. In terms of the expressions of WQC value, they can be expressed numerically and narratively according to different indicator categories and extrapolation methods of WQC. Most of the numerical values are expressed as the concentrations of pollutants in water environments. For those criteria that cannot give specific numerical indicators, such as color and turbidity, narrative criteria are often used [12,13].

Various relevant factors are comprehensively considered in the derivation process of WQC, and the determination of criteria values is influenced by many water environmental factors, such as water hardness, temperature, pH and dissolved organic matter, and so on [14–17]. Especially in the WQC studies of heavy metals, many countries will calibrate the toxicity values with water environment parameters. At present, the common correction methods for specific regional WQC in different countries and regions include single water quality parameter correction (i.e., hardness correction or pH correction), dual water quality parameter correction, and multiple water quality parameter correction [18–20]. In the 1980s, when it was fully realized that the toxicity of metals is determined by their interactions with other components in the water, the US EPA began to derive the WQC of metals based on a function of hardness [21,22]. This is a big improvement over the previous methods, but the method does not take into account the effects of other factors such as temperature, pH, dissolved organic carbon, sulfides, and alkalinity on metal toxicity. The WQC of pentachlorophenol in the United States takes pH as an important consideration, and the WQC is finally expressed as a function of pH. The most representative case of two-parameter correction is Canada. When formulating long-term WQC for manganese, Canada has conducted hardness correction for fish and invertebrate toxicity data and pH correction for plant toxicity data [23]. The most representative method for multiple water quality parameter correction is the biological ligand model (BLM) [24,25]. At present, Cu-BLM is the most mature BLM applied to heavy metals in freshwater environments. In addition, BLM models for heavy metals such as silver, cadmium, zinc, nickel, cobalt, and lead are also being established and developed. The criteria for copper published in the United States are derived using the BLM [26]. Meanwhile, some studies have also used multiple linear regression methods to study the toxicity and WQC of heavy metals [17,27,28].

Based on the mentioned above, the main purpose of the present study is to systematically summarize and integrate the current technical methods of WQC, deeply analyze the key issues and technical systems of WQC research methods and summarize the current case studies of WQC. Finally, some new ideas of the derivation method of WQC for emerging pollutants are also explored and prospected.

2. Interpretation of WQC Guidelines for the Protection of Aquatic Organisms in Different Countries

The pioneering study on WQC began in the early 20th century. The United States was the earliest country to conduct WQC research. The development process of water quality standards is presented in the form of a series of WQC papers, reports, and monographs. Additionally, the United States finally issued its national water quality guideline for the protection of aquatic organisms in 1985. In addition, the European Union, the Netherlands, Canada, Australia and New Zealand, Japan, China, and other countries have also formulated their own relevant technical guidelines for WQC and established their own WQC and ecological risk assessment research systems. With the deepening of WQC research, many countries have revised, improved, and updated their guidelines in recent years (Table 1).

Table 1. Technical guidelines for water quality criteria for the protection of aquatic life in different countries/organizations.

Country /Organization	Technical Guidelines	Release Time	Description
the United States	Guidelines for deriving numerical national WQC for the protection of aquatic organisms and their uses	1985 (first release)	CMC, CCC
European Union (EU)	Technical guidance document on risk assessment-partII Common implementation strategy for the water framework directive (2000/60/EC) guidance document No. 27 technical guidance for deriving environmental quality standards	2003 (first release) 2018 (latest edition)	PNEC
OECD	Guidance Document for Aquatic Effects Assessment	1995 (first release)	HC_5
Canada	A protocol for the derivation of water quality guidelines for the protection of aquatic life A protocol for the derivation of water quality guidelines for the protection of aquatic life	1999 (first release) 2007 (latest edition)	WQG
Australia and New Zealand	Australian and New Zealand guidelines for fresh and marine water quality Revised method for deriving Australian and New Zealand water quality guideline values for toxicants	2000 (first release) 2018 (latest edition)	TV GV
The Netherlands	Guidance document on deriving environmental risk limits Guidance for the derivation of environmental risk limits within the framework of 'International and national environmental quality standards for substances in the Netherlands' (INS) Revision 2007	2001 (first release) 2007 (latest edition)	ERL
Japan	Environmental Quality Standards for Water Pollution Environmental Quality Standards for Water Pollution	1971 (first release) 2021 (latest edition)	EQS
China	Technical guideline for deriving WQC for the protection of freshwater aquatic organisms Technical guideline for deriving WQC for freshwater organisms	2017 (first release) 2022 (latest edition)	WQC

Note: CMC is criteria continuous concentration; CCC is criteria maximum concentration; PNEC is predicted no-effect concentration; HC_5 is 5 percent hazardous concentration; WQG is water quality guideline; TV is trigger value; GV is guideline value; ERL is environmental risk limit; EQS is environmental quality standard; WQC is water quality criteria.

For example, the technical guidance of the United States was issued in 1985 [9], and the main method used is the toxicity percentage ranking method. The main core of this method is to arrange the average toxicity values of the species from small to large, to select the four most sensitive genera, and to use a series of formulas to calculate the final criteria values. The criteria values obtained by use of this method include the criteria maximum concentration (CMC) and criteria continuous concentration (CCC). CMC considers the acute toxic effect of pollutants on aquatic animals, which is equal to half of the final acute value; CCC considers the chronic toxic effect of pollutants on aquatic animals, which is equal to the minimum of final chronic value, final plant value and final residual value.

The European Union promulgated the Water Framework Directive (WFD) in 2000, which played an important role in developing and promoting the setting of water environmental quality standards. The WFD is a legal framework designed to protect freshwater and marine ecosystems from the adverse effects of pollutants and to protect human health. The European Union (EU) uses environmental risk assessment technology to derive the predicted no-effect concentration (PNEC) of pollutants as water quality protective objectives for environmental management and issued the "*Technical Guidance Document for Risk Assessment*" (TGD) in 2003 [8]. In 2004, at the request of the European Commission for the Environment, the Fraunhofer Institute (FHI), based on TGD Guidelines, prepared a technical guidance manual on the establishment of WQC for priority pollutants of the WFD. In 2007, under the framework of the EU's WFD, the expert group on environmental quality criteria conducted the compiling work of environmental quality criteria in the field of water environments. The event is led and organized by the United Kingdom and the joint research center with support from a working group of experts from EU member states. In 2011, the EU issued the technical guidelines for the derivation of environmental quality criteria, which were updated in 2018 [29]. This is also a programmatic document that has been used to guide WQC research.

In addition to the two mainstream countries of the United States and the European Union, some other countries and organizations have also conducted relevant research on WQC. For example, the Organization for Economic Cooperation and Development (OECD) issued the *Guidance Document for Aquatic Effects Assessment* in 1995 to evaluate the hazardous effect of pollutants on water environments [30]. In 1999, the Canadian Council of Ministers of the Environment (CCME) first published "*A protocol for the derivation of water quality guidelines for the protection of aquatic life*" [31], which was revised and improved in 2007 [32]. The Australian and New Zealand Environmental Protection Council and Agriculture and Resource Management Council (ANZECC/ARMCANZ) issued "*Australian and New Zealand guidelines for fresh and marine water quality*" in 2000 [33], which was revised and updated in 2018 [34]. The National Institute of Public Health and the Environment (RIVM) of the Netherlands published *the Guidance document on deriving environmental risk limits* in 2001 [35], which was revised and improved in 2007 [36]. The Ministry of Environment of Japan published *Environmental Quality Standards for Water Pollution* in 1971 [37], which was revised and improved in 2021 [38].

The research on WQC in China started relatively late, dating back to the 1980s. The initial research was only the collection and collation of relative references and data and integrated the research methods of WQC in different countries and organizations. Subsequently, on the basis of a large number of theoretical explorations and WQC case studies, some monographs and books related to WQC were published one after another [4,5,39]. The technical guide document at the national level is "*Technical guideline for deriving WQC for freshwater organisms*", which was first released by the Ministry of Ecology and Environment in 2017 [40]. It is also the first technical guideline in the field of environmental criteria in China. In 2022, some details of the guidelines were improved and revised for the first time [3]. The methodology of China's WQC guideline is to fully absorb and refer to the latest international WQC research methods, and recommend the SSD method as the extrapolation method of WQC. Meanwhile, a supporting national standardization software named "EEC-SSD" was also developed for WQC derivation [41], which is also an important innovation and highlight in China's research on WQC.

3. Key Points of Theoretical Methodology of WQC for the Protection of Aquatic Organisms

There are many key steps involved in the extrapolation of WQC for the protection of aquatic organisms. In general, the factors affecting WQC mainly include the reliability of basic data, standardization of toxicity data and the scientificity of statistical analysis methods. All these elements play a very important role in the study of WQC.

Firstly, in terms of species selection, the selection of regional native species is crucial for WQC. Many studies have shown that, for the same pollutant, biota differences will lead to the ultimate difference in the WQC value of the pollutant, which further highlights the importance of site-specific or native species [4,42,43]. Therefore, the desired species need to be selected according to regional characteristics and differences in biota. Different countries have defined the quantity requirements for species selection. For example, the US EPA, the European Union, Canada, Australia and New Zealand have requirements on the number of species selected, including at least three trophic levels of fish, invertebrates and plants, and at least five species before subsequent data fitting and WQC extrapolation [11]. After the completion of a species screening process, the acquisition and screening of species toxicity data are also important parts. The quality of the obtained data should also be evaluated according to the reliability and relevance of the data. Klimisch et al. [44] divided toxicity data into four levels according to reliability, correlation and appropriateness and generally screened the data of level 1 and level 2 for WQC research. For example, the European Union evaluates the quality of data according to the principles of reliability and relevance [10], while the UK, the Netherlands, Canada, Australia, New Zealand and China all adopted similar data screening principles [5,11,32,33,36].

Secondly, in terms of WQC derivation methods, the derivation system based on the SSD method and supplemented by the assessment factor (AF) method has been basically formed in the world at present. The main steps of WQC derivation mainly include data collection, data screening and evaluation, and WQC derivation (Figure 1). The basic idea of the AF method is to divide the most sensitive toxicity data of the pollutant by the assessment factor to obtain the WQC value of the pollutant. The AF method requires less basic data, and the calculation method is simple. The disadvantage is that this method is empirical and depends on the toxicity value of sensitive organisms, with high uncertainty. In addition, the AF method does not consider the relationship between species and the biological enrichment effect of pollutants. Therefore, it is only used when the data are difficult to obtain or for comparative verification. In contrast, the advantage of the SSD method is that it makes full use of the toxicity data available for all species and assumes that a limited number of species are randomly sampled from the ecosystem and can be representative of the entire ecosystem. It uses all the toxicity data of known pollutants to fit the sensitivity distribution curve of species and then extrapolates the WQC value. The criteria value is the concentration corresponding to a specified percentage point on the curve, usually expressed as HC_5, which is the concentration at which 5% of species are at risk [3,45,46]. In terms of the scope of use, when the toxicity data of pollutants are relatively sufficient, the SSD method is used to fit the model and extrapolate the criteria, which will statistically reduce the uncertainty. It can be seen that the SSD method has become the international trend of WQC methods and will also be the main method of WQC research in the future. The guidelines for WQC issued by most countries are mainly based on the SSD method. Only the guidelines issued by Canada in 1999 [31] only used the AF method for WQC derivation. However, after the revision in 2007 [32], the SSD method was added. In addition, the principle of giving priority to the SSD followed by the AF method was also proposed.

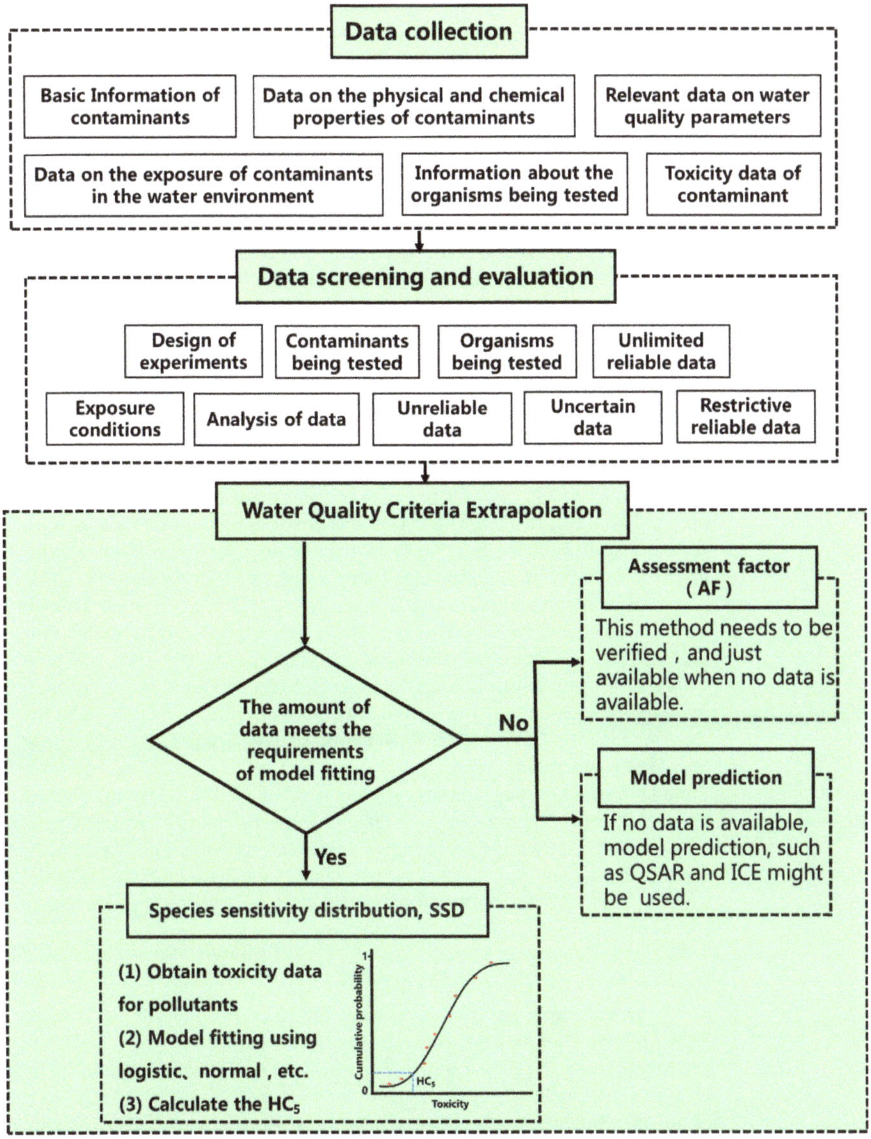

Figure 1. The main process and method of water quality criteria derivation.

The SSD method is an international mainstream method used to establish ecological environment criteria. Model selection is the core and key of this method. When building SSD models, the commonly used derivation methods according to the amount of toxicity data include the normal distribution model, logistic distribution model, etc., and the mainstream trend is to use the best-fit curve to derive the final WQC value [47–49]. Some mathematical statistics software, such as Origin, Matlab, Sigmaplot and other software commonly used in mathematical statistics are mostly used. The fitting results of different models and different statistical software will have certain differences, which may lead to differences in model selection and calculation results of different scholars [50]. In order to solve this problem, some countries have developed or stipulated their own national environmental criteria calculation software. For example, the Netherlands recommended using EcoToX software. Burrlioz software was recommended in Australia and New Zealand [34]. The United States has developed an SSD toolbox for fitting SSD using Fortran language programming [51]. Based on the experience of other countries and combined with empirical data, China developed and first released national ecological environment criteria calculation software–species sensitivity distribution method, EEC-SSD in 2022 [41], which provided standardized technical support for the setting of national ecological and environmental criteria. The release of standardized software at the national level mentioned above provides a guarantee for the fine management of the environmental criteria. With the continuous deepening of model research methods, some relatively mature models have gradually been recommended by WQC researchers and recommended as mainstream derivation models, such as the Burr III model [34].

Thirdly, in terms of the description and value of the WQC for specific pollutants, the double-value criteria system is mainly used internationally at present. The United States, Canada, China, and other countries adopt double-value criteria. Double-value criteria generally refer to long-term criteria and short-term criteria (or chronic criteria and acute criteria). Long-term criteria (chronic criteria) are designed to protect aquatic life (all species and their life stages) from the negative effects caused by the indefinite long-term effects of pollutants. In contrast, short-term criteria are designed to protect aquatic organisms from serious negative effects (such as death) caused by short-term effects. Some countries and organizations also use single-value criteria such as the European Union [8]. There are also countries where the criteria value is neither double-value nor single-value but is classified according to the different protection levels. For example, in Australia and New Zealand, the WQC were divided into four level criteria, i.e., 99%, 95%, 90% and 80% of the criteria values according to the scope of the protected objects (Table 1). In addition, there are also several differences in the description or naming of WQC. For example, in the United States, they were named CMC and CCC. In China, the WQC were divided into short-term WQC for aquatic organisms (SWQC) and long-term WQC for aquatic organisms (LWQC). In Canada, they were used as water quality guidelines (WQG) for the protection of aquatic life, which were divided into long-term concentration and short-term concentration. The European Union directly uses PNEC to represent its WQC. Additionally, in Australia and New Zealand, they were termed trigger values (TVs) in the 2000 Guidelines [33] and named water quality guideline values (GVs) in the updated 2018 Guidelines [34].

4. Research Progress of Case Studies of WQC for Environmental Pollutants

4.1. Bibliometric Analysis of WQC Research

In recent years, great progress has been made in international research focusing on WQC. Many scholars have carried out a large number of WQC case studies of typical pollutants and published a series of research papers, monographs, etc. These studies not only summarized and explored the theoretical methodology of WQC for the protection of aquatic organisms but also provided case studies on WQC, including conventional pollutants, some physical and chemical parameters of water bodies, and new emerging pollutants [4,5,11,16,39,45,52]. Bibliometrics uses statistical methods to conduct quantitative analysis of scientific papers, which can describe the research status and emerging trends in this field and explore future research hotspots and directions [53]. CiteSpace is a citation visualization analysis software based on scientific metrology data and information visualization technology for analyzing potential information in the literature [54,55]. This software can not only analyze the co-citation of literature and mining clustering information, but also analyze the cooperation information of authors, institutions, and countries/regions. Based on this, bibliometrics was used to statistically analyze the previously published papers. The retrieval deadline was September 2022. With the theme of "WQC", "water quality guidelines" and "water quality standards", relevant articles were retrieved on Web of Science (WoS) and China National Knowledge Infrastructure (CNKI), respectively. The results showed that a total of 501 and 203 relevant articles (704 articles in total) were retrieved on WoS and CNKI, respectively (Figure 2). The first article on WQC was published in 1953 and was retrieved in WoS [56]. Since then, the annual average number of documents issued has been less than 10. Since 2010, the number of articles related to WQC has increased rapidly, reaching a peak in 2015. The research on WQC in CNKI started late. The first article related to WQC was published in 1984 [57], and the number of articles published gradually increased from 2010 but showed a downward trend in recent years.

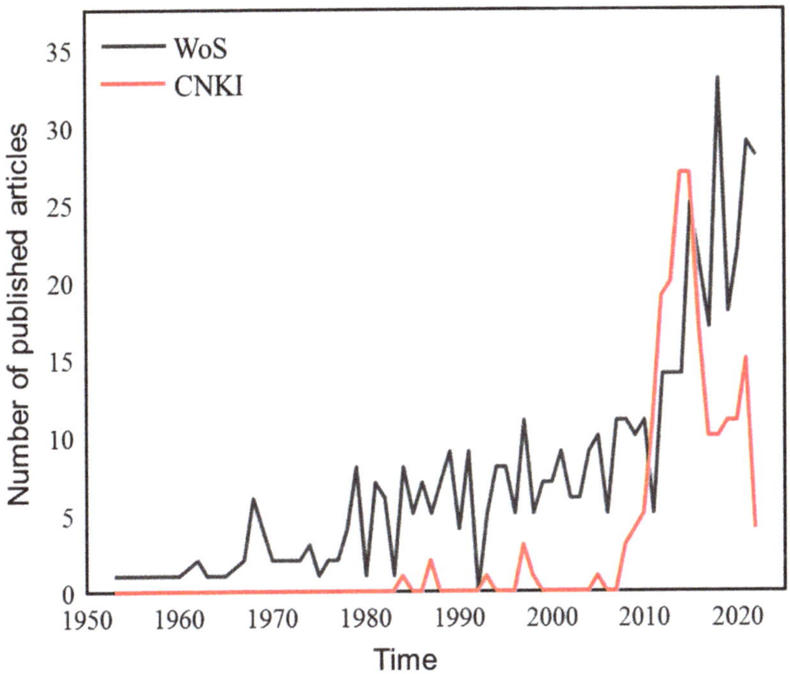

Figure 2. Schematic diagram of the trend of the annual average number of articles published by WoS and CNKI on water quality criteria (Note: WoS is Web of Science; CNKI is China National Knowledge Infrastructure).

At the same time, the countries and organizations studying WQC were screened, and it was found that China and the United States were significantly ahead of other countries in terms of the number of publications on WQC, with 284 and 134 publications (41% and 19% of the total, respectively). In China, the Chinese research academy of environmental sciences published the most articles, and its secondary institution, the state key laboratory of environmental criteria and risk assessment, contributed more than 118 articles. The top five countries in terms of the number of articles issued were China, the United States, Canada, the United Kingdom, and Italy. Additionally, the number of WQC articles issued by these five countries accounted for 60% of all the published articles (Figure 3).

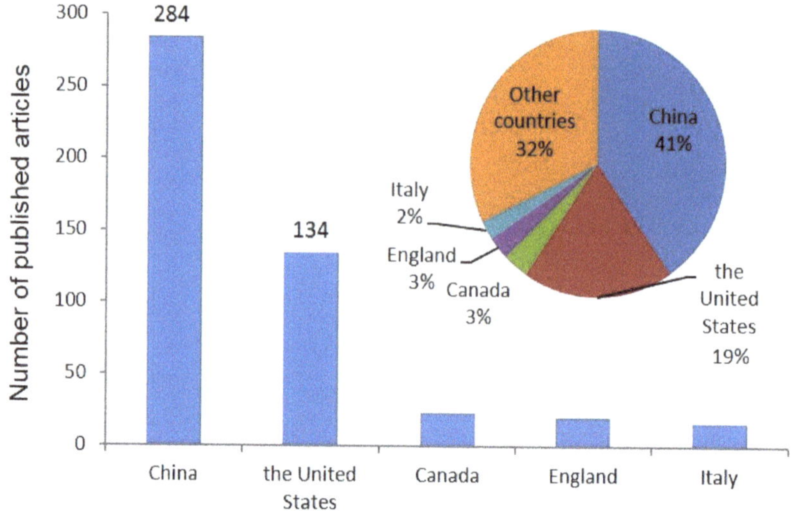

Figure 3. Number of articles published on water quality criteria in different countries.

4.2. WQC Values of Toxic Substances Published at the National Level

In terms of the WQC values of toxic substances published at the national level, some countries have also published WQC for the protection of aquatic organisms for some toxic substances at the national level. For example, since the release of the guidelines for WQC in the United States, a number of WQC for toxic substances have been published, and the WQC are updated almost every 2–3 years in combination with the latest research. EPA's compilation of nationally recommended WQC is presented as a summary table containing recommended WQC for the protection of aquatic life and human health in surface waters. These criteria are published pursuant to the Clean Water Act and provide guidance for states and tribes to use to establish water quality standards and ultimately provide a basis for controlling discharges or releases of pollutants. At present, the latest WQC published by the US EPA contains 186 indicators [21,22], including 61 indicators to protect aquatic life (including 31 indicators of organic matter, 21 indicators of inorganic matter and 9 other indicators, such as pH, temperature, dissolved oxygen, etc.) and 125 indicators to protect human health (Table 2).

In addition to the United States, Australia and New Zealand [58], Canada [59], Japan [60] and China [61–63] have also published WQC for some toxic substances at the national level. Based on the water quality guideline, Australia and New Zealand also published the default guideline values (DGVs) of the pollutants in fresh water and seawater, which were updated in 2023. DGVs were derived using toxicity data for at least three species from at least three taxonomic groups. The listed criteria indicators included a total of 159 freshwater WQC. The WQC value published by Canada contains 194 toxic substance items. In Japan, there are only three indicators in the criteria for the protection of aquatic life. That is, total zinc, nonylphenol and straight-chain alkylbenzene sulfonate, and their criteria values are determined according to the type of water and the type of organism. Up to now, China has only published WQC for toxic substances, namely cadmium, ammonia nitrogen and phenol, and all of them were published in 2020 [61–63].

Table 2. Summary of WQC for aquatic life in the United States.

Indicator Categories	The Specific Name	Number
Organic substances	4,4′-DDT, Acrolein, Aldrin, alpha-Endosulfan, Atrazine, beta-Endosulfan, Carbaryl, Chlordane, Chlorpyrifos, Cyanide, Demeton, Diazinon, Dieldrin, Endrin, gamma-BHC (Lindane), Guthion, Heptachlor Epoxide, Heptachlor, Malathion, Methoxychlor, Methyl Tertiary-Butyl Ether (MTBE), Mirex, Nonylphenol, Oil and Grease, Parathion, Pentachlorophenol, Perfluorooctane Sulfonate (PFOS), Perfluorooctanoic Acid (PFOA), Polychlorinated Biphenyls (PCBs), Toxaphene, Tributyltin	31
Inorganic substances	Alkalinity, Aluminum, Ammonia, Arsenic, Boron, Cadmium, Chloride, Chlorine, Chromium (III), Chromium (VI), Copper, Hardness, Iron, Lead, Mercury, Nickel, Phosphorus Elemental, Selenium, Silver, Sulfide-Hydrogen Sulfide, Zinc	21
Other indicators	Aesthetic Qualities; Color;Gases, Total Dissolved; Nutrients; Oxygen, Dissolved Freshwater; pH; Solids Suspended and Turbidity; Tainting Substances; Temperature	9
	Total	61

5. Exploration of New Theories and Methods of WQC Research

5.1. Exploration of WQC Research Methods Based on Model Prediction

The acquisition of toxicity data is the key issue of WQC research. Different types of pollutants have different influencing factors in the study of WQC. For example, the toxic effect of organic matter on organisms is relatively complex, so the endpoint of the toxic effect of organic pollutants on organisms should be clarified in the WQC study, and then the criteria are deduced. The toxic effects of heavy metals are greatly influenced by environmental factors. Another problem encountered in WQC studies is the lack of data on the toxicity of the selected pollutants. The toxicity data could not meet the requirements of fitting the SSD curve, nor could they meet the requirements of the AF method. However, when environmental management is in urgent need, it is necessary to predict the toxicity data of pollutants by means of model prediction. Model prediction includes the following levels. First, how to use laboratory experiments to predict actual toxicity effects in the field? The Biotic Ligand Models (BLM) approach can be used. The BLM is a mechanistic approach that greatly improves our ability to generate site-specific ambient water quality criteria for metals in the natural environment [26]. Water environmental factors (such as hardness, organic matter content, pH, etc.) have a great influence on metal toxicity. The BLM allows metal–organism interactions to be taken into account and given site-specific information on actual water chemistry to evaluate the dissolved metal concentration associated with a critical level of metal accumulation that is toxic to an organism. It serves as a powerful tool for predicting metal toxicity because it accounts for the concurrent influences of several environmental factors that alter site-specific metal bioavailability in an organism. In contrast to full BLMs, which require the input of field data on upwards of 10 parameters, simplified BLMs integrate data on dissolved metal concentrations, pH, DOC, and calcium concentrations to predict the amount of metal available for uptake by organisms on a site-specific basis, such as Cu, Zn, Ni, Pb (BIO-MET, M-BAT and PNEC-pro) and Mn (M-BAT). At present, most toxicity data are based on laboratory toxicity tests, which mainly consider the total concentration of metal and do not reflect the concept of bioavailability. In fact, the toxicity of metals is closely related to the form of metals, on the one hand, and also closely related to water environmental factors such as organic matter content, hardness, pH value and so on. That is, the concentration of the effective state can better reflect the actual toxicity of the metal. Therefore, the impact of water environmental factors on their bioavailability should be taken into consideration when studying the criteria of such pollutants [4,64,65]. For example, in the WQC for the protection of aquatic organisms published in the United States [26], the copper criteria adopted the method of BLM method, which fully considered the impact of water environmental conditions on acute copper toxicity and its criteria. The European Union (EU) has launched a European Union Water Framework Directive. WQC for copper and other metals that have been developed for use within this regulatory framework are typically based on site-specific effect levels that are evaluated by means of chronic BLMs. In addition to the above-mentioned models, there are also some Multiple Linear Regression (MLR) models [17,66,67], normalized SSD models, etc., which some scholars are trying to adopt for the study of WQC. Second, how to use the toxicity of known species to predict the toxicity of unknown species? Some scholars proposed the use of Interspecies Correlation Estimation (ICE) models and verified the applicability of this model through case studies [68,69]. ICE was first developed by the US EPA, with the surrogate species toxicity data to predict the toxicity data of other unknown species [70]. The basic principle is that there is a linear relationship between the toxicity of the replacement species and the predicted species so that the toxicity of other species can be predicted based on the known toxicity data of the replacement species. At present, it is only used to predict acute toxicity. Thus, the purpose of supplementing the toxicity data of unknown species and then deriving the WQC can be realized. Third, how can the chemical structure of a pollutant be used to predict its toxic effects? Quantitative structural activity correlation (QSAR) models can be used for experiments, and it has been shown that metal toxicity and WQC can also be derived using QSAR models [16,71].

5.2. New Ideas on WQC Research for Emerging Pollutants

Emerging pollutants refer to the pollutants produced in production, construction or other activities, which are caused by human activities, that clearly exist but have not been regulated by laws, regulations and standards and harm the living and ecological environment. With the continuous detection of emerging pollutants in the environment, they are gradually receiving widespread attention [72–74]. The first characteristic of emerging pollutants is that they are "new". There are many kinds of emerging pollutants. At present, there are more than 20 categories of emerging pollutants of global concern, and each category contains dozens or hundreds of chemical substances. This mainly includes persistent organic pollutants (POPs), endocrine-disrupting chemicals (EDCs), antibiotics and microplastic, etc. The second characteristic is "high environmental risk", which is mainly reflected in the severity of the hazard, the hidden risk, the persistence of the environment, the extensive source, and the complexity of the management. Therefore, it is very important to study the WQC and risk assessment of emerging pollutants.

Compared with conventional pollutants, emerging pollutants have the following differences. Taking EDC as an example, first, the dose–response relationship of conventional pollutants generally follows the principle of "low dose, low toxicity", and its toxicity value has a certain threshold. While for EDCs, it is also toxic at low doses, with the characteristics of "low dose, high toxicity". In addition, EDCs may exhibit certain biological effects, including delayed and multigenerational effects, and may exhibit non-monotonic dose–response relationships [75,76]. Secondly, in terms of the selection of sensitive species, the requirements for the derivation of conventional pollutant WQC include fish, amphibians, invertebrates, aquatic plants, algae, etc. However, aquatic plants, algae and some lower invertebrates do not have endocrine systems, which is not suitable for the derivation of WQC for EDCs. Therefore, toxicity data of aquatic plants, algae and lower invertebrates should not be considered when deriving WQC for EDCs [43,55]. Moreover, the toxic effects of conventional pollutants are generally growth and death. While, for EDCs, their toxic effects are complex and changeable, including delayed effect, non-monotonic effect, and more generation effect, which has a great deal of uncertainty. It is necessary to identify its sensitive toxic effect endpoint and select the most sensitive toxic effect endpoint suitable for such substances, such as development, reproduction, and other sensitive effects, to derive the WQC.

Furthermore, in terms of WQC values, when deriving WQC for conventional pollutants, double criteria values, such as short-term criteria values and long-term criteria values, are generally formulated. Nonetheless, EDCs in a short time exposure generally will not cause serious toxicity on aquatic organisms, while, under trace concentrations, will have irreversible toxicity effects on aquatic organisms. The acute-to-chronic ratio of EDCs is generally large; the largest may reach 10,000 or 100,000 times. Therefore, it is recommended to use single-value criteria when deriving WQC for EDCs, and only the values obtained from chronic toxicity are used to derive the long-term criteria value. In addition, it is also important to establish an effective correlation between toxic effects and endpoints [77]. Many effects, from the molecular level to the individual level, are used to evaluate the endocrine mechanism of different groups, but the relationship between these effects and the group-level endpoint is often unclear. If new methods are applied, such as adverse outcome pathways (AOPs) and population modeling, the relationship between EDCs at the level of low-level biological tissues and adverse endpoints at the population level can be better understood. An AOP framework should be built by fully utilizing the existing toxicity data of EDCs based on in vivo and in vitro to completely and deeply understand their toxic mechanisms. Through molecular initiation events (MIEs), key events (KEs) and adverse outcomes (AOs), AOP can organically arrange the existing pollutants' toxic effect mechanisms and toxicity endpoints and provide a theoretical basis for WQC research and predict the toxicity of new pollutants in the near future [74]. At present, most of the toxicity data of chemical pollutants are based on single-species toxicity tests at the individual or organizational level. However, the safety threshold derived from individual-level toxicity testing cannot guarantee population safety. The ecological threshold based on population modeling is of great significance for the protection of aquatic biodiversity and the structure and function of the entire aquatic ecosystem [78]. In Japan, population modeling approaches that have been applied to the threshold of some high-priority chemicals such as nonylphenol, polychlorinated biphenyls, and tributyltin included the derivation of the predicted-no-effect concentration (PNEC) for medaka population-level impact based on population growth rate, as well as the lowest observed-effect concentration (LOEC), the no-observed-effect concentration (NOEC), and the maximum-acceptable-toxic concentration (MATC) [79].

In addition, a non-monotonic dose–effect relationship may appear in both in vitro and short-term in vivo studies of the EDCs, but it might not be able to widely predict the toxicity endpoint in long-term in vivo studies. In the absence of the toxicity threshold of EDCs, probabilistic methods should be used to predict the threshold, which is also a new idea for the future study of WQC of emerging pollutants.

Author Contributions: Conceptualization, C.F.; methodology, C.F.; validation, H.L. and Y.Q.; formal analysis, W.H.; investigation, D.L.; writing—original draft preparation, C.F.; writing—review and editing, C.F. and H.L.; supervision, H.L.; project administration, C.F.; funding acquisition, C.F. All authors have read and agreed to the published version of the manuscript.

Funding: This work was supported by the National Natural Science Foundation of China (42277274) and the National Key Research and Development Program of China (2021YFC3200104).

Institutional Review Board Statement: Not applicable.

Informed Consent Statement: Not applicable.

Data Availability Statement: Not applicable.

Conflicts of Interest: The authors declare no conflict of interest.

References

1. Wu, F.C.; Meng, W.; Zhao, X.L.; Li, H.X.; Zhang, R.Q.; Cao, Y.J.; Liao, H.Q. China Embarking on Development of its Own National Water Quality Criteria System. *Environ. Sci. Technol.* **2010**, *44*, 7992–7993. [CrossRef]
2. Wu, F.C.; Mu, Y.S.; Chang, H.; Zhao, X.L.; Giesy, J.P.; Wu, K.B. Predicting Water Quality Criteria for Protecting Aquatic Life from Physicochemical Properties of Metals or Metalloids. *Environ. Sci. Technol.* **2013**, *47*, 446–453. [CrossRef]
3. *MEE Technical Guideline for Deriving Water Quality Criteria for Freshwater Organisms (HJ 831-2022)*; China Environmental Science Press: Beijing, China, 2022.
4. Feng, C.L.; Wu, F.C.; Zhao, X.L.; Li, H.X.; Chang, H. Water quality criteria research and progress. *Sci. China Earth. Sci.* **2012**, *55*, 882–891. [CrossRef]
5. Wu, F.C. *Theory and Methodology of Water Quality Criteria and Case Studies (2012)*; Science Prezss: Beijing, China, 2012.
6. Leung KM, Y.; Merrington, G.; Warne, M.S.J.; Wenning, R. Scientific derivation of environmental quality benchmarks for the protection of aquatic ecosystems: Challenges and opportunities. *Environ. Sci. Pollut. R* **2014**, *21*, 1–5. [CrossRef] [PubMed]
7. Merrington, G.; Van Sprang, P. Deriving environmental quality standards in European surface waters: When are there too few data? Environ. *Sci. Pollut. R* **2014**, *21*, 67–76. [CrossRef] [PubMed]
8. Lu, L.; Yanjiao, H.; Kang, S.; Fazhi, X.; Huixian, L.; Fuhong, S. Derivation of water quality criteria of zinc to protect aquatic life in Taihu Lake and the associated risk assessment. *J. Environ. Manag.* **2021**, *296*, 113175.
9. *USEPA Guidelines for Deriving Numerical National Water Quality Criteria for the Protection of Aquatic Organisms and Their Uses: PB85-227049 (440-9-76-023) (1985)*; United States Environmental Protection Agency: Washington, DC, USA, 1985.
10. EC. *Technical Guidance Document in Support of the Commission Directive 93/67/EEC on Risk Assessment for New Notified Substances and the Commission Regulation EC 1488/94 on Risk Assessment for Existing Substances (2003)*; European Commission Joint Research Centre: Helsinki, Finland, 2003.
11. Feng, C.L.; Li, H.; Yan, Z.F.; Wang, Y.J.; Wang, C.; Fu, Z.Y.; Liao, W.; Giesy, J.P.; Bai, Y.C. Technical study on national mandatory guideline for deriving water quality criteria for the protection of freshwater aquatic organisms in China. *J. Environ. Manag.* **2019**, *250*, 109539. [CrossRef]
12. USEPA. *Quality Criteria for Water (1976)*; Office of Water Regulations and Standards: Washington, DC, USA, 1976.
13. Feng, C.L.; Wu, F.C.; Zheng, B.H.; Meng, W.; Paquin, P.R.; Wu, K.B. Biotic Ligand Models for Metals-A Practical Application in the Revision of Water Quality Standards in China. *Environ. Sci. Technol.* **2012**, *46*, 10877–10878. [CrossRef]

14. Yung, M.M.N.; Wong, S.W.Y.; Kwok, K.W.H.; Liu, F.Z.; Leung, Y.H.; Chan, W.T.; Li, X.Y.; Djurisic, A.B.; Leung, K.M.Y. Salinity-dependent toxicities of zinc oxide nanoparticles to the marine diatom Thalassiosira pseudonana. *Aquat. Toxicol.* **2015**, *165*, 31–40. [CrossRef]
15. Michael, K.; Kreiss, C.M.; Hu, M.Y.; Koschnick, N.; Bickmeyer, U.; Dupont, S.; Portner, H.O.; Lucassen, M. Adjustments of molecular key components of branchial ion and pH regulation in Atlantic cod (*Gadus morhua*) in response to ocean acidification and warming. *Comp. Biocgem. Phys. B* **2016**, *193*, 33–46. [CrossRef]
16. Mu, Y.S.; Wang, Z.; Wu, F.C.; Zhong, B.Q.; Yang, M.R.; Sun, F.H.; Feng, C.L.; Jin, X.W.; Leung, K.M.Y.; Giesy, J.P. Model for Predicting Toxicities of Metals and Metalloids in Coastal Marine Environments Worldwide. *Environ. Sci. Technol.* **2018**, *52*, 4199–4206. [CrossRef] [PubMed]
17. Liao, W.; Zhu, Z.W.; Feng, C.L.; Yan, Z.F.; Hong, Y.J.; Liu, D.Q.; Jin, X.W. Toxicity mechanisms and bioavailability of copper to fish based on an adverse outcome pathway analysis. *J. Environ. Sci.* **2023**, *127*, 495–507. [CrossRef]
18. Stauber, J.; Golding, L.; Peters, A.; Merrington, G.; Adams, M.; Binet, M.; Batley, G.; Gissi, F.; McKnight, K.; Garman, E.; et al. Application of Bioavailability Models to Derive Chronic Guideline Values for Nickel in Freshwaters of Australia and New Zealand. *Environ. Toxicol. Chem.* **2021**, *40*, 100–112. [CrossRef] [PubMed]
19. Merrington, G.; Peters, A.; Wilson, I.; Cooper, C.; Van Assche, F.; Ryan, A. Deriving a bioavailability-based zinc environmental quality standard for France. *Environ. Sci. Pollut. R* **2021**, *28*, 1789–1800. [CrossRef]
20. Iain, W.; Adam, P.; Graham, M.; Stijn, B. Following the evidence and using the appropriate regulatory tools: A European-wide risk assessment of copper in freshwaters. *Integr. Environ. Assess. Manag.* **2023**, *20*, 4768–4778.
21. USEPA. *National Recommended Water Quality Criteria (2018)*; Office of Water, Office of Science and Technology: Washington, DC, USA, 2018.
22. USEPA. *National Recommended Water Quality Criteria-Correction (2018)*; Office of Water Office of Science and Technology: Washington, DC, USA, 2018.
23. CCME. *Scientific Criteria Document for the Development of the Canadian Water Quality Guidelines for the Protection of Aquatic Life—MANGANESE (2019)*; Canadian Council of Ministers of the Environment: Winnipeg, MB, Canada, 2019.
24. Niyogi, S.; Wood, C.M. Biotic ligand model, a flexible tool for developing site-specific water quality guidelines for metals. *Environ. Sci. Technol.* **2004**, *38*, 6177–6192. [CrossRef]
25. Chung, J.; Hwang, D.; Park, D.H.; Dong, H.Y.; Tae, J.P.; Jinhee, C.; Jong, H.L. Derivation of acute copper biotic ligand model-based predicted no-effect concentrations and acute-chronic ratio. *Sci. Total Environ.* **2021**, *780*, 146425. [CrossRef]
26. USEPA. *Aquatic Life Ambient Freshwater Quality Criteria-Copper (2007)*; Office of Water Regulations and Srandards Criteria Division: Washington, DC, USA, 2007.
27. Peters, A.; Wilson, I.; Merrington, G.; Heijerick, D.; Baken, S. Assessing compliance of European fresh waters for copper: Accounting for bioavailability. *Bull. Environ. Contam. Tox.* **2019**, *102*, 153–159. [CrossRef]
28. Van, G.E.; Stauber, J.L.; Delos, C.; Eignor, D.; Gensemer, R.W.; McGeer, J.; Merrington, G.; Whitehouse, P. Best Practices for Derivation and Application of Thresholds for Metals Using Bioavailability-Based Approaches. *Environ. Toxicol. Chem.* **2020**, *39*, 118–130.
29. EC. *Common Implementation Strategy for the Water Framework Directive (2000/60/ec) Guidance Document No. 27 Technical Guidance for Deriving Environmental Quality Standards (2018)*; European Commission: Copenhagen, Denmark, 2018.
30. OECD. *Guidance Document for Aquatic Effects Assessment (1995)*; Organization for Economic Co-operation and Development: Paris, France, 1995.
31. CCME. *A Protocol for the Derivation of Water Quality Guidelines for the Protection of Aquatic Life (1999)*; Canadian Council of Ministers of the Environment: Ottawa, ON, Canada, 1999.
32. CCME. *A Protocol for the Derivation of Water Quality Guidelines for the Protection of Aquatic Life (2007)*; Canadian Council of Ministers of the Environment: Ottawa, ON, Canada, 2007.
33. ANZECC and ARMCANZ. *Australian and New Zealand Guidelines for Fresh and Marine Water Quality (2000)*; Australian and New Zealand Environment and Conservation Council and Agriculture and Resource Management Council of Australia and New Zealand: Canberra, ACT, Australia, 2000.
34. ANZG. *Australian and New Zealand Guidelines for Fresh and Marine Water Quality (2018)*; Australian and New Zealand Governments and Australian State and Territory Governments: Canberra, ACT, Australia, 2018.
35. RIVM. *Guidance Document on Deriving Environmental Risk Limits: 601501 012 (2001)*; National Institute for Public Health and the Environment: Bilthoven, The Netherlands, 2001.
36. RIVM. *Guidance for the Derivation of Environmental Risk Limits within the Framework of 'International and National Environmental Quality Standards for Substances in the Netherlands' (INS) Revision 2007: 601782001 (2007)*; National Institute for Public Health and the Environment: Bilthoven, The Netherlands, 2007.
37. ME. *Environmental Quality Standards for Water Pollution (1971)*; Ministry of the Environment: Helsinki, Finland, 1971.
38. ME. *Environmental Quality Standards for Water Pollution (2021)*; Ministry of the Environment: Helsinki, Finland, 2021.
39. Wu, F.C. *Introduction to Theory and Methodology of Water Quality Criteria (2021)*; Science Press: Beijing, China, 2021.
40. MEE. *Technical Guideline for Deriving Water Quality Criteria for the Protection of Freshwater Aquatic Organisms (HJ 831-2017)*; China Environmental Science Press: Beijing, China, 2017.
41. Available online: http://www.mee.gov.cn/ywgz/fgbz/ (accessed on 12 January 2023).

42. Jin, X.; Wang, Z.; Wang, Y.; Lv, Y.; Rao, K.; Jin, W.; Giesy, J.P.; Leung, K.M.Y. Do water quality criteria based on nonnative species provide appropriate protection for native species? *Environ. Toxicol. Chem.* **2015**, *34*, 1793–1798. [CrossRef]
43. Jin, X.W.; Zha, J.M.; Xu, Y.P.; Wang, Z.J.; Kumaran, S.S. Derivation of aquatic predicted no-effect concentration (PNEC) for 2,4-dichlorophenol: Comparing native species data with non-native species data. *Chemosphere* **2011**, *84*, 1506–1511. [CrossRef] [PubMed]
44. Klimisch, H.J.; Andreae, M.; Tillmann, U. A systematic approach for evaluating the quality of experimental toxicological and ecotoxicological data. *Regul. Toxicol. Pharmacol.* **1997**, *25*, 1–5. [CrossRef] [PubMed]
45. Hong, Y.; Li, H.; Feng, C.; Liu, D.; Yan, Z.; Qiao, Y.; Bai, Y.; Wu, F. A Review on the Water Quality Criteria of Nonylphenol and the Methodological Construction for Reproduction Toxicity Endocrine Disrupting Chemicals. *Rev. Environ. Contam. T.* **2022**, *260*, 5. [CrossRef]
46. Feng, C.L.; Wu, F.C.; Mu, Y.S.; Meng, W.; Dyer, S.D.; Fan, M.; Raimondo, S.; Barron, M.G. Interspecies Correlation Estimation-Applications in Water Quality Criteria and Ecological Risk Assessment. *Environ. Sci. Technol.* **2013**, *47*, 11382–11383. [CrossRef]
47. Newman, M.C.; Ownby, D.R.; Mezin, L.C.A.; Powell, D.C.; Christensen, T.R.L.; Lerberg, S.B.; Anderson, B.A. Applying species-sensitivity distributions in ecological risk assessment: Assumptions of distribution type and sufficient numbers of species. *Environ. Toxicol. Chem.* **2000**, *19*, 508–515. [CrossRef]
48. Liu, Y.D.; Wu, F.C.; Mu, Y.S.; Feng, C.L.; Fang, Y.X.; Chen, L.L.; Giesy, J.P. Setting Water Quality Criteria in China: Approaches for Developing Species Sensitivity Distributions for Metals and Metalloids. In *Reviews of Environmental Contamination and Toxicology*; Whitacre, D.M., Ed.; Springer Science and Business Media LLC: New York, NY, USA, 2014; Volume 230, pp. 35–57.
49. Wang, Y.; Feng, C.L.; Liu, Y.D.; Zhao, Y.J.; Li, H.X.; Zhao, T.H.; Guo, W.J. Comparative study of species sensitivity distributions based on non-parametric kernel density estimation for some transition metals. *Environ. Pollut.* **2014**, *221*, 343–350. [CrossRef] [PubMed]
50. NCEEC. *User Manual for "National Ecological Environment Criteria Calculation Software—Species Sensitivity Distribution Method (Version 1.0)"(2022)*; National Committee of Expert on Environmental Criteria: Deer Park, NY, USA, 2022.
51. Species Sensitivity Distribution (SSD) Toolbox. Available online: https://www.epa.gov/chemical-research/species-sensitivity-distribution-ssd-toolbox (accessed on 12 February 2023).
52. Wang, Y.Y.; Xiong, J.J.; Ohore, O.E.; Cai, Y.E.; Fan, H.L.; Sanganyado, E.; Li, P.; You, J.; Liu, W.H.; Wang, Z. Deriving freshwater guideline values for neonicotinoid insecticides: Implications for water quality guidelines and ecological risk assessment. *Sci. Total. Environ.* **2022**, *828*, 154569. [CrossRef] [PubMed]
53. Niu, L.; Zhao, X.; Wu, F.; Tang, Z.; Lv, H.; Wang, J.; Fang, M.; Giesy, J.P. Hotpots and trends of covalent organic frameworks (COFs) in the environmental and energy field: Bibliometric analysis. *Sci. Total. Environ.* **2021**, *783*, 146838. [CrossRef] [PubMed]
54. Chen, C. CiteSpace II: Detecting and visualizing emerging trends and transient patterns in scientific literature. *J. Am. Soc. Inf. Tec.* **2006**, *57*, 359–377. [CrossRef]
55. Hong, Y.J.; Feng, C.L.; Jin, X.W.; Xie, H.Y.; Liu, N.; Bai, Y.C.; Wu, F.C.; Raimondo, S. A QSAR-ICE-SSD model prediction of the PNECs for alkylphenol substances and application in ecological risk assessment for rivers of a megacity. *Environ. Int.* **2022**, *167*, 107367. [CrossRef]
56. Borgeson, E.C.; Bacon, V.W.; Gilman, R.H.; McKee, J.E. Judicial Aspects of Water Quality Criteria. *Sew. Ind. Wastes* **1953**, *25*, 325–334.
57. Chen, X. The Establishment Method of the Latest Water Quality Standard in the United States. *Res. Environ. Sci.* **1984**, *6*, 21–25.
58. Australian & New Zealand Toxicant Default Guideline Values for Water Quality in Aquatic Ecosystems. Available online: https://www.waterquality.gov.au/anz-guidelines/guideline-values/default/water-quality-toxicants/search (accessed on 12 March 2022).
59. Canadian Council of Ministers of the Environment Water Quality Guideline Values. Available online: https://ccme.ca/en/summary-table (accessed on 12 April 2023).
60. Japanese Environmental Quality Standards for Water Pollution. Available online: https://www.env.go.jp/en/water/wq/wp.pdf (accessed on 2 June 2023).
61. MEE. *Technical Report on Water Quality Criteria for Freshwater Aquatic Organisms—Cadmium (2020)*; Ministry of Ecology and Environment of the People's Republic of China: Beijing, China, 2020.
62. MEE. *Technical Report on Water Quality Criteria for Freshwater Aquatic Organisms—Ammonia Nitrogen (2020)*; Ministry of Ecology and Environment of the People's Republic of China: Beijing, China, 2020.
63. MEE. *Technical Report on Water Quality Criteria for Freshwater Aquatic Organisms—Phenol (2020)*; Ministry of Ecology and Environment of the People's Republic of China: Beijing, China, 2020.
64. Villavicencio, G.; Urrestarazu, P.; Arbildua, J.; Rodriguez, P.H. Application of an acute biotic ligand model to predict chronic copper toxicity to daphnia magna in natural waters of chile and reconstituted synthetic waters. *Environ. Toxicol. Chem.* **2011**, *30*, 2319–2325. [CrossRef] [PubMed]
65. Liao, W.; Feng, C.L.; Liu, N.; Liu, D.Q.; Yan, Z.F.; Bai, Y.C.; Xie, H.W.; Shi, H.; Wu, D.S. Influence of Hardness and Dissolved Organic Carbon on the Acute Toxicity of Copper to Zebrafish (Danio rerio) at Different Life Stages. *Bull. Environ. Contam. Tox.* **2019**, *103*, 789–795. [CrossRef] [PubMed]

66. Brix, K.V.; DeForest, D.K.; Tear, L.; Grosell, M.; Adams, W.J. Use of Multiple Linear Regression Models for Setting Water Quality Criteria for Copper: A Complementary Approach to the Biotic Ligand Model. *Environ. Sci. Technol.* **2017**, *51*, 5182–5192. [CrossRef]
67. DeForest, D.K.; Brix, K.V.; Tear, L.M.; Cardwell, A.S.; Stubblefield, W.A.; Nordheim, E.; Adams, W.J. Updated Multiple Linear Regression Models for Predicting Chronic Aluminum Toxicity to Freshwater Aquatic Organisms and Developing Water Quality Guidelines. *Environ. Toxicol. Chem.* **2020**, *39*, 1724–1736. [CrossRef]
68. Dyer, S.D.; Versteeg, D.J.; Belanger, S.E.; Chaney, J.G.; Mayer, F.L. Interspecies correlation estimates predict protective environmental concentrations. *Environ. Sci. Technol.* **2006**, *40*, 3102–3111. [CrossRef]
69. Feng, C.L.; Wu, F.C.; Dyer, S.D.; Chang, H.; Zhao, X.L. Derivation of freshwater quality criteria for zinc using interspecies correlation estimation models to protect aquatic life in China. *Chemosphere* **2013**, *90*, 1177–1183. [CrossRef]
70. Raimondo, S.; Jackson, C.R.; Barron, M.G. *Web-Based Interspecies Correlation Estimation (Web-ICE) for Acute Toxicity: User Manual Version 3.2*; U.S Environmental Protection Agency, Office of Research and Development, National Health and Environmental Effects Research Laboratory, Gulf Ecology Division: Gulf Breeze, FL, USA, 2013.
71. Wu, F.C.; Feng, C.L.; Zhang, R.Q.; Li, Y.S.; Du, D.Y. Derivation of water quality criteria for representative water-body pollutants in China. *Sci. China. Earth. Sci.* **2012**, *55*, 900–906. [CrossRef]
72. Liu, N.; Jin, X.W.; Yan, Z.; Luo, Y.; Feng, C.L.; Fu, Z.Y.; Tang, Z.; Wu, F.C.; Giesy, J.P. Occurrence and multiple-level ecological risk assessment of pharmaceuticals and personal care products (PPCPs) in two shallow lakes of China. *Environ. Sci. Eur.* **2020**, *32*, 69. [CrossRef]
73. Yan, Z.F.; Feng, C.L.; Jin, X.W.; Liu, D.Q.; Hong, Y.J.; Qiao, Y.; Bai, Y.C.; Moon, H.B.; Qadeer, A.; Wu, F.C. In vitro metabolic kinetics of cresyl diphenyl phosphate (CDP) in liver microsomes of crucian carp (*Carassius carassius*). *Environ Pollu.* **2021**, *274*, 116586. [CrossRef] [PubMed]
74. Yan, Z.F.; Jin, X.W.; Liu, D.Q.; Hong, Y.J.; Liao, W.; Feng, C.L.; Bai, Y.C. The potential connections of adverse outcome pathways with the hazard identifications of typical organophosphate esters based on toxicity mechanisms. *Chemosphere* **2021**, *266*, 128989. [CrossRef] [PubMed]
75. Hong, Y.J.; Feng, C.L.; Yan, Z.F.; Wang, Y.; Liu, D.Q.; Liao, W.; Bai, Y.C. Nonylphenol occurrence, distribution, toxicity and analytical methods in freshwater. *Environ. Chem. Lett.* **2020**, *18*, 2095–2106. [CrossRef]
76. Yan, Z.; Feng, C.; Jin, X.; Wang, F.; Liu, C.; Li, N.; Qiao, Y.; Bai, Y.; Wu, F.; Giesy, J.P. Organophosphate esters cause thyroid dysfunction via multiple signaling pathways in zebrafish brain. *Env. Sci. Ecotechnol.* **2022**, *12*, 100198. [CrossRef] [PubMed]
77. Liu, D.Q.; Hong, Y.J.; Feng, C.L.; Yan, Z.F.; Bai, Y.C.; Xu, Y.P. General Challenges and Recommendations for the Water Quality Criteria of Endocrine Disrupting Chemicals (EDCs). *Bull. Environ. Contam. Toxicol.* **2022**, *108*, 995–1000. [CrossRef]
78. Bechmann, R.K. Use of life-tables and LC_{50} tests to evaluate chronic and acute toxicity effects of copper on the marine coperpod Tisbe furcata (Baird). *Environ. Toxicol. Chem.* **1994**, *13*, 1509–1517. [CrossRef]
79. Hansen, F.T.; Forbes, V.E.; Forbes, T.L. Effects of 4-n-nonylphenol on life -history traits and population dynamis of a polychaete. *Ecol. Appl.* **1999**, *9*, 482–495. [CrossRef]

toxics

MDPI

Article

Molecular Toxicity Mechanism Induced by the Antibacterial Agent Triclosan in Freshwater *Euglena gracilis* Based on the Transcriptome

Ting Lu [1], Tong Zhang [1], Weishu Yang [1], Bin Yang [1], Jing Cao [1], Yang Yang [2] and Mei Li [1,*]

[1] State Key Laboratory of Pollution Control and Resource Reuse, School of Environment, Nanjing University, Nanjing 210023, China
[2] School of Resources and Environmental Sciences, Nanjing Agricultural University, Nanjing 210095, China
* Correspondence: meili@nju.edu.cn

Abstract: Triclosan (TCS), a commonly used antibacterial preservative, has been demonstrated to have high toxicological potential and adversely affects the water bodies. Since algae are one of the most significant primary producers on the planet, understanding the toxicological processes of TCS is critical for determining its risk in aquatic ecosystems and managing the water environment. The physiological and transcriptome changes in *Euglena gracilis* were studied in this study after 7 days of TCS treatment. A distinct inhibition ratio for the photosynthetic pigment content in *E. gracilis* was observed from 2.64% to 37.42% at 0.3–1.2 mg/L, with TCS inhibiting photosynthesis and growth of the algae by up to 38.62%. Superoxide dismutase and glutathione reductase significantly changed after exposure to TCS, compared to the control, indicating that the cellular antioxidant defense responses were induced. Based on transcriptomics, the differentially expressed genes were mainly enriched in biological processes involved in metabolism pathways and microbial metabolism in diverse environments. Integrating transcriptomics and biochemical indicators found that changed reactive oxygen species and antioxidant enzyme activities stimulating algal cell damage and the inhibition of metabolic pathways controlled by the down-regulation of differentially expressed genes were the main toxic mechanisms of TCS exposure to *E. gracilis*. These findings establish the groundwork for future research into the molecular toxicity to microalgae induced by aquatic pollutants, as well as provide fundamental data and recommendations for TCS ecological risk assessment.

Keywords: triclosan; *Euglena gracilis*; oxidative stress; photosynthetic pigment; molecular toxicity

Citation: Lu, T.; Zhang, T.; Yang, W.; Yang, B.; Cao, J.; Yang, Y.; Li, M. Molecular Toxicity Mechanism Induced by the Antibacterial Agent Triclosan in Freshwater *Euglena gracilis* Based on the Transcriptome. *Toxics* **2023**, *11*, 414. https://doi.org/10.3390/toxics11050414

Academic Editors: Zhen-Guang Yan, Zhi-Gang Li and Jinzhe Du

Received: 28 March 2023
Revised: 23 April 2023
Accepted: 25 April 2023
Published: 27 April 2023

1. Introduction

Water environmental management issues resulting from the increased wastewater generation associated with antimicrobial agents are a major challenge for countries [1]. Triclosan (TCS), an effective polychlorinated aromatic antibacterial agent, is broadly added to many medical and personal care products to achieve concentrations of 0.1 to 0.3%. In the last few decades, its use has grown exponentially in many products, such as TCS-coated antibacterial sutures, TCS-contained composite materials, hand sanitizer and detergent, and cosmetics [2–4]. TCS pollution is widely monitored around the world, due to mass consumption of these and other products. Research shows that TCS continuously enters the aquatic environment, consequently accumulating in water bodies, especially freshwater environments [4,5]. A United States Geological Survey (USGS) report indicates that TCS is one of the top 10 river pollutants in the United States [6]. The global maximum measured TCS concentrations in water are as high as 86 μg/L, 5.3 μg/L, 40μg/L, and 0.1 μg/L for influent, wastewater, surface water, and seawater, respectively [7]. Additionally, the concentration ranges of TCS in the tributary of the Yangtze River, Nanjing, China reached 0.25–0.43 μg/L [8]. More seriously, TCS has been found in urine, breast milk, and blood samples. Specifically, the reported TCS concentration in the urine of Chinese children ranged from none detectable to 681.38 μg/L, and the concentration in blood samples was 0.126–0.161 μg/L [3,8,9]. Considering its potential risk to human health, TCS has been banned in human hygiene products, including household soaps in the United States, since 2017 [10,11]. However, TCS is still widely used in many countries, resulting in high residual levels in aquatic environments, which is potentially harmful for the health of aquatic organisms [4].

As a typical hydrophobic organic compound, TCS exhibits further environmental persistence, based on its high octanol water partition coefficient ($\log_{Know}$ = 4.8) and long half-life [12,13]. Compared to TCS, a variety of conversion products of TCS are more persistent, due to its higher hydrophobicity and lower potential for photodegradation, such as chlorophenols, methyl-triclosan, and dioxins [14]. Research found evidence that an abundance of TCS and its degradation products exist in the environment, especially in the aquatic environment [15,16]. Due to its hydrophobicity, TCS shows high bioaccumulation in organs far exceeding its environmental water concentration [17], which may cause strong toxicity to aquatic organisms, such as algae [4], protozoa [18], insects ([19], crustaceans [20], fish [21,22], and amphibians [23]. Among these aquatic organisms, there are many related studies on the 96 h half-lethal concentration (LC_{50}) values of TCS for fish and microalgae. For fish, the LC_{50}–96 h was 600.0 μg/L for *Poecilia vivipara* and 1700.0 μg/L for *Oryzias latipes* [24]. For microalgae, TCS induced the median effective concentrations (EC_{50}) of 27.1 and 93 μg/L for *Pseudokirchneriella subcapitata* and *Dunaliella tertiolecta*, respectively, while freshwater algae are more sensitive than marine algae and bacteria [25]. Toxicity studies for different algae (excluding *P. subcapitata*) at TCS concentrations ranging from 20.0 to 4000.0 μg/L report that this biocide promotes a reduction in the chlorophyll concentration [26], increases cell membrane activity and permeability [8,27], and interferes with photosynthesis [28].

However, most of the current assessments addressing the toxic effects of TCS focus on the typical phenotypic-based endpoints, such as growth, antioxidant activity, and other indicators. Little information is available on the algal responses to TCS on the molecular level [29,30]. Recently, the transcriptomic has often been used to explore the mechanism of toxic effects by detecting the whole gene expression of organisms and providing insight into the cellular biochemistry. Currently, transcriptomic analysis is used to reveal the toxicity mechanisms of nanoparticles, heavy metals, and organic pollutants in algae [31–33]. This technique is being used to comprehensively analyze transcriptomic changes and reveal the molecular mechanism caused by TCS toxicity in organisms. Transcriptomics and biochemical research in *Labeo rohita* indicated that TCS caused liver and kidney damage, abnormal metabolic processes, and digestive system disorders [34,35]. Transcriptome analysis of zebrafish revealed the role of the liver as a target organ for TCS toxicity, with liver steatosis mainly resulting from increased fatty acid synthetase activity, and the uptake and suppression of β-oxidation [8]. Moreover, in the green alga *Raphidocelis subcapitata,* TCS suppressed molecular signaling pathways, including porphyrin and chlorophyll metabolism, photosynthesis–antenna proteins, and photosynthesis [36].

Euglena gracilis, a secondary green alga that mostly lives in fresh water, has the dual characteristics of flora and fauna [37]. Meanwhile, due to its lack of cell wall and sensitivity to environmental pressure, it is often used as a model organism to evaluate the ecotoxicity of various chemicals [38–40]. The purpose of this study was to investigate the harmful effects of various TCS concentrations on *E. gracilis*. The broad transcriptome and metabolic pathway alterations in *E. gracilis* generated by TCS were studied to better understand the underlying toxicological processes. The major genes and molecular pathways reacting to TCS were screened using the enrichment analysis of biological functions and signal pathways. As a result, the alterations in metabolism and gene information processing caused by TCS exposure were explained. These findings contribute to a better understanding of the toxic mechanism of TCS on *E. gracilis* and give new insight into future aquatic environmental toxicological investigations and assessments.

2. Materials and Methods

2.1. Triclosan and Euglena Gracilis Cultivation

TCS (purity > 97%) was purchased from Aladdin Reagent Co., Ltd. (Shanghai, China). *Euglena gracilis* was obtained from the Freshwater Algae Culture Collection at the Institute of Hydrobiology (FACHB-Collection, Wuhan, China). Checcucci culture medium was used to cultivate the microalgae at 25 ± 1 °C under a 12 h/12 h light/dark cycle, with an illumination intensity of 3000 lux and three replicates [41]. The conical flasks were shaken, and their positions were randomly changed every day to reduce differences in growth among different algal flasks.

2.2. Triclosan Exposure

The pretreatment of algae was conducted as per our previous research [39]. The algal inoculum was resuspended 3 days prior to the toxicity tests. The supernatant was removed after centrifugation at 3500× *g* for 15 min. Then, 5 mL of phosphate buffer solution (PBS, Sevier, Wuhan, China) was used to resuspend the algal cells while ensuring that the original microalgal density was in the exponential growth phase at approximately 1×10^5 cells/mL. Briefly, to calculate EC_{50}, *E. gracilis* was exposed to different concentrations of TCS (0.3, 0.6, 0.9, and 1.2 mg/L), and blank and acetone solvent control groups (0.15%) were performed. The cell density was measured every 24 h at a wavelength of 680 nm, which had a linear correlation with the cell number, according to our previous study using Multiscan spectroscopy (INFINITE M200, Beijing, China), allowing the establishment of a 96 h growth curve for *E. gracilis* [39]. To further explore the toxicity mechanism, 0.30 mg/L (minimum effective concentration, TCS-E) and 1.20 mg/L (maximum effective concentration, TCS-H) TCS were selected for subsequent transmission electron microscopy (TEM) and transcriptome analysis after 96 h of exposure.

2.3. Transmission Electron Microscopy Analysis

After 96 h of exposure, 15 mL of algal solution (>10^7 cells/mL) from the TCS-E and TCS-H treatments and the acetone solvent control group were centrifuged at 3500× *g* for 15 min in a 1.5 mL centrifuge tube. Subsequently, 1 mL of room temperature 2.5% glutaraldehyde (Aladdin Reagent Co., Ltd., Shanghai, China) was added to the pellet. The cells were fixed at room temperature for 2 h in the dark and then stored at 4 °C prior to analysis. Ultrastructure images were taken by TEM (HITACHI HT7700, Tokyo, Japan).

2.4. Pigments Content

After 96 h of exposure to TCS, 1 mL of microalgal suspension from each treatment group was centrifuged at 3500× *g* for 15 min. The supernatant was removed; 1 mL of 80% acetone(Aladdin Reagent Co., Ltd., Shanghai, China) was added to samples and mixed well, and the suspension was then placed at room temperature in the dark for 24 h. After extracting the pigments, the mixture was centrifuged again at 3500× *g* for 15 min, and 80% acetone was used for the control group. The collected supernatant was subjected to Multiscan spectrum analysis at wavelengths of 663, 645, and 470 nm. Chlorophyll a (*Chl a*), chlorophyll *b* (*Chl b*), and carotenoid (*CAR*) contents and the inhibition rate were calculated according to the following equations [42]:

$$Chl\ a = 12.21A_{663} - A_{645} \tag{1}$$

$$Chl\ b = 20.13A_{645} - 5.03A_{663} \tag{2}$$

$$CAR = (1000A_{470} - 3.27Chl\ a - 104Chl\ b)/229 \tag{3}$$

$$\text{The inhibition rate} = \frac{\text{Sample value} - \text{Control value}}{\text{Control value}} \times 100\% \tag{4}$$

A663, A645, and A470 are the fluorescence values at wavelengths of 663, 645, and 470 nm measured by the microplate reader (Synergy H1, Bio Tek, Winooski, VT, USA).

2.5. Oxidative Stress

After 96 h of exposure to TCS, algal cells were centrifuged and rinsed three times with phosphate buffer saline (PBS). The activities of superoxide dismutase (SOD, No. A001-3), glutathione (GSH, No. A006-2), and reactive oxygen species (ROS, No. E004-1) were determined using commercial kits (Jiancheng Bioeng. Inst., Nanjing, China) for the estimation of oxidative stress [43]. The protein content (No. A045-4) was measured by the Coomassie brilliant blue method using a kit (Jiancheng, Nanjing, China) to standardize the enzyme activities. All enzyme activity results were expressed by fluorescence values directly or after calculation via the instructions.

2.6. RNA Extraction and Sequencing

After 96 h of exposure to TCS-E and TCS-H, total RNA was extracted from 10 mL of algal cultures (>5 $\times$ 10^6 cells) via TRIzol extraction (Takara, Maebashi, Japan). A NanoDrop instrument was used to determine the RNA concentration, while the RNA integrity was checked by automated electrophoresis on an Agilent 4150 Tapestation system. After meeting the requirements of sequencing and library construction, the library was constructed.

2.7. De Novo Transcriptome Assembly

Clean reads were assembled using Trinity software v2.13.2 (Broad Institute and the Hebrew University of Jerusalem, Jerusalem, Israel, 2022), and unigenes were generated by TGICL purification [44]. The assembled unigenes were annotated by comparing the five functional databases, namely the RefSeq nonredundant protein (ftp://ftp.ncbi.nlm.nih.gov/blast/db (accessed on 15 August 2022)) (NR), Pfam (http://pfam.xfam.org/ (accessed on 15 August 2022)), Gene Ontology (GO), Kyoto Encyclopedia of Gene Genotype (KEGG), and SwissProt (http://ftp.ebi.ac.uk/pub/databases/swissprot (accessed on 15 August 2022)) databases. Subsequently, the unigenes were annotated to the corresponding classification by NCBI BLAST 2.6.0 + software [45] to obtain the corresponding functional annotation. GO and KEGG analyses were used to determine the gene function and key biological pathways.

2.8. Differential Gene Expression

The FPKM (fragments per kilobase of exon model per million mapped fragments) value of each gene in each sample was calculated by featureCounts software v2.0.3 (Shi Lab, Austin, Australia, 2021) to compare the differential gene expression between samples. Using the read-out data as the input data for horizontal analysis, the differentially expressed genes (DEGs) were analyzed according to DEseq2 ($|\log_2$foldchange$| \geq 1$, p value ≤ 0.05). According to the differential gene detection, this experiment classified and enriched the gene ontology (GO) function and analyzed the KEGG biological pathways and enrichment of the obtained DEGs.

2.9. Statistical Analysis

Statistical analysis considered the mean value $\pm$ standard deviation (mean $\pm$ SD) and employed one-way analysis of variance (ANOVA) using SPSS 24.0 software (IBM, Armonk, NY, USA). Statistical differences between treatments were considered significant at $p < 0.05$. The FPKM value of gene expression in the samples was calculated using the featureCounts software, and the DEGs between groups were analyzed using Deseq2 ($p < 0.05$ and $|\log_2$foldchange$| > 1$). All samples were performed in triplicate.

3. Results

3.1. Dose–Effect Relationship between Triclosan and Euglena Gracilis Growth

The effects on *E. gracilis* growth were explored under exposure to different concentrations of TCS (0, 0.3, 0.6, 0.9, and 1.2 mg/L). During the 96 h exposure, the *E. gracilis* cell density began to decline, compared with the controls (Figure 1a). With prolonged TCS exposure, the algal cell density in all treatments showed an upwards trend within 24 h, except for the highest concentration group. After 96 h of exposure, a dramatic decrease in the algal growth rate was observed with the incremental TCS concentrations. The growth inhibition rate of algae exposed to 0.6, 0.9, and 1.2 mg/L TCS for 96 h decreased to 11.52%, 24.11%, and 38.62%, respectively, of the control (Figure 1b). The higher the TCS concentration, the more severe the *E. gracilis* growth inhibition, forming a significant dose–effect relationship (Figure 1b). According to the linear interpolation, the 96 h EC_{50} value was calculated to be 1.82 mg/L. The acute toxicity to *E. gracilis* due to TCS exposure was moderately toxic (0.3 mg/L < EC_{50} ≤ 3.0 mg/L), according to the toxicity classification of aquatic organisms (GB/T 31270.14-2014).

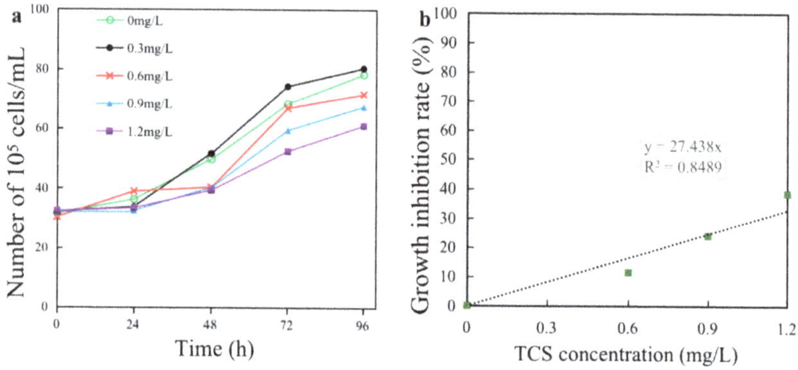

Figure 1. Effect of triclosan (TCS) on the growth of *Euglena gracilis*. (**a**) Cell density; (**b**) growth inhibition rate.

3.2. *Effects on Cell Morphology and Ultrastructure*

After 96 h of exposure, the morphology of the algae was observed under a 40× optical microscope. The results implied that the algae gradually tended to become distorted, and there was an increasing number of deformed algae with an increasing TCS concentration (Figure 2a–e). All *E. gracilis* cells after exposure to TCS had morphological deformities. TEM images were used to observe the microalgal ultrastructure after TCS exposure for 96 h (Figure 2f–h). After exposure to TCS, the number of vacuoles in the *E. gracilis* cells increased notably. Morphological changes in the algal cells were observed after treatment with TCS, including the fragmentation of cells around the chloroplasts. Compared with the control, the chloroplast membranes of *E. gracilis* cells exposed to 0.3 and 1.2 mg/L TCS were slightly abnormal and loosely arranged, which was consistent with the optical microscopy observations. This implies that the growth of *E. gracilis* may be inhibited through toxicity caused by the exposure of chloroplasts to TCS.

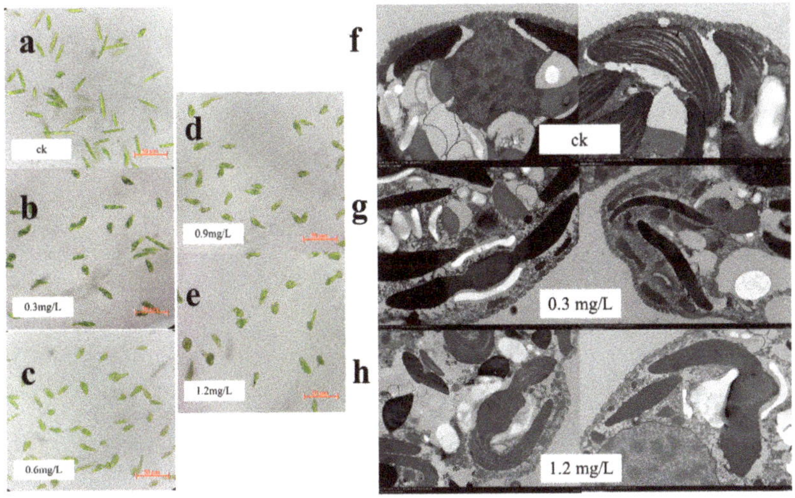

Figure 2. Observations of algal cell morphology. Optical microscope image of the control group (**a**), 0.3 mg/L (**b**), 0.6 mg/L (**c**), 0.9 mg/L (**d**), and 1.2 mg/L (**e**) TCS: the morphology changes in *E. gracilis*; TEM images of the control group (**f**), 0.3 mg/L (**g**), and 1.2 mg/L (**h**) TCS: the ultrastructure of *E. gracilis*, such as the number of vacuoles and chloroplast arrangement and damage.

3.3. Physiological Index Changes Induced by Triclosan Exposure

The photosynthetic pigment content is an important toxicological index of algae and is widely used to indicate the effects of pollutants on photosynthesis. For *E. gracilis*, all photosynthetic pigment indices (Chl-*a*, Chl-*b* and *CAR*) significantly decreased at high TCS concentrations (1.2 mg/L), compared to the control ($p < 0.01$) (Figure 3a), and the inhibition rates reached 37.42%, 32.23%, and 35.80%, respectively, for Chl-*a*, Chl-*b*, and *CAR*. Subsequently, the activities of SOD and GSH, which are essential enzymes of the antioxidative system, were determined. TCS generally changed the activities of SOD and GSH in a concentration-independent manner. SOD activity was inhibited by 24% at 1.2 mg/L TCS (Figure 3b). In contrast, the activities of the GSH level were significantly higher than those of the control (Figure 3c), reaching 303.47%, which indicates that oxidative damage occurred in the algae exposed to a high TCS concentration. Meanwhile, the ROS level declined with the increasing concentration of TCS, although it increased at 1.2 mg/L (Figure 3d). Thus, 1.2 mg/L was considered the critical TCS concentration for algae, after which the ROS levels might exceed its own regulation/detoxification ability. These findings show that exposure to high concentrations of TCS (1.2 mg/L) can induce oxidative stress and reduce photosynthetic pigment in the microalgae, leading to an increase in GSH and ROS production and a decrease in SOD activity.

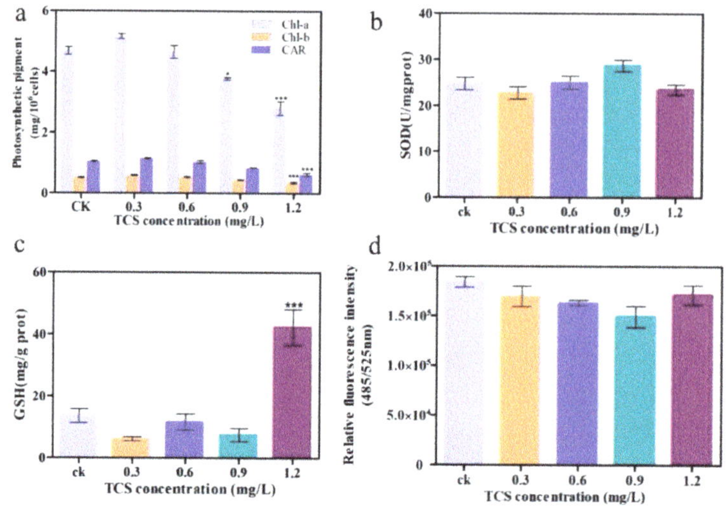

Figure 3. Changes in physiological indicators induced by triclosan (TCS). (**a**) Photosynthetic pigment content; (**b**) the activities of superoxide dismutase (SOD); (**c**) the activities of glutathione (GSH); and (**d**) the activities of relative fluorescence intensity. * $p < 0.05$, *** $p < 0.001$, compared with the control group.

3.4. Transcriptome Analysis

To explore the molecular toxicity mechanism of TCS to *E. gracilis*, TCS-E and TCS-H (Figure 3) were selected for transcriptomics analysis, according to their toxic effects. A total of 285,020 unigenes were obtained from the *E. gracilis* cells. Among the Venn diagrams (Figure 4a), 63 DEGs were shared among all TCS treatments. The histogram showed 87 and 406 DEGs in the TCS-E and TCS-H exposures, respectively, compared to the control (Figure 4c), suggesting that TCS-H induced more severe transcriptional changes. These results were further supported by the principal component analysis (PCA) loading plot (Figure 4b), which exhibited a clear separation between the control and contaminant treatments, especially for the TCS-H exposure. Furthermore, the hierarchical clustering heat map showing the abundances of the top 20 DEGs for TCS-E and TCS-H exhibited downward response trends, with 11 shared DEGs (Figure 4d,e), indicating that the transcriptome response patterns were similar. Compared to TCS-E, there were 65 DEGs in TCS-H, with the top 20 DEGs significantly decreased (Figure 4f). The adverse effect of TCS-H exposure on *E. gracilis* might be greater from the perspective of the transcriptome, and TCS exposure mainly interfered with its growth viability by inhibiting gene expression.

According to the GO functional analysis, biological processes were the most abundant functional gene-encoding products, followed by cellular components and molecular functions (Figure 5a). Cellular and metabolic processes were significantly affected within the biological processes. A total of 10,963 unigenes were annotated into 128 KEGG pathways. Among them, the most significant 34 pathways were enriched (Figure 5b) and divided into 5 KEGG classifications, including metabolism (43.53%) and organic systems (20.99%). Thus, TCS exposure mainly altered the metabolic processes of *E. gracilis*.

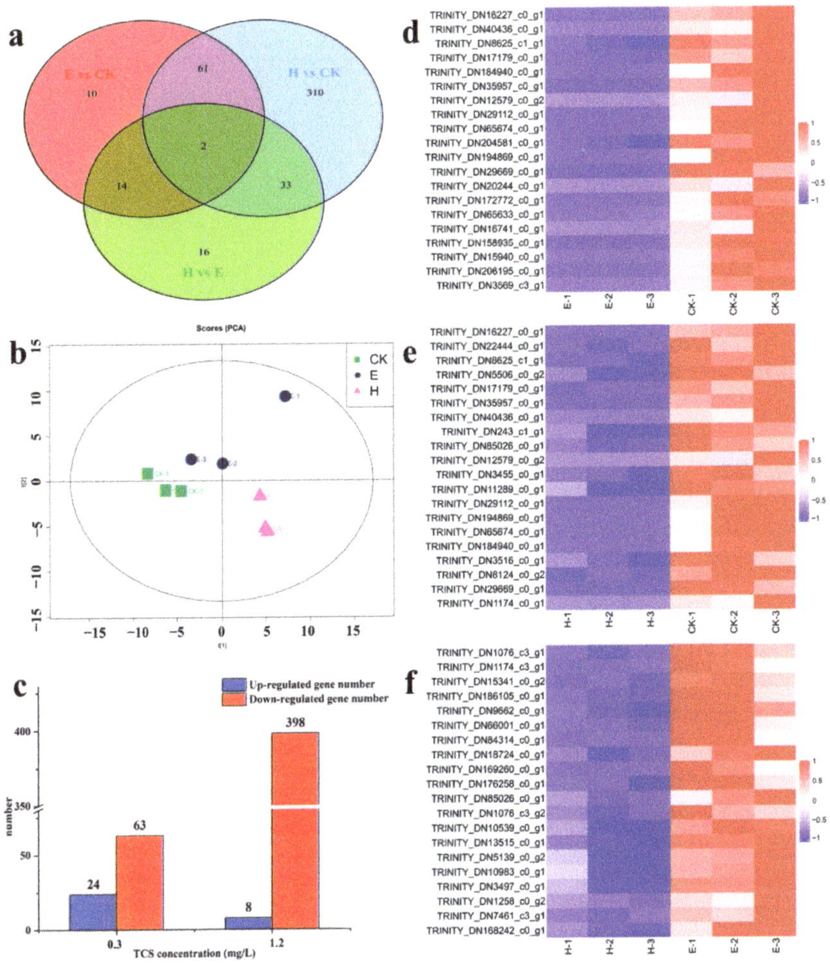

Figure 4. (**a**) Venn diagram of differentially expressed genes (DEGs); (**b**) principal component analysis (PCA) for *Euglena gracilis*; (**c**) differentially expressed gene (DEG) numbers due to triclosan (TCS) exposure (FDR < 0.05); clustering heat map of the top 20 DEGs between groups; (**d**) TCS-E (0.3 mg/L) vs. control; (**e**) TCS-H (1.2 mg/L) vs. control; and (**f**) TCS-H (1.2 mg/L) vs. TCS-E (0.3 mg/L).

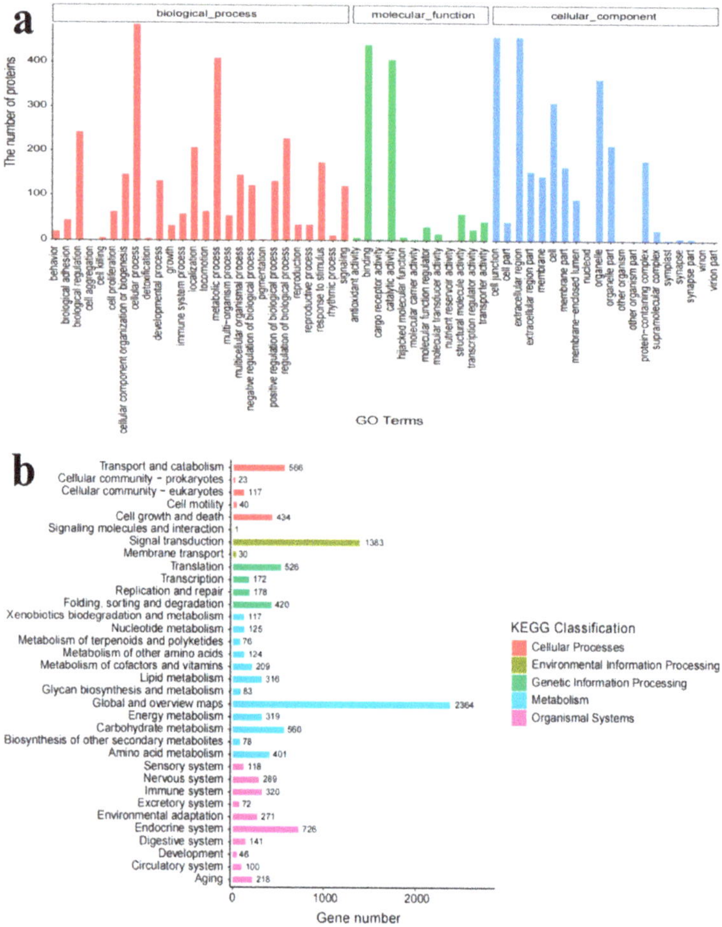

Figure 5. (**a**) Gene classification of *E. gracilis* according to the GO database; (**b**) gene pathway classification of *E. gracilis* according to the KEGG database.

The top 20 pathways were selected by functional analysis of DEGs in the secondary KEGG pathways (Figure 6a,b), with 19 pathways shared between TCS-E and TCS-H. Almost all significantly altered pathways were shared, such as metabolic pathways, microbial metabolism in diverse environments, carbon metabolism, and biosynthesis of secondary metabolites. This demonstrates the similar toxic mechanism over the range of TCS concentrations (0.3–1.2 mg/L), while TCS-H had more DEGs in the KEGG-enriched pathways, which indicates its greater toxicity. Furthermore, the KEGG pathway showed that carbon and nitrogen metabolism in algae were significantly altered, which is important for the growth and development of *E. gracilis*, especially the carbon fixation of this photosynthetic organism [46].

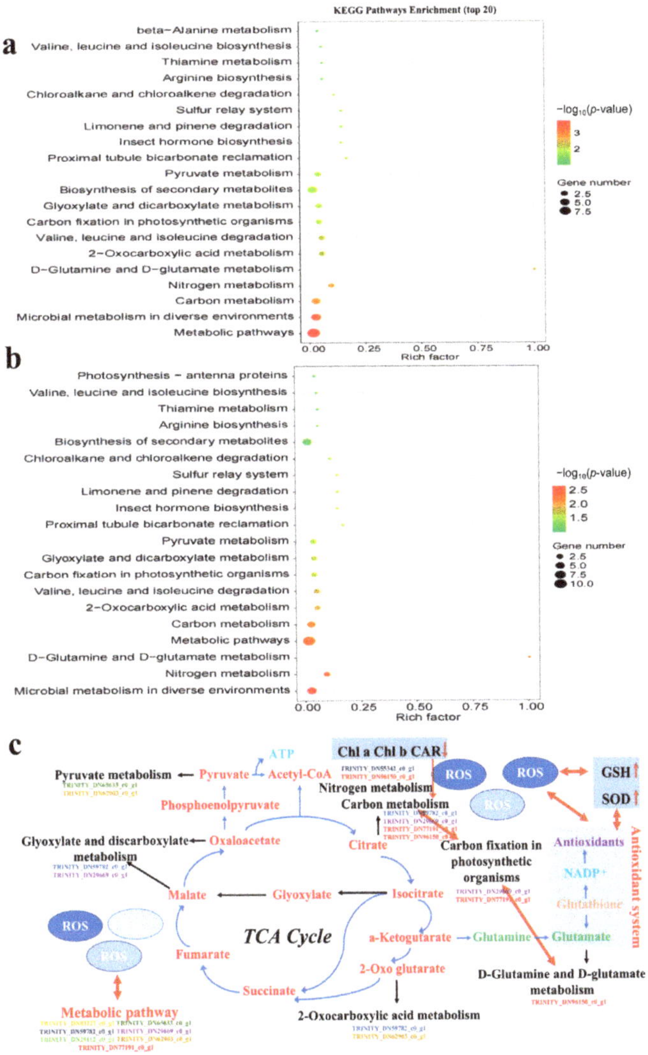

Figure 6. KEGG pathway analysis of *Euglena gracilis* induced by triclosan (TCS). (**a**) TCS-E; (**b**) TCS-H; (**c**) schematic of the proposed metabolic pathways of *Euglena gracilis*.

As shown in Figure 6c, the tricarboxylic acid (TCA) cycle is a pathway related to energy and metabolism, and DEGs cause damage to the related pathway and affect the up- and down-regulations of metabolites, ultimately resulting in metabolization-related toxicity to algae. All DEGs were down-regulated in the KEGG-enriched pathways of the glyoxylate and dicarboxylate metabolisms; pyruvate metabolism; carbon fixation in photosynthetic organisms; 2-oxocarboxylic acid metabolism; D-glutamine and D-glutamate metabolism; and nitrogen metabolism. The metabolic and carbon metabolism pathways, representing 7 and 4 DEGs, respectively, exhibited more disturbances than other pathways. Each DEG could control multiple metabolic pathways, indicating that these metabolic pathways were interrelated. For example, carbon fixation in photosynthetic organisms and carbon metabolism shared two DEGs (*TRINITY_DN29669_c0_g1*, *TRINITY_DN77191_c0_g1*) that were significantly down-regulated, resulting in the significant inhibition of photosynthesis in *E. gracilis*, and all the other enrichment pathways shared genes with metabolic pathways.

4. Discussion

In order to explore the toxic effect and mechanism of TCS on *E. gracilis*, an acute toxicity test was carried out. The results showed that TCS concentration formed a significant dose–effect relationship with the growth inhibition rate of *E. gracilis* (Figure 1b), which is consistent with the study of *Chlamydomonas reinhardtii* [26]. The sensitivity of different types of algae to TCS was variable. Compared with the EC_{50}–96 h values of the green algae *Microcystis aeruginosa* (9.2 μg/L) and *Scenedesmus* subspicatus (2.8 μg/L) and the EC_{50}–72 h value of the diatom *Navicula* sp. (145.6 μg/L), *E. gracilis* (1820 μg/L) displayed stronger tolerance to TCS. This phenomenon may be related to the nutrients, water quality, and algal morphology required for the growth of the different algae. In addition, as a heterotroph, *E. gracilis* can survive in the dark, indicating that it may still survive after chloroplast damage, which may be the reason for the high resistance of *E. gracilis* to TCS, compared with other algae [47].

Electron microscope and TEM images were used to observe the microalgal cell damage after TCS exposure for 96 h (Figure 2a–h). In the results, an increased TCS dose induced the morphology of the algae to be more spherical, rather than a slender strip, due to enhanced osmotic stress, indicating algal adaptation to external pressure [48,49]. All *E. gracilis* cells after exposure to TCS had morphological deformities, which may be related to the occurrence of cell pellicule rupture and degradation [50]. After exposure to TCS, the number of vacuoles in the *E. gracilis* cells increased notably, indicating that toxic substances in the algal cell may be transferred into the vacuoles, where antioxidant molecules, such as glutathione S-transferase (GST), can protect or detoxify the algae [51]. Compared with the control, the chloroplast membranes of *E. gracilis* cells exposed to 0.3 and 1.2 mg/L TCS were slightly abnormal and loosely arranged, indicating that the growth of *E. gracilis* may be inhibited through toxicity caused by the exposure of chloroplasts to TCS.

Photosynthetic pigment content is an important toxicological test index of algae and is widely used to indicate the effect of pollutants on photosynthesis [42]. As shown in Figure 3a, a decline in the algal photosynthetic pigment content after exposure to TCS caused a reduction in photosynthetic activity. Meanwhile, research shows that the exposure of *P. subcapitata* to TCS also induced a reduction in photosynthetic pigments *Chl a*, *Chl b*, and photosynthetic activity [4]. Combined with the conclusions in Figure 2, it was shown that TCS exposure caused cell damage and chloroplast damage in *E. gracilis*, resulting in the hindrance of chlorophyll synthesis, which is the main reason for the reduction in photosynthetic pigments and the inhibition of photosynthesis.

Furthermore, the oxidative stress response of *E. gracilis* under TCS stress was studied to further understand its tolerance and adaptability to TCS stress. The activities of SOD and GSH, which are essential enzymes of the antioxidative system, were determined. An increase in SOD levels and an increase in GSH levels were caused by the elimination of excess oxygen free radicals and hydrogen peroxide in cells [52]. Meanwhile, the ROS level increased at 1.2 mg/L (Figure 3d). To avoid oxidative damage caused by excessive ROS, the balance between the production and elimination of ROS by enzymatic antioxidants is critical for microalgae [53]. Moreover, lipid peroxidation and membrane structure damage of algal cells exposed to high concentrations of TCS may be one of the explanations for the hindrance of chlorophyll synthesis, which results in cell photosynthesis and growth inhibition. Dioxins, one of the transformation products of TCS, significantly enhances the oxidative stress response of organisms, thereby causing irreversible oxidative damage to organisms, indicating that dioxins produced by TCS transformation may also cause oxidative damage in *E. gracilis* [54]. In conclusion, the exposure of TCS and its transformation products resulted in oxidative stress, triggering cell structure damage and the impairment of cell function in *E. gracilis*, which can further threaten the growth and reproduction of the population.

From the perspective of the transcriptome, the higher number of DEGs from TCS-H exposure suggested that the adverse effect of high TCS exposure on *E. gracilis* might be greater

(Figure 4a), indicating that TCS-H induced more severe transcriptional changes. To reflect an in-depth understanding of the toxicity mechanism, analyses were performed on the GO- and KEGG-enriched pathways. The enrichment results of GO verified that metabolic processes were the most affected, which are closely related to the growth, development, and reproduction of the population. In addition, there were enriched pathways related to oxidative stress. This is because ROS, such as superoxide (O^{2-}), hydroxyl radicals ($OH^{\bullet}$), and hydrogen peroxide (H_2O_2), are generated in cells when exposed to chemical stresses, such as metals, nanomaterials, and organic chemicals [51]. This result suggests that TCS exposure caused serious damage to the antioxidant system of *E. gracilis*, which is related to the changes in the antioxidant enzyme activity, and in addition, ROS production stimulated enriched pathways related to oxidative stress, which is the main mechanism of the toxic effects of TCS exposure in *E. gracilis*. In addition, through KEGG secondary pathway function and enrichment analysis, it was shown that both TCS-E and TCS-H exposure affected genes related to *E. gracilis* metabolism, suggesting that the metabolic mechanism of algae was blocked under greater toxicity, resulting in a corresponding stress response. This result was consistent with Liao et al. (2020) [42], who found that the combined exposure group of cadmium and microplastics mainly enriched the DEGs of *E. gracilis* in the gene pathways related to metabolism, suggesting that the metabolic mechanism of algae was blocked under the action of greater toxicity, resulting in a corresponding stress response. Studies have shown that organochlorines can induce inflammatory processes in organisms, stimulate oxidative stress responses, and have a bidirectional relationship with endocrine disorders, eventually leading to a variety of metabolic diseases [55]. As a ubiquitous organochlorine, TCS also has the potential to cause metabolic disorders. The KEGG pathway showed that highly coordinated carbon and nitrogen metabolism in the unicellular algae was significantly altered and resulted in the significant inhibition of algal photosynthesis [46]. A greater carbon source was required for nitrogen metabolism in the TCS-H group to enhance the D-glutamine and D-glutamate metabolism cycles, which may be the main reason for the difference in the carbon fixation pathway, compared to TCS-E. Herein, TCS exposure may further affect metabolic pathways by stimulating oxidative stress responses, thereby inhibiting the growth and development of *E. gracilis*, especially the inhibition of photosynthesis by carbon and nitrogen metabolism.

The TCA cycle is a central pathway of primary metabolism for energy production [56], as shown in Figure 6c; it is difficult to speculate which metabolic processes are most influenced by it. The TCA cycle likely serves to make carbon available from amino acids, fatty acids, and other carbon-containing molecules for energy generation [57]. When genes involved in the TCA cycle are suppressed, a reduction in energy-consuming sugar biosynthesis is expected, resulting in photosynthesis inhibition. Meanwhile, algal cells exposed to pollutants require more essential nutrients to survive because the pollutants can accumulate in the algal cell, impeding the absorption/uptake of essential nutrients, thereby inhibiting growth. When exposed to environmental pollutants, the balance between endogenous and exogenous ROS and antioxidant enzyme activity in organisms may be interrupted, and the change in them can subsequently result in oxidative damage to and metabolic disorders in organisms [58]. In Figure 6c, ROS production damaged the antioxidant system; it caused the change in the metabolite glutamate, affected the D-glutamine and D-glutamate metabolism pathways, and down-regulated related the genes in *TRINITY_DN96150_c0_g1* existing in multiple pathways (ko00220, ko00250, ko00471, and so on), which could lead to the disorder of the TCA cycle metabolic system, including carbon and nitrogen metabolism. Herein, the down-regulation of metabolic pathways, as indicated by DEGs, and the change in ROS and the antioxidant enzyme activity mainly caused oxidative stress and photosynthesis inhibition in *E. gracilis*.

Based on the observed responses of physiological biomarkers and transcriptomic analysis, *E. gracilis* exposed to TCS exhibited an inhibition of population growth, with oxidative damage and metabolic pathways significantly altered. As the primary producer, the pollutants consumed by microalgae are easily transferred to larger organisms along the food chain, such as zooplankton, and is potentially harmful to the entire ecological environment and humans [59]. The inhibition of microalgal population growth indicates an increasing ecological risk to other aquatic species, which may lead to a population decline for higher organisms. Furthermore, TCS has a high adsorption potential, allowing it to adsorb to sedimented sewage sludge and migrate to the soil environment [12], which may cause similar ecological risks to the soil environment, thereby affecting soil microorganisms, plants, and animals. Therefore, exploring the toxic mechanism of TCS on *E. gracilis* not only brings more attention to the harm of pollutants in daily necessities, it also lays the foundation for adequate water environment management and studying the toxic effects of pollutants on other aquatic and terrestrial organisms.

5. Conclusions

In this study, the adverse effects of TCS on freshwater microalgae (*E. gracilis*), including morphological alterations, reduced photosynthesis, and oxidative stress, were investigated. Additionally, the cell's own capacity for detoxification was surpassed by the ongoing stress of greater TCS concentrations (1.2 mg/L). It was shown that TCS, to some extent, interfered with the metabolism and gene information processing of *E. gracilis*, leading to neuronal death brought on by oxidative stress damage from functional analysis of DEGs utilizing the GO and KEGG pathways. Therefore, the main toxic mechanisms of TCS exposure to *E. gracilis* were the changes in ROS and antioxidant enzyme activities to stimulate algal cell damage and the inhibition of the TCA cycle metabolic system, including carbon metabolism, nitrogen metabolism, and D-glutamine and D-glutamate metabolism pathways controlled by the down-regulation of DEGs, which were further manifested as oxidative stress and photosynthesis inhibition effects. These results serve as a starting point for further investigation into the specific molecular pathways in microalgae that are affected negatively by the toxic effects of aquatic pollutants. At the same time, it provides theoretical guidance for the application of antibacterial agents in aquatic environments and promotes water environment management.

Author Contributions: Conceptualization, M.L. and J.C.; data curation, T.L.; formal analysis, T.L.; funding acquisition, M.L.; investigation, T.Z., W.Y. and B.Y.; supervision, M.L.; writing—original draft, T.L.; writing—review and editing: B.Y., J.C., Y.Y. and M.L. All authors have read and agreed to the published version of the manuscript.

Funding: This work was supported by the Science and Technology Support Program of Jiangsu Province (No. BZ2022006) and the National Natural Science Foundation of China (No. 52192682, 22176094, 41773115) of Professor M.L.

Institutional Review Board Statement: Not applicable.

Informed Consent Statement: Not applicable.

Data Availability Statement: Not applicable.

Conflicts of Interest: All authors of this article declare that they have no conflict of interest regarding the publication of this study.

References

1. Eren, G.; Egemen, A.; Emel, T.; Elif, P.-M. Effect of triclosan and its photolysis products on marine bacterium *V. fischeri* and freshwater alga *R. subcapitata*. *J. Environ. Manag.* **2018**, *211*, 218–224.
2. Yueh, M.-F.; Tukey, R.-H. Triclosan: A widespread environmental toxicant with many biological effects. *Annu. Rev. Pharmacol. Toxicol.* **2016**, *56*, 251–272. [CrossRef]
3. Li, X.-Q.; Shang, Y.; Yao, W.-W. Comparison of transcriptomics changes induced by TCS and MTCS exposure in human hepatoma HepG2 cells. *ACS Omega* **2020**, *5*, 10715–10724. [CrossRef]
4. Machado, M.-D.; Soares, E.-V. Toxicological effects induced by the biocide triclosan on *Pseudokirchneriella subcapitata*. *Aquat. Toxicol.* **2021**, *230*, 105706. [CrossRef]
5. Xie, J.-H.; Zhao, N.; Zhang, Y.-Y.; Hu, H.-M.; Zhao, M.-R.; Jin, H.-B. Occurrence and partitioning of bisphenol analogues triclocarban and triclosan in seawater and sediment from East China Sea. *Chemosphere* **2022**, *287*, 132218. [CrossRef] [PubMed]
6. Halden, R.-U. On the need and speed of regulating triclosan and triclocarban in the United States. *Environ. Sci. Technol.* **2014**, *48*, 3603–3611. [CrossRef]

7. Montaseri, H.; Forbes, P.-B.-C. A review of monitoring methods for triclosan and its occurrence in aquatic environments. *TrAC Trends Anal. Chem.* **2016**, *85*, 221–231. [CrossRef]
8. Wei, M.; Yang, X.; Watson, P.; Yang, F.; Liu, H. A cyclodextrin polymer membrane-based passive sampler for measuring triclocarban triclosan and methyl triclosan in rivers. *Sci. Total Environ.* **2019**, *648*, 109–115. [CrossRef] [PubMed]
9. Wang, L.; Mao, B.; He, H.; Shang, Y.; Zhong, Y.; Yu, Z.; Yang, Y.; Li, H.; An, J. Comparison of hepatotoxicity and mechanisms induced by triclosan (TCS) and methyl-triclosan (MTCS) in human liver hepatocellular HepG2 cells. *Toxicol. Res.* **2019**, *8*, 38–45. [CrossRef]
10. European Comission. *Commission Implementing Decision (EU) 2016/110 of 27 January 2016 Not Approving Triclosan as an Existing Active Substance for Use in Biocidal Products for Product Type 1*; European Union: Maastricht, The Netherlands, 2016.
11. US-FDA, Food and Drug Administration. *USA, Safety and Effectiveness of Health Care Antiseptics; Topical Antimicrobial Drug Products for Overthe-Counter Human Use*; Final rule; US-FDA, Food and Drug Administration: Silver Spring, MD, USA, 2017.
12. Dhillon, G.-S.; Kaur, S.; Pulicharla, R.; Brar, S.-K.; Cledón, M.; Verma, M. Triclosan: Current status occurrence environmental risks and bioaccumulation potential. *Int. J. Environ. Res. Public. Health* **2015**, *12*, 5657–5684. [CrossRef]
13. Ramires, P.-F.; Tavella, R.-A.; Escarrone, A.-L.; Volcão, L.-M.; Honscha, L.-C. Ecotoxicity of triclosan in soil: An approach using different species. *Environ. Sci. Pollut. Res.* **2021**, *28*, 41233–41241. [CrossRef] [PubMed]
14. Ma, Y.-J.; Zhang, T.-L.; Zhu, P.; Cai, H.-T.; Jin, Y.; Gao, K.-G.; Li, J. Fabrication of Ag_3PO_4/polyaniline-activated biochar photocatalyst for efficient triclosan degradation process and toxicity assessment. *Sci. Total Environ.* **2022**, *821*, 153453. [CrossRef]
15. Kookana, R.-S.; Shareef, A.; Fernandes, M.-B.; Hoare, S.; Gaylard, S.; Kumar, A. Bioconcentration of triclosan and methyl-triclosan in marine mussels (Mytilus gallo-provincialis) under laboratory conditions and in metropolitan waters of Gulf St Vin-cent South Australia. *Mar. Pollut. Bull.* **2013**, *74*, 66–72. [CrossRef] [PubMed]
16. Solá-Gutiérrez, C.; Schröder, S.; San-Román, M.-F.; Ortiz, I. Critical review on the mechanistic photolytic and photocatalytic degradation of triclosan. *J. Environ. Manag.* **2020**, *260*, 110101. [CrossRef] [PubMed]
17. Schebb, N.-H.; Flores, I.; Kurobe, T.; Franze, B.; Ranganathan, A.; Hammock, B.-D. Bioconcentration metabolism and excretion of triclocarban in larval Qurt medaka (*Oryzias latipes*). *Aquat. Toxicol.* **2011**, *105*, 448–454. [CrossRef]
18. Gao, L.; Yuan, T.; Cheng, P.; Bai, Q.; Zhou, C.; Wang, W. Effects of triclosan and triclocarban on the growth inhibition cell viability genotoxicity and multixenobiotic resistance responses of *Tetrahymena thermophila*. *Chemosphere* **2015**, *139*, 434–440. [CrossRef]
19. Martínez-Paz, P. Response of detoxification system genes on *Chironomus riparius* aquatic larvae after antibacterial agent triclosan exposures. *Sci. Total Environ.* **2018**, *624*, 1–8. [CrossRef]
20. Rowett, C.-J.; Hutchinson, T.-H.; Comber, S.-D.-W. The impact of natural and anthropogenic dissolved organic carbon (DOC) and pH on the toxicity of triclosan to the crustacean *Gammarus pulex* (L.). *Sci. Total Environ.* **2016**, *565*, 222–231. [CrossRef]
21. Dar, O.I.; Sharma, S.; Singh, K.; Sharma, A.; Bhardwaj, R.; Kaur, A. Biochemical markers for prolongation of the acute stress of triclosan in the early life stages of four food fishes. *Chemosphere* **2020**, *247*, 125914. [CrossRef]
22. Paul, T.; Kumar, S.; Shukla, S.-P.; Pal, P.; Kumar, K.; Poojary, N. A multibiomarker approach using integrated biomarker response to assess the effect of pH on triclosan toxicity in Pangasianodon hypophthalmus (Sauvage 1878). *Environ. Pollut.* **2020**, *260*, 114001. [CrossRef]
23. Martins, D.; Monteiro, M.-S.; Soares, A.-M.-V.-M.; Quintaneiro, C. Effects of 4-MBC and triclosan in embryos of the frog *Pelophylax perezi*. *Chemosphere* **2017**, *178*, 325–332. [CrossRef]
24. Sager, E.; Scarcia, P.; Marino, D.; Mac, L.-T.; Rossi, A. Oxidative stress responses after exposure to triclosan sublethal concentrations: An integrated biomarker approach with a native (*Corydoras paleatus*) and a model fish species (*Danio rerio*). *J. Toxicol. Environ. Health* **2021**, *85*, 291–306. [CrossRef]
25. Machado, M.-D.; Soares, E.-V. Sensitivity of freshwater and marine green algae to three compounds of emerging concern. *J. Appl. Phycol.* **2019**, *31*, 399–408. [CrossRef]
26. Pan, C.-G.; Peng, F.-J.; Shi, W.-J.; Hu, L.-X.; Wei, X.-D.; Ying, G.-G. Triclosan-induced transcriptional and biochemical alterations in the freshwater green algae *Chlamydomonas reinhardtii*. *Ecotoxicol. Environ. Saf.* **2018**, *148*, 393–401. [CrossRef]
27. Gonzalez, P.-M.; Rioboo, C.; Reguera, M.; Abreu, I.; Leganes, F.; Cid, A. Calcium mediates the cellular response of *Chlamydomonas reinhardtii* to the emerging aquatic pollutant triclosan. *Aquat. Toxicol.* **2017**, *186*, 50–66. [CrossRef]
28. Xin, X.; Huang, G.; An, C.; Raina-Fulton, R.; Weger, H. Insights into long-term toxicity of triclosan to freshwater green algae in Lake Erie. *Environ. Sci. Technol.* **2019**, *53*, 2189–2198. [CrossRef] [PubMed]
29. Yan, J.; Zou, Y.; Zhang, F.-R.; Zhang, S.-H.; Huang, X.-Y.; Benoit, G. Growth ROS accumulation site and photosynthesis inhibition mechanism of *Chlorella vulgaris* by triclosan. *Environ. Sci. Pollut. Res.* **2022**, *30*, 12125–12137. [CrossRef] [PubMed]
30. Ding, T.-D.; Wei, L.-Y.; Hou, Z.-M.; Lin, S.-Q.; Li, J.-Y. Biological responses of alga *Euglena gracilis* to triclosan and galaxolide and the regulation of humic acid. *Chemosphere* **2022**, *307*, P1135667. [CrossRef]
31. Fernando, P.-S.; Silvia, D.; Vanessa, P.; Angeles, A.; Sanna, O. Basis of genetic adaptation to heavy metal stress in the acidophilic green alga *Chlamydomonas acidophila*. *Aquat. Toxicol.* **2018**, *200*, 62–72.
32. Wang, W.-J.; Sheng, Y.-Q. Effects and mechanisms of decabromodiphenyl ethane on *Chlorella sorokiniana*: Transcriptomics proteins and fatty acid production. *Mar. Environ. Res.* **2022**, *181*, 105764. [CrossRef]
33. Gao, L.; Xie, Y.; Su, Y.-Y.; Mehmood, T.; Bao, R.-Q.; Fan, H.-J. Elucidating the negatively influential and potentially toxic mechanism of single and combined micro-sized polyethylene and petroleum to *Chlorella vulgaris* at the cellular and molecular levels. *Ecotoxicol. Environ. Saf.* **2022**, *245*, 245114102. [CrossRef]

34. Sharma, S.; Dar, O.-I.; Singh, K.; Kaur, A.; Faggio, C. Triclosan elicited biochemical and transcriptomic alterations in *Labeo rohita* larvae. *Environ. Toxicol. Pharmacol.* **2021**, *88*, 103748. [CrossRef] [PubMed]
35. Sharma, S.; Dar, O.-I.; Singh, K.; Thakur, S.; Kesavan, A.-K.; Kaur, A. Genomic markers for the biological responses of Triclosan stressed hatchlings of *Labeo rohita*. *Environ. Sci. Pollut. Res.* **2021**, *28*, 67370–67384. [CrossRef] [PubMed]
36. Mo, J.-Z.; Qi, Q.-J.; Hao, Y.-G.; Lei, Y.; Guo, J.-H. Transcriptional response of a green alga (*Raphidocelis subcapitata*) exposed to triclosan: Photosynthetic systems and DNA repair. *J. Environ. Sci.* **2022**, *111*, 400–411. [CrossRef] [PubMed]
37. Cordoba, J.; Perez, E.; Van, V.-M.; Bertrand, A.-R.; Lupo, V.; Cardol, P. *De Novo* transcriptome meta-assembly of the mixotrophic freshwater microalga *Euglena gracilis*. *Genes* **2021**, *12*, 842. [CrossRef]
38. Zakryś, B.; Milanowski, R.; Karnkowska, A. Evolutionary origin of *Euglena*. *Adv. Exp. Med. Biol.* **2017**, *979*. [CrossRef]
39. Yu, S.-P.; Cole, M.; Chan, B. Review: Effects of microplastic on zooplankton survival and sublethal responses. *Oceanogr. Mar. Biol.* **2020**, *58*, 351–394.
40. He, J.-Y.; Liu, C.-C.; Du, M.-Z.; Zhou, X.-Y.; Hu, Z.-L.; Lei, A. Metabolic responses of a model green microalga *Euglena gracilis* to different environmental stresses. *Front. Bioeng. Biotechnol.* **2021**, *9*, 662655. [CrossRef]
41. Li, M.; Gao, X.-Y.; Wu, B.; Qian, X.; Giesy, J.-P.; Cui, Y.-B. Microalga *Euglena* as a bioindicator for testing genotoxic potentials of organic pollutants in Taihu Lake China. *Ecotoxicology* **2014**, *23*, 633–640. [CrossRef] [PubMed]
42. Liao, Y.-C.; Jiang, X.-F.; Xiao, Y. Exposure of microalgae *Euglena gracilis* to polystyrene microbeads and cadmium: Perspective from the physiological and transcriptional responses. *Aquat. Toxicol.* **2020**, *228*, 105650. [CrossRef]
43. Deng, X.-Y.; Cheng, J.; Hu, X.-L.; Wang, L.; Li, D.; Gao, K. Biological effects of TiO_2 and CeO_2 nanoparticles on the growth photosynthetic activity and cellular components of a marine diatom *Phaeodactylum tricornutum*. *Sci. Total Environ.* **2017**, *575*, 87–96. [CrossRef]
44. Pertea, G.; Huang, X.-Q.; Liang, F.; Antonescu, V.; Sultana, R.; Karamycheva, S. TIGR Gene Indices clustering tools (TGICL): A software system for fast clustering of large EST datasets. *Bioinformatics* **2003**, *19*, 651–652. [CrossRef] [PubMed]
45. Altschul, S.-F.; Gish, W.; Miller, W.; Myers, E.-W.; Lipman, D.-J. Basic local alignment search tool. *J. Mol. Biol.* **1990**, *215*, 403–410. [CrossRef] [PubMed]
46. Prado, R.; Rioboo, C.; Herrero, C. The herbicide paraquat induces alterations in the elemental and biochemical composition of nontarget microalgal species. *Chemosphere* **2009**, *76*, 1440–1444. [CrossRef] [PubMed]
47. Wang, Y.-M.; Seppänen-Laakso, T.; Rischer, H.; Wiebe, M.a.-G. Euglena gracilis growth and cell composition under different temperature, light and trophic conditions. *PLoS ONE* **2018**, *13*, e0195329. [CrossRef] [PubMed]
48. Azizullah, A.; Richter, P.; Haeder, D.-P. Comparative toxicity of the pesticides carbofuran and malathion to the freshwater flagellate *Euglena gracilis*. *Ecotoxicology* **2011**, *20*, 1442–1454. [CrossRef]
49. Mao, Y.; Ai, H.; Chen, Y. Phytoplankton response to polystyrene microplastics: Perspective from an entire growth period. *Chemosphere* **2018**, *208*, 59–68. [CrossRef]
50. Xin, X.; Huang, G.; An, C.; Feng, R. Interactive toxicity of triclosan and nanoTiO_2 to green alga *Eremosphaera viridis* in Lake Erie: A new perspective based on Fourier Transform Infrared spectromicroscopy and synchrotron-based X-ray fluorescence imaging. *Environ. Sci. Technol.* **2019**, *53*, 9884–9894. [CrossRef]
51. Kim, H.; Wang, H.; Ki, J.-S. Chloroacetanilides inhibit photosynthesis and disrupt the thylakoid membranes of the dinoflagellate *Prorocentrum minimum* as revealed with metazachlor treatment. *Ecotoxicol. Environ. Saf.* **2021**, *211*, 111928. [CrossRef]
52. Peng, D.-L.; Wang, W.-J.; Liu, A.-R.; Zhang, Y.; Li, X.-Z.; Wang, G. Comparative transcriptome combined with transgenic analysis reveal the involvement of salicylic acid pathway in the response of *Nicotiana tabacum* to triclosan stress. *Chemosphere* **2021**, *270*, 129456. [CrossRef]
53. Caverzan, A.; Casassola, A.; Brammer, S.-P. Antioxidant responses of wheat plants under stress. *Genet. Mol. Biol.* **2016**, *39*, 1–6. [CrossRef]
54. VanEtten, S.-L.; Bonner, M.-R.; Ren, X.-F.; Birnbaum, L.-S.; Kostyniak, P.-J.; Wang, J.; Olson, J.-R. Effect of Exposure to 2,3,7,8-Tetrachlorodibenzo-p-dioxin (TCDD) and Polychlorinated Biphenyls (PCBs) on Mitochondrial DNA (mtDNA) Copy Number in Rats. *Toxicology* **2021**, *454*, 152744. [CrossRef] [PubMed]
55. Peinado, F.-M.; Artacho-Cordón, F.; Barrios-Rodríguez, R.; Arrebola, J.-P. Influence of polychlorinated biphenyls and organochlorine pesticides on the inflammatory milieu. A systematic review of in vitro, in vivo and epidemiological studies. *Environ. Res.* **2020**, *186*, 109561. [CrossRef]
56. Du, C.-L.; Zhang, B.; He, Y.-L.; Hu, C.-Y. Biological effect of aqueous C 60 aggregates on *Scenedesmus obliquus* revealed by transcriptomics and non-targeted metabolomics. *J. Hazard. Mater.* **2017**, *324*, 221–229. [CrossRef]
57. Matthijs, M.; Fabris, M.; Obata, T.; Foubert, I. The transcription factor bZIP14 regulates the TCA cycle in the diatom *Phaeodactylum tricornutum*. *EMBO J.* **2017**, *36*, 1559–1576. [CrossRef]

58. Jin, Y.-X.; Zheng, S.-S.; Fu, Z.-W. Embryonic exposure to cypermethrin induces apoptosis and immunotoxicity in zebrafish (*Danio rerio*). *Fish. Shellfish. Immunol.* **2011**, *30*, 1049–1054. [CrossRef]
59. Chae, Y.; Kim, D.; Kim, S.-W.; An, Y. Trophic transfer and individual impact of nano-sized polystyrene in a four-species freshwater food chain. *Sci. Rep.* **2018**, *8*, 284. [CrossRef]

toxics

MDPI

Article

Distribution and Characterization of Typical Antibiotics in Water Bodies of the Yellow River Estuary and Their Ecological Risks

Jindong Wang [1], Zhenfei Yan [1], Yu Qiao [1], Daqing Liu [1,2], Chenglian Feng [1,*] and Yingchen Bai [1]

1 State Key Laboratory of Environmental Criteria and Risk Assessment, Chinese Research Academy of Environmental Sciences, Beijing 100012, China
2 College of Water Science, Beijing Normal University, No. 19, Outer Street, Xinjiekou, Beijing 100875, China
* Correspondence: fengcl@craes.org.cn

Abstract: A total of 34 antibiotics from five major classes of antibiotics, including macrolides, sulfonamides, quinolones, tetracyclines and chloramphenicol, were considered as contaminants, considering the Yellow River Estuary as the study area. The distribution, sources and ecological risks of typical antibiotics in the Yellow River Estuary were investigated using an optimized solid-phase extraction pre-treatment and an Agilent 6410B tandem triple-quadrupole liquid chromatography–mass spectrometer for antibiotic detection. The results show that antibiotics were widely present in the water bodies of the Yellow River Estuary, with 14 antibiotics detected to varying degrees, including a high detection rate for lincomycin hydrochloride. Farming wastewater and domestic sewage were the primary sources of antibiotics in the Yellow River Estuary. The distribution characteristics of antibiotics in the study area were linked to the development of farming and social activities. The ecological risk evaluation of 14 antibiotics in the Yellow River Estuary watershed showed that clarithromycin and doxycycline hydrochloride were present at medium-risk levels, and lincomycin hydrochloride, sulfamethoxazole, methomyl, oxifloxacin, enrofloxacin, sulfadiazine, roxithromycin, sulfapyridine, sulfadiazine and ciprofloxacin were present at low-risk levels in the samples collected from water bodies of the Yellow River Estuary. This study provides novel, beneficial information for the assessment of the ecological risk presented by antibiotics in the Yellow River Estuary water bodies and provides a scientific basis for future antibiotic pollution control in the Yellow River Basin.

Keywords: antibiotics; Yellow River Estuary; ecological risk assessment

Citation: Wang, J.; Yan, Z.; Qiao, Y.; Liu, D.; Feng, C.; Bai, Y. Distribution and Characterization of Typical Antibiotics in Water Bodies of the Yellow River Estuary and Their Ecological Risks. *Toxics* **2023**, *11*, 400. https://doi.org/10.3390/toxics11050400

Academic Editor: Antonia Concetta Elia

Received: 28 February 2023
Revised: 15 April 2023
Accepted: 18 April 2023
Published: 23 April 2023

1. Introduction

The Yellow River is the mother river of the Chinese nation. As the second largest river in China, it plays an essential role as the country's northern drinking water supply and feeds the agricultural system, but it also receives natural or treated effluent from urban centers. A large amount of wastewater (up to 4.4 billion tonnes/year) is generated from industrial production, livestock farming and agricultural surface sources. Introducing effluent from multiple sources has led to a deterioration of the Yellow River's water quality in localized sections, with large amounts of antibiotics detected frequently. In the 18th Party Congress, a national strategy was formed to promote the ecological protection and high quality of water in the Yellow River Basin [1]. In 2019, the proportion of Class I-II surface water quality sections in the Yellow River Basin was on the rise, and the balance between IV and poor V categories of water quality was on the decline. However, poor V sections still account for 8.8%, and water pollution in tributaries is still relatively severe [2]. In October 2021, the Ministry of Ecology and Environment issued the "Action Plan for the Treatment of New Pollutants (Draft for Public Comments)", proposing specific targets and visions for the treatment of new pollutants [3]. In September 2022, the General Office of the Ministry of Ecology and the Environment issued the "List of New Pollutants for Priority Control

(2022 Version) (Draft for Public Comments)", proposing that antibiotic residues should be managed following hazardous waste protocols, and that the battle against pollution should be fought head on [4].

Antibiotics are organic substances synthesized naturally by microorganisms through secondary metabolism or synthesized artificially by industry. They can inhibit the growth or metabolic activity of other microorganisms and can even cause their metabolism and death [5,6]. As a new contaminant, antibiotics are used by humans and animals. Furthermore, due to incomplete intestinal absorption and incomplete metabolism, antibiotics can be excreted through feces and urine and enter water bodies [7], soil [8] and sediments [9]. Antibiotics have played an essential role in the development of modern medicine. Still, with their extensive clinical use, over 50,000 tonnes of antibiotic residues are "released" into the water environment each year [10], making many rivers major reservoirs of these pollutants [11,12]. Antibiotic contamination not only poses a severe risk to aquatic ecosystems [13,14], but can also induce microbial resistance, posing a severe threat to public health [15].

According to statistics, in 1999, 65% of the 13,216 tons of antibiotics used in Germany were applied to treat human diseases, and in Denmark, 20.8%, 27.4% and 51.8% were used for human, veterinary and growth purposes. The annual use of antibiotics in the United States is stipulated as 22,700 tons, 50% for humans and 50% for animals, agriculture and aquaculture [16]. China produces 75% of the total antibiotics, and the abuse of antibiotics is a serious situation. In recent years, most studies on antibiotic pollution in the Yellow River basin have focused on the Jinan section of the lower Yellow River [17], the Yellow River delta wetlands [18,19] and the lower Yellow River [20], and relevant studies on the Yellow River Estuary basin are still insufficient.

In the present study, typical antibiotics were detected in the surface waters of the Yellow River Estuary, and the distribution and characterization of antibiotics were also analyzed. At the same time, the ecological risks of antibiotics were assessed via the use of risk quotient methods (RQs). This study deepens the understanding of the concentration levels of antibiotics in the water bodies of the Yellow River Estuary, provides theoretical support for environmental protection of the water in the basin and provides a reference for maintaining the health of the ecosystem and drinking water safety in the Yellow River Estuary.

2. Materials and Methods

2.1. Sample Collection

The Yellow River Estuary is located in Dongying, Shandong Province, which is close to Bohai Bay and Laizhou Bay. It forms the fan-shaped accumulation plain of the modern-day Yellow River Estuary. The terrain is low in elevation with an average altitude below 15 m, and the range is between 37°34′–37°53′ N and 118°53′–119°21′ E (Figure 1). It has a temperate continental monsoon climate, four distinct seasons, an average yearly temperature of 12.1 °C, an average yearly precipitation of 530–630 mm with concurrent high rain and humidity.

Combined with field investigation and a review of relevant data, eight water samples were collected at eight points in the Yellow River Estuary in September 2022, as shown in Figure 1.

A total of 2 L of water was collected in a brown glass bottle, and the growth of bacteria in the water sample was inhibited by adding 50 mL of methanol to slow down the degradation of antibiotics by microorganisms. To improve antibiotic recovery, 100 μL of concentrated H_2SO_4 was added to the water sample, and the pH was adjusted to around 3. The sample was stored at a low temperature and transported back to the laboratory, where the samples were processed within 24 h.

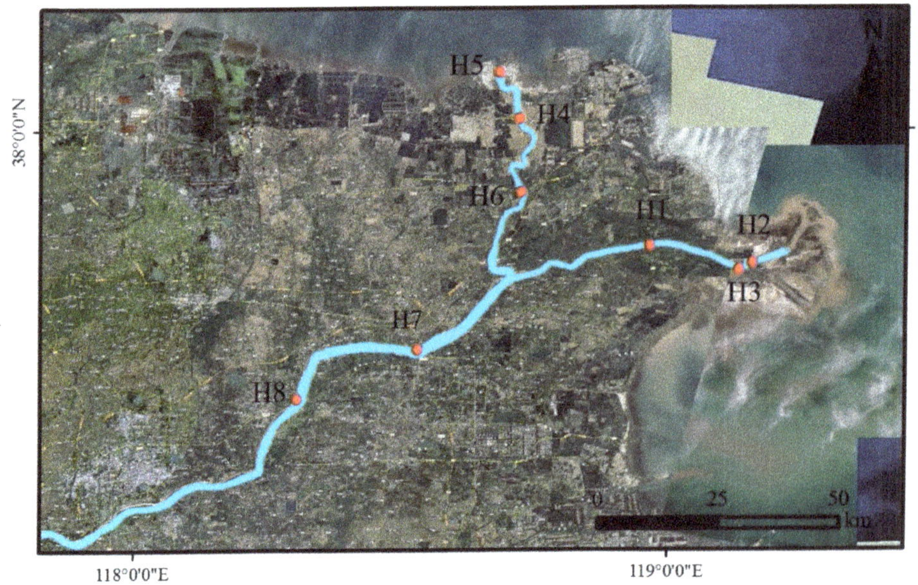

Figure 1. Sample points in the study area of the Yellow River.

2.2. Advanced Analysis Instruments and Reagents

An Agilent 6410B triple-quadrupole liquid chromatography–mass spectrometer (1290-6460, Agilent Technologies, USA); Extend-C18 Column (2.1 mm × 100 mm × 3.5 μm, Agilent Technologies, USA); Vortex (UVS-3, Beijing Yousheng United Technology Co., Ltd., China); electronic balance (AR224CN, OHAUS Instruments (Changzhou Co., Ltd., China); and CNC ultrasonic cleaner (KQ-250DE, Kunshan Ultrasonic Instrument Co., Ltd., China) were the main advanced analysis instruments employed.

A total of 34 antibiotic standards were grouped into five categories, i.e., (1) macrolides, including erythromycin, roxithromycin, hythromycin, azithromycin and tylosycin, clindamycin; (2) sulfonamides, including sulfaacetic amide, sulfaclodazine, sulfadimethoxypyrimidine, sulfapyridine, sulfathiazole, sulfamethiodiazole, sulfadiazine, sulfamethazine, sulfamethoxazole, sulfadimethazole, sulfadimethylpyrimidine, trimethoprim and sulfaquinoxaline; (3) quinolones, including ofloxacin, norfloxacin, ciprofloxacin, enrofloxacin, salafloxacin, lomefloxacin, flurofloxacin and difloxacin; (4) tetracyclines, including doxycycline hydrochloride, tetracycline (hydrochloride), oxytetracycline (oxytetracycline) and chlortetracycline (chlortetracycline); (5) chloramphenicol, including chloramphenicol, florfenicol, thiamphenicol and rifampicin. The above reagents were imported from the German company Dr. Ehrenstorfer.

Four internal standards, sulfamerazine-D4 (SMR-D4), ciprofloxacin-D8 (CIPROFLOXACIN-D8, CIP-D8), normeclocycline (DTC) and erythromycin-^{13}C, D3 (erythromycin-^{13}C, D3, ERY-^{13}C, D3), were imported from Dr. Ehrenstorfer in Germany.

A Water Oasis HLB (6 mL, 200 mg) solid-phase extraction cartridge, methanol, acetonitrile, hydrochloric acid, Na_2EDTA, ethyl acetate, dichloromethane, ammonium acetate, formic acid, disodium hydrogen phosphate and citric acid were purchased from Shanghai Anpu Experimental Technology Co., Ltd., and a 0.7μm (70 mm) GF/F filter membrane was purchased from Whatman Company in the United Kingdom.

2.3. Sample Treatment

The water sample was filtered through a 0.45 μm pore glass-fiber membrane, and weighed 1.0 L of water accurately. Eight samples were taken in two replicates for a total of sixteen samples. An amount of 0.2 g of Na_2EDTA was added to reduce the chelation of antibiotics and metal ions in the water sample, about 300 μL of hydrochloric acid was added to the water sample to adjust the pH of the water sample to 3.0~4.0, 25 ng of antibiotic internal standard was added, and then the cartridge was extracted using solid-phase Oasis HLB (200 mg/6 cc) at a rate of 5 mL/min. The HLB cartridge was activated with 10 mL of methanol, 10 mL of purified water and 10 mL of pure water with a pH of 4.0. After the sampling, the column was cleaned with 10 mL of pure water, drained, dried under the protection of nitrogen for 30 min, eluted in 3 times using 6 mL of methanol, nitrogen-blown until nearly dry and reconstituted with the initial mobile phase (0.1% formic acid–ammonium formate aqueous solution/acetonitrile) to be measured.

2.4. Instrumental Analysis

The HPLC-MS/MS used the Agilent 6410B tandem triple-quadrupole LC-MS/MS, Waters Xterra C18 separation column (100 mm × 2.1 mm, 3.5 μm) ESI ionization source. Mobile phase: A phase, 0.1% formic acid–ammonium formate; B: acetonitrile. Linear gradient: 0 min, 5% B; 0.1~10 min, 10~60% B; 10~12 min, 60%; 12.1~22 min 10% B. The flow rate was 0.25 mL/min. The column temperature was maintained at 25 °C and the injection volume was 200 μL. MS conditions: gas temperature of 350 °C, gas flow rate of 8 mL, nebulizer pressure of 25 psi, capillary voltage of 4000 V.

2.5. Ecological Risk Assessment

Risk quotient methods (*RQs*) are one of the most effective methods for assessing the environmental risks of aquatic biochemicals [21]. This study used data on the antibiotic concentrations in the water of the Yellow River Estuary for ecological risk assessment. According to the methodology for environmental risk assessment presented in the EU's technical guidance document, ecological risks can be assessed using the risk quotient (*RQ*).

The *RQ* is calculated as follows:

$$RQ = \frac{MEC}{PNEC} \tag{1}$$

$$PNEC = \frac{EC_{50}(LC_{50})}{AF} \tag{2}$$

where *MEC* represents the measured environmental concentration and *PNEC* indicates the predicted non-effect concentration for each contaminant. *PNEC* is the quotient of the toxicological relevant concentration with an appropriate assessment factor (*AF*). LC_{50} represents the median lethal concentration and EC_{50} represents the half maximal effective concentration. LC_{50} or EC_{50} represent the lowest maximal effective concentration value according to the available literature. According to the RQ, it can be divided into three risk levels: high risk ($RQ > 1$), medium risk ($0.1 < RQ < 1$) and low risk ($RQ < 0.1$) [22].

2.6. Data Analysis

The sampling sites were mapped using ArcGIS software and Bigemap Gis Office software. OriginPro 2023 software was also used to produce box plots and bar charts to visualize and clearly show the distribution of antibiotics in the Yellow River Estuary waters at the eight sampling sites.

3. Results and Discussion

3.1. Concentration Levels of Antibiotics in the Water Bodies of the Yellow River Estuary

The results of the antibiotic monitoring experiment in the Yellow River Estuary are shown in Figure 2, with the maximum, 75th percentile, mean, median, 25th percentile and minimum values shown in order from highest to lowest in the box plot.

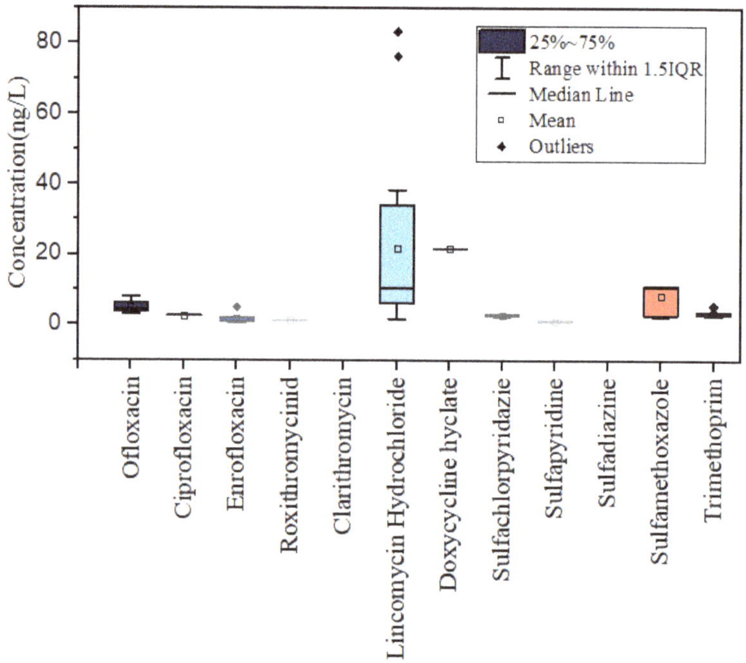

Figure 2. Box plots of measured concentrations of twelve antibiotics in water samples from the Yellow River Estuary.

Sulfonamides had the highest average concentration in the waters of the Yellow River Estuary, including all sulfonamide derivatives and sulfa analogs based on the chemical synthesis of p-aminobenzenesulfonamide, whose structures are connected to a free amino and sulfonamide group in the para-position of the benzene ring. These are broad-spectrum synthetic antibacterial agents with the advantages of low price, stable performance and good therapeutic effect. They are commonly used in the medical, agricultural, aquaculture and livestock industries for the prevention and treatment of bacterial and protozoan infections [23]. Their average concentration in the water bodies of the Yellow River Estuary reached 11.80 ng·L^{-1}; this was followed by macrolides at 8.40 ng·L^{-1}, quinolones at 4.40 ng·L^{-1}, tetracyclines at 1.37 ng·L^{-1} and chloramphenicol at the lowest level, that is, not detected. The highest concentration detected at any site was 83.31 ng·L^{-1}, found at sampling site H2. As a macrolide, lincomycin hydrochloride has similar effects to erythromycin and has a better impact on Gram-positive cocci. The antibiotics with the highest mean concentrations seen at each site were, in descending order, lincomycin hydrochloride (8.36 ng·L^{-1}), sulfamethoxazole (8.91 ng·L^{-1}), ofloxacin (5.08 ng·L^{-1}), methicillin (4.27 ng·L^{-1}), sulfamonomethoxazole (3.54 ng·L^{-1}) and sulfadiazine (2.85 ng·L^{-1}), with the rest of the antibiotics having mean concentrations below 1.7 ng·L^{-1}.

Sulfonamides accounted for 45.47% of the antibiotics detected in the water samples, with a detection rate of 50%. The average concentration of sulfamethoxazole was 8.91 ng·L^{-1}, which is much higher than that detected for any other sulfonamide. The proportion of macrolides was 32.34%, of which lincomycin hydrochloride had the highest detection rate of 93.75% with an average concentration of 8.36 ng·L^{-1}. Quinolones accounted for 16.93% of detected antibiotics, of which ofloxacin had the highest detection rate of 68.75% with an average concentration of 5.08 ng·L^{-1}. Norfloxacin, salafloxacin, lomefloxacin, fleroxacin and diflufenacin were all detected. This is due to the fact that most of the quinolone antibiotics have a strong adsorption capacity and are better able to adsorb sediment or suspended matter in rivers, making their detection rate low [24]. The proportion of tetracyclines was 5.26%, and their concentration in the sediment was relatively high due to the strong adsorption of hygromycin [25]. Chloramphenicol antibiotics were not detected at any of the eight sampling sites.

Comparing the Yellow River Estuary with other sections of the Yellow River Basin, a total of 14 antibiotics were detected in the Yellow River Estuary, as shown in Table 1, with concentrations starting from ND~415.53 ng·L^{-1} and the average concentration of the 34 antibiotics being 25.97 ng·L^{-1}. In the Jinan section of the lower Yellow River [17], a total of 36 of the target antibiotics were detected in 35 sampling locations, and the concentrations of detected antibiotics starting from ND~13.462 ng·L^{-1}, with an average concentration of 373.94 ng·L^{-1}. Sulfonamides and macrolides were seen at a high rate; the total concentration of antibiotics seen in the Yellow River Delta section [18] during an abundant water period was ND~256.6 ng·L^{-1}, with an average concentration of 15.09 ng·L^{-1}; the total concentration of antibiotics detected in the intertidal zone of the Yellow River Delta [19] was ND~82.94 ng·L^{-1} with an average concentration of 10.37 ng·L^{-1}; the total antibiotic concentration in surface waters such as canals, rivers and fish ponds in Kaifeng [20], a key city in the lower reaches of the Yellow River, Henan Province, was ND~12,224.99 ng·L^{-1}, with an average concentration of 815.00 ng·L^{-1}; meanwhile the total concentration of antibiotics in the Wei River [26], the largest tributary of the Yellow River Basin, was ND~573.26 ng·L^{-1} with an average concentration of 13.98 ng·L^{-1}. In summary, the current level of antibiotic concentrations in the Yellow River Estuary is moderate.

Table 1. Comparison of antibiotic concentration levels in surface water in the Yellow River Estuary and other sections of the Yellow River Basin ① ng·L^{-1}.

Lakes (Year of Survey)	Antibiotic Concentration	Average Concentration
Yellow River Estuary (2022)	ND~415.53	25.97
Jinan section of the lower Yellow River (2022)	ND~13,462	373.94
Yellow River Delta Section (2019)	ND~256.6	15.09
Yellow River Delta intertidal zone (2016)	ND~82.94	10.37
Canal in Kaifeng, Henan, a key city on the lower reaches of the Yellow River (2022)	ND~12,224.99	815.00
Weihe River (2018)	ND~573.26	13.98

Note: ① ND stands for not detected.

3.2. Spatial Distribution of Antibiotics

The point distribution of antibiotics in the Yellow River Estuary is shown in Figure 3. As can be seen, the total number of antibiotics detected was highest at points H2 and H8 and lowest at points H3 and H5. Two substances were detected at all sampling locations, including one macrolide and one quinolone. Erythromycin, roxithromycin, telithromycin, sulfadimethoxypyrimidine, sulfathiazole, sulfamethoxazole, sulfamethoxypyrimidine, sulfadimethoxypyrimidine, sulfoquinoxaline, norfloxacin, salafloxacin, lomefloxacin, fleroxacin, difluoxacin, tetracycline, oxytetracycline, chlortetracycline, chloramphenicol, fluphenazole, methomycin and rifampicin were not detected at any of the eight sampling locations. Lincomycin hydrochloride had the highest detected mass concentration of 83.31 ng·L^{-1} at site H2 and doxycycline hydrochloride had the highest detected mass concentration of 21.86 ng·L^{-1} at site H6; sulfamethoxazole's highest mass concentration was 11.32 ng·L^{-1}, at site H8.

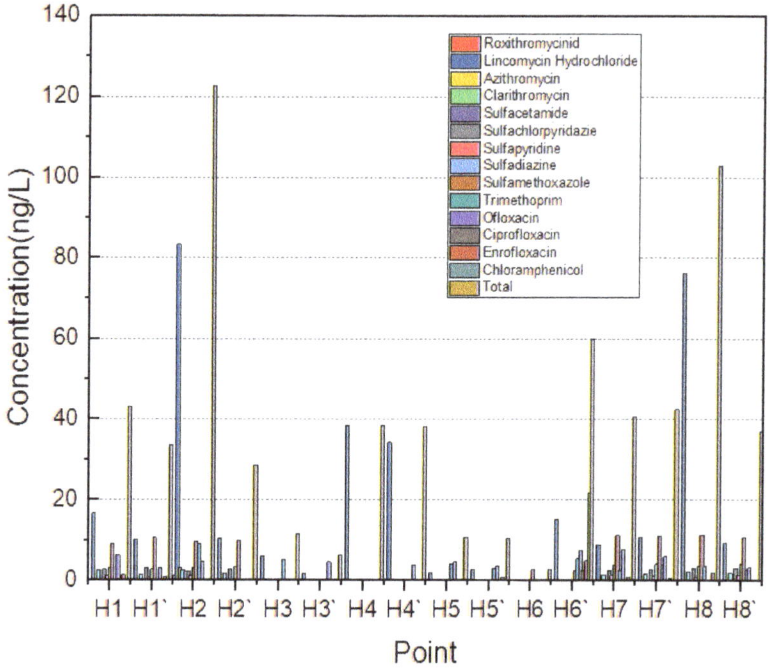

Figure 3. Distribution of antibiotic concentrations at the eight sampling sites.

Overall, more than half of the eight sampling sites had higher concentrations of sulphonamide antibiotics than the other four categories, which are widely used in medical, agricultural, aquatic and livestock industries for the prevention and treatment of bacterial and protozoal infections because of their broad-spectrum antibacterial strength, low price, stable performance and sound therapeutic effects [26]. Sulfadiazine and sulfamethoxazole are commonly used to treat human diseases such as urinary tract infections and respiratory tract infections, and are the most widely used classes of sulfonamide antibiotics. Sulfamethoxazole is frequently used in farming to promote animal growth and increase production, while sulfadiazine is highly toxic. Sulfamethoxazole is quickly oxidized when exposed to light and is often used to suppress intestinal and soft skin tissue infections caused by sensitive bacteria, among other things. Sulphonamide antibiotics have a stable structure, degrade slowly in the environment and persist in the aqueous environment for an extended period of time. The sampling period coincided with the rainy season, with many cloudy days, which weakened natural degradation processes such as photodegradation, thus making the concentration of sulfonamide antibiotics significantly higher. Point H8 is the Lijin Hydrological Station, with a section width of 598 m. The main channel is 355 m wide and the beach area is 243 m wide. The beach is full of crops, which has a specific deterrent effect on the flooding of the beach. The Lijin Hydrological Station is part of the Yellow River Delta National Nature Reserve. Point H2 and point H3 are ecological tourist zones in the Yellow River Estuary. The general flow direction of rivers in China is from west to east; as point H2 is to the east of point H3, the water flows from point H3 to point H2, so the concentration of antibiotics at point H2 is higher than at point H3. Point H5 is the Feiyan Beach, which is one of the best areas in the Yellow River Estuary in terms of water quality due to the low level of human activities in the area.

3.3. Ecological Risk Assessment

To better evaluate the risk level of antibiotics in the waters of the Yellow River Estuary, this study used the risk quotient method to perform a preliminary analysis of the

14 antibiotic-like substances detected. Toxicity data for each compound were screened from the literature, and the PNEC values (Table 2), as well as the risk quotient values (Figure 4), were calculated using the evaluation factor method. Of the 14 compounds detected, azithromycin and sulphonamide acetate were not evaluated here for ecological risk due to the lack of toxicity data from which to derive PNEC values.

Table 2. PNEC for common antibiotics ng·L^{-1}.

Antibiotics	PNEC	References
Ofloxacin	21~17,400	[27–29]
Ciprofloxacin	2~30,000	[30,31]
Enrofloxacin	28.8~49	[32,33]
Roxithromycinid	4.3~10,000	[34]
Clarithromycin	2	[22,35]
Lincomycin Hydrochloride	50~50,000	[28,29,36]
Doxycycline hyclate	131	[37]
Sulfachlorpyridazie	2330~1,720,000	[38,39]
Sulfapyridine	460~5280	[38,39]
Sulfadiazine	107.394~135	[28,29,38,40]
Sulfamethoxazole	27~4674	[28,29,38,40]
Trimethoprim	29~255.516	[28,29,38,40]

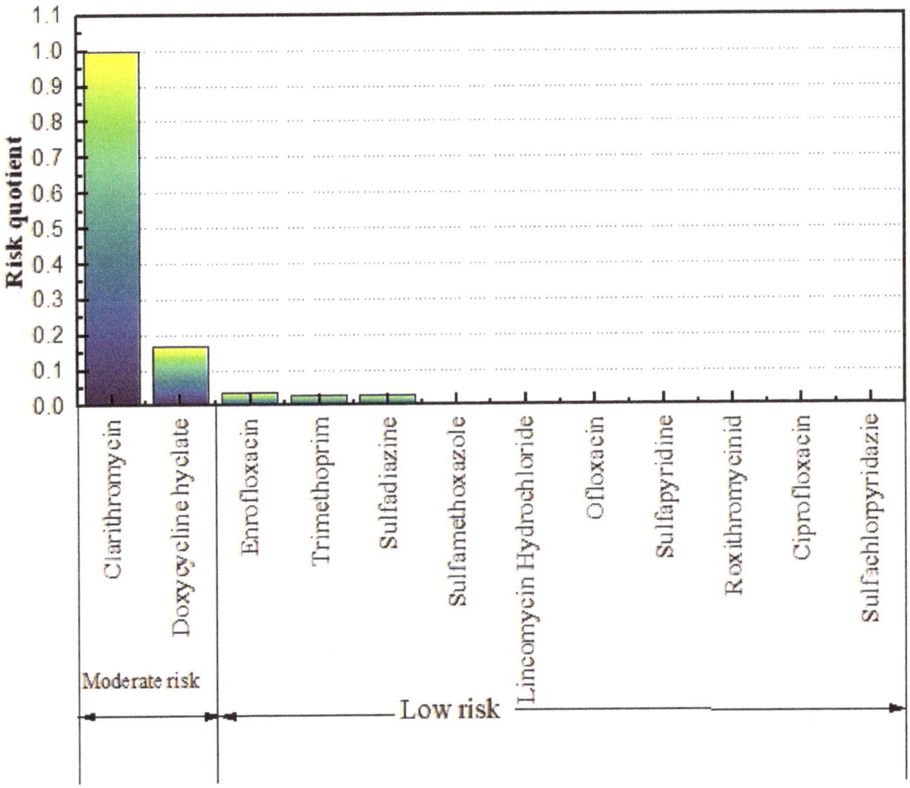

Figure 4. Risk quotient (RQ) values for antibiotics in water bodies of the Yellow River Estuary.

4. Conclusions

The following conclusions were drawn from a survey of the pollution status and ecological risk assessment of 34 antibiotics at eight sites in the Yellow River Estuary. A total of 14 antibiotics were detected, with concentrations in the following descending order: sulphonamides, macrolides, quinolones, tetracyclines and chloramphenicol. The detection rate of sulfa antibiotics reached 45.47%, and the highest concentration detected at a single site was for lincomycin hydrochloride, with a concentration of 83.31 $ng \cdot L^{-1}$; the concentration levels of antibiotics at sites near villages, fishing grounds and hospitals were significantly higher than those around scenic areas, confirming that the concentrations of antibiotics in urban water bodies are closely related to human activities. The ecological risk assessment of the detected antibiotics using the risk quotient method showed that clarithromycin and doxycycline hydrochloride pose a medium risk, while lincomycin hydrochloride, sulfamethoxazole, meperidine, ofloxacin, enrofloxacin, sulfadiazine, roxithromycin, sulfapyridine, sulfadiazine and ciprofloxacin pose a low risk according to their concentrations in the water bodies of the Yellow River Estuary.

Author Contributions: Conceptualization, J.W. and C.F.; Methodology, J.W. and C.F.; Software, Y.Q.; Formal analysis, C.F.; Investigation, J.W. and Z.Y.; Resources, D.L.; Data curation, J.W.; Writing—original draft, J.W.; Writing—review & editing, C.F.; Visualization, Y.B.; Supervision, C.F.; Project administration, C.F. All authors have read and agreed to the published version of the manuscript.

Funding: This research was funded by the Beijing Natural Science Foundation (Grant No. 8222077) and the National Key Research and Development Program of China (Grant No. 2021YFC3200104).

Institutional Review Board Statement: Not applicable.

Informed Consent Statement: Not applicable.

Data Availability Statement: Not applicable.

Conflicts of Interest: The authors declare no conflict of interest.

References

1. Xi, J.P. Speech at the symposium on ecological protection and high: Quality development of the Yellow River Basin. *China Water Resour.* **2019**, *20*, 1–3. (In Chinese)
2. Li, J.; Peng, K.; Hao, G. Research progress on quantification and control of non-point source pollution load in the Yellow River Basin. *Water Resour. Prot.* **2021**, *37*, 90–102. (In Chinese)
3. Ministry of Ecology and Environment. The 14th Five-Year Plan for Ecological and Environmental Monitoring. Available online: https://www.mee.gov.cn/xxgk2018/xxgk/xxgk03/202201/t20220121_967927.html (accessed on 28 December 2021).
4. General Office of the Ministry of Ecology and Environment. Public Consultation on the List of Key Control New Pollutants (2022 Version) (Draft for Comments). Available online: https://www.mee.gov.cn/xxgk2018/xxgk/xxgk06/202209/t20220927_995054.html (accessed on 27 September 2022).
5. Kim, K.-R.; Owens, G.; Kwon, S.-I.; So, K.-H.; Lee, D.-B.; Ok, Y.S. Occurrence and Environmental Fate of Veterinary Antibiotics in the Terrestrial Environment. *Water Air Soil Pollut.* **2011**, *214*, 163–174. [CrossRef]
6. Zhang, J.; Ge, H.; Shi, J.; Tao, H.; Li, B.; Yu, X.; Zhang, M.; Xu, Z.; Xiao, R.; Li, X. A tiered probabilistic approach to assess antibiotic ecological and resistance development risks in the fresh surface waters of China. *Ecotoxicol. Environ. Saf.* **2022**, *243*, 114018. [CrossRef]
7. Zhang, Q.; Xin, Q.; Zhu, J. Progress of research on the status of antibiotic pollution in major waters of China and its ecological effects. *Environ. Chem.* **2014**, *33*, 1075–1083.
8. Peng, Q.; Wang, W.Z.; Xu, W.H. Ecological Risk Assessment of Tetracycline Antibiotics in Livestock Manure and Vegetable Soil of Chongqing. *Huan Jing Ke Xue* **2020**, *41*, 4757–4766. [PubMed]
9. Wang, T.; Zhang, W.; Li, L.; Zhang, M.; Liao, G.; Wang, D. Distribution characteristics and risk assessment of antibiotics and polycyclic aromatic hydrocarbons in sediments from Baiyangdian dredging demonstration area. *Environ. Sci.* **2021**, *42*, 5303–5311. [CrossRef]
10. Zhang, Q.Q.; Ying, G.G.; Pan, C.G.; Liu, Y.S.; Zhao, J.L. Comprehensive evaluation of antibiotics emission and fate in the river basins of China: Source analysis, multimedia modeling, and linkage to bacterial resistance. *Environ. Sci. Technol.* **2015**, *49*, 6772–6782. [CrossRef]
11. Lei, K.; Zhu, Y.; Chen, W.; Pan, H.Y.; Cao, Y.X.; Zhang, X.; Guo, B.B.; Sweetman, A.; Lin, C.Y.; Ouyang, W.; et al. Spatial and seasonal variations of antibiotics in river waters in the Haihe River Catchment in China and ecotoxicological risk assessment. *Environ. Int.* **2019**, *130*, 104919. [CrossRef]

12. Huang, F.; An, Z.; Moran, M.J.; Liu, F. Recognition of typical antibiotic residues in environmental media related to groundwater in China (2009–2019). *J. Hazard. Mater.* **2020**, *399*, 122813. [CrossRef]
13. Ting, R.U.A.N.; Guibin, J.I.A.N.G. Basic theory and methods for the discovery of novel environmental organic pollutants. *Proc. Chin. Acad. Sci.* **2020**, *35*, 1328–1336. [CrossRef]
14. Li, X.; Zhang, R.; Tian, T.; Shang, X.; Du, X.; He, Y.; Matsuura, N.; Luo, T.; Wang, Y.; Chen, J.; et al. Screening and ecological risk of 1200 organic micropollutants in Yangtze Estuary water. *Water Res.* **2021**, *201*, 117341. [CrossRef]
15. Bielen, A.; Šimatović, A.; Kosić-Vukšić, J.; Senta, I.; Ahel, M.; Babić, S.; Jurina, T.; Plaza, J.J.G.; Milaković, M.; Udiković-Kolić, N. Negative environmental impacts of antibiotic-contaminated effluents from pharmaceutical industries. *Water Res.* **2017**, *126*, 79–87. [CrossRef]
16. Klaus, K. Antibiotics in the aquatic environment—A review—Part I. *Chemosphere* **2008**, *75*, 417–434.
17. Zhao, K.; Wang, Q.; Qian, S.; Li, F. Spatial and temporal distribution characteristics of antibiotics and heavy metals in the Yitong River basin and ecological risk assessment. *Sci. Rep.* **2023**, *13*, 4202. [CrossRef] [PubMed]
18. Liu, X.; Wang, Z.; Wang, X.L.; Li, Z.; Yang, C.; Li, E.H.; Wei, H.M. Evaluation of antibiotic pollution in surface water environment and its ecological risk in a typical region of China. *Environ. Sci.* **2019**, *40*, 2094–2100. [CrossRef]
19. Zhao, S.; Liu, X.; Cheng, D.; Liu, G.; Liang, B.; Cui, B.; Bai, J. Temporal–spatial variation and partitioning prediction of antibiotics in surface water and sediments from the intertidal zones of the Yellow River Delta, China. *Sci. Total Environ.* **2016**, *569–570*, 1350–1358. [CrossRef] [PubMed]
20. Lü, D.Y.; Yu, C.; Zhuo, Z.J.; Meng, S.R.; Liu, S.B. The distribution and ecological risks of antibiotics in surface water in key cities along the lower reaches of the Yellow River: A case study of Kaifeng City, China. *China Geol.* **2022**, *5*, 411–420. (In Chinese) [CrossRef]
21. Shen, L.; Zhang, L.; Qin, S.; Yao, B.; Cui, J.S. Characteristics of quinolone antibiotic contamination in Baiyangdian and its correlation with environmental factors. *J. Environ. Sci.* **2019**, *39*, 3888–3897. [CrossRef]
22. Li, J.; Wang, Y.; Dong, Y.; Wang, M.; Zhao, Q.; Zhou, Y.; Ding, H.; Yin, L.; Cao, Y.; Xie, J.; et al. Distribution characteristics and ecological risk assessment of typical antibiotics in Yuan River water bodies in Poyang Lake basin. *J. Ecotoxicol.* **2022**, *17*, 563–574. (In Chinese)
23. Wu, W.; Qi, M.; Zhang, Z.; Lan, Q.; Hu, H. Study on the current situation and detection methods of sulfonamide antibiotics contamination. *Environ. Sci. Manag.* **2022**, *47*, 121–126. (In Chinese)
24. Duong, H.A.; Pham, N.H.; Nguyen, H.T.; Hoang, T.T.; Pham, H.V.; Pham, V.C.; Berg, M.; Giger, W.; Alder, A.C. Occurrence, fate and antibiotic resistance of fluoroquinolone antibacterials in hospital wastewaters in Hanoi, Vietnam. *Chemosphere* **2008**, *72*, 968–973. [CrossRef] [PubMed]
25. Figueroa, R.A.; MacKay, A.A. Sorption of oxytetracycline to iron oxides and iron oxide-rich soils. *Environ. Sci. Technol.* **2005**, *39*, 6664–6671. [CrossRef]
26. Zhu, T.; Zhou, M.; Yang, S.K.; Wang, Z.Z.; Wang, R.Z.; Wang, W.K.; Zhao, Y.Q. Characteristics of antibiotic distribution and ecological risk assessment in the Shaanxi section of the Weihe River. *People's Yellow River* **2018**, *40*, 85–91. (In Chinese)
27. Backhaus, T.; Scholze, M.; Grimme, L.H. The single substance and mixture toxicity of quinolones to the bioluminescent bacterium Vibrio fischeri. *Aquat. Toxicol.* **2000**, *49*, 49–61. [CrossRef]
28. Huang, F.; Zou, S.; Deng, D.; Lang, H.; Liu, F. Antibiotics in a typical karst river system in China: Spatiotemporal variation and environmental risks. *Sci. Total Environ.* **2019**, *650 Pt 1*, 1348–1355. [CrossRef]
29. Goyne, K.W.; Chorover, J.; Kubicki, J.D.; Zimmerman, A.R.; Brantley, S.L. Sorption of the antibiotic ofloxacin to mesoporous and nonporous alumina and silica. *J. Colloid Interface Sci.* **2005**, *283*, 160–170. [CrossRef]
30. Brain, R.A.; Johnson, D.J.; Richards, S.M.; Sanderson, H.; Sibley, P.K.; Solomon, K.R. Effects of 25 pharmaceutical compounds to Lemna gibba using a seven-day static-renewal test. *Environ. Toxicol. Chem.* **2004**, *23*, 371–382. [CrossRef]
31. Wu, T.Y.; Li, J.; Yang, A.J.; Li, Y.C.; Chen, Y.; He, Q.; Ma, K.; Hu, X.; Wang, B.; Ai, J.; et al. Characteristics and Risk Assessment of Antibiotic Contamination in Chishui River Basin, Guizhou Province, China. *Huan Jing Ke Xue* **2022**, *43*, 210–219.
32. Robinson, A.A.; Belden, J.B.; Lydy, M.J. Toxicity of fluoroquinolone antibiotics to aquatic organisms. *Environ. Toxicol. Chem.* **2005**, *24*, 423–430. [CrossRef]
33. Xu, L.; Ye, X.; Hao, G. Characterization and ecological risk evaluation of typical antibiotic pollution in surface waters of Campsis River. *Mod. Agric. Sci. Technol.* **2020**, *765*, 180–183+187. (In Chinese)
34. Ferrari, B.; Mons, R.; Vollat, B.; Fraysse, B.; Paxēaus, N.; Giudice, R.L.; Pollio, A.; Garric, J. Environmental risk assessment of six human pharmaceuticals: Are the current environmental risk assessment procedures sufficient for the protection of the aquatic environment? *Environ. Toxicol. Chem.* **2004**, *23*, 1344–1354. [CrossRef] [PubMed]
35. Lützhøft, H.C.H.; Halling-Sørensen, B.; Jørgensen, S.E. Algal toxicity of antibacterial agents applied in Danish fish farming. *Arch. Environ. Contam. Toxicol.* **1999**, *36*, 1–6. [CrossRef]
36. Zhao, T.; Chen, Y.; Han, W.; He, Y. The contamination characteristics and ecological risk assessment of typical antibiotics in the upper reaches of the Dongjiang River. *Ecol. Environ. Sci.* **2016**, *25*, 1707–1713.
37. Zhao, F.; Gao, H.; Zhang, K. Fugacity and risk assessment of antibiotics in typical river waters in China. *Environ. Pollut. Prev.* **2021**, *43*, 94–102. [CrossRef]
38. Li, Y.; Fang, J.; Yuan, X.; Chen, Y.; Yang, H.; Fei, X. Distribution characteristics and ecological risk assessment of tetracyclines pollution in the Weihe river, China. *Int. J. Environ. Res. Public Health* **2018**, *15*, 1803. [CrossRef]

39. Wu, X.Y.; Zou, H.; Zhu, R.; Jingguo, W. Characterization and ecological risk assessment of antibiotic pollution in the waters of Lake Taiko Gonghu Bay. *Environ. Sci.* **2016**, *37*, 4596–4604. (In Chinese) [CrossRef]
40. Wang, T.; Yang, Z.F.; Chen, Y.H.; Yaoyao, Z.; Ranran, S.; Ying, X.; Mengting, Z. Ecological risk assessment of sulfonamide antibiotics in surface water. *J. Ecol. Environ.* **2016**, *25*, 1508–1514. [CrossRef]

toxics

MDPI

Article

The Application of Reference Dose Prediction Model to Human Health Water Quality Criteria and Risk Assessment

Shu-Hui Men [1,2], Xin Xie [3], Xin Zhao [4], Quan Zhou [1], Jing-Yi Chen [1], Cong-Ying Jiao [3,*] and Zhen-Guang Yan [1,*]

1 State Key Laboratory of Environmental Criteria and Risk Assessment, Chinese Research Academy of Environmental Sciences, Beijing 100012, China
2 College of Water Sciences, Beijing Normal University, Beijing 100875, China
3 China National Environmental Monitoring Center, Beijing 100012, China
4 Vehicle Emission Control Center, Chinese Research Academy of Environmental Sciences, Beijing 100012, China
* Correspondence: jiaocy@cnemc.cn (C.-Y.J.); zgyan@craes.org.cn (Z.-G.Y.)

Abstract: Oral reference dose (RfD) is a key parameter for deriving the human health ambient water quality criteria (AWQC) for non-carcinogenic substances. In this study, a non-experimental approach was used to calculate the RfD values, which explore the potential correlation between toxicity and physicochemical characteristics and the chemical structure of pesticides. The molecular descriptors of contaminants were calculated using T.E.S.T software from EPA, and a prediction model was developed using a stepwise multiple linear regression (MLR) approaches. Approximately 95% and 85% of the data points differ by less than 10-fold and 5-fold between predicted values and true values, respectively, which improves the efficiency of RfD calculation. The model prediction values have certain reference values in the absence of experimental data, which is beneficial to the advancement of contaminant health risk assessment. In addition, using the prediction model constructed in this manuscript, the RfD values of two pesticide substances in the list of priority pollutants are calculated to derive human health water quality criteria. Furthermore, an initial assessment of the health risk was performed by the quotient value method based on the human health water quality criteria calculated by the prediction model.

Keywords: reference dose; molecular descriptor; multiple liner stepwise regression; ambient water quality criteria; health risk assessment

Citation: Men, S.-H.; Xie, X.; Zhao, X.; Zhou, Q.; Chen, J.-Y.; Jiao, C.-Y.; Yan, Z.-G. The Application of Reference Dose Prediction Model to Human Health Water Quality Criteria and Risk Assessment. *Toxics* **2023**, *11*, 318. https://doi.org/10.3390/toxics11040318

Academic Editors: Oriana Motta and Lisa Truong

Received: 7 January 2023
Revised: 15 March 2023
Accepted: 25 March 2023
Published: 28 March 2023

1. Introduction

Water environmental quality criteria are the maximum dose or level of pollutants or harmful factors in the water environment that do not have harmful effects on human health and water ecosystems [1]. The oral reference dose (RfD) is an evaluation metric presented by the US Environmental Protection Agency (EPA) to evaluate the risk of non-carcinogens [2]. It is the estimate of the mean daily dose of exogenous compounds, which is commonly defined as the amount of a chemical to which a person can be exposed on a daily basis over an extended period of time (usually a lifetime) without suffering a deleterious effect. It is an important component of the risk characterization of chemical substances and is also one of the important parameters for the development of water quality criteria for the non-carcinogenic effects of pollutants. RfD was first proposed in a report published by the US Environmental Protection Agency (EPA) in 1988, before which the acceptable daily intake (ADI) was more widely used in the field of toxicology and risk management [3]. Since ADI has some limitations in the field of risk assessment and control, the concept of RfD was introduced to promote consistency in the risk assessment of non-carcinogenic chemicals. The threshold of toxicological concern (TTC) is another important parameter in the field of chemical substance risk assessment. However, the TTC method is not suitable for assessing the safety of chemicals for which toxicological data are required [4].

RfD is an estimate of the average daily exposure dose of exogenous chemicals in environmental media. The two main traditional methods for calculating RfD are the NOAEL/LOAEL method and the benchmark dose method (BMDL) [5,6], and the RfD value is derived by the uncertainty factor UF and the correction factor MF by these traditional methods [7]. These derivation methods require a large investment of time for exposure experiments on mammals [8–10]. Since the U.S. EPA issued risk assessment guidelines in the 1980s, RfD values have been obtained for only a few hundred chemical substances [2], so the traditional methods for obtaining RfD are inefficient and constrain the health risk assessment studies of chemical substances. Additionally, the National Science Board proposed in its 2007 report that the study of hazards and risks of contaminants in the environment should make greater use of modern scientific tools and systematic data integration, replacing traditional toxicological methods based on animal experiments [11].

In recent years, a number of studies have used modeling approaches to predict the toxic effects of contaminants [12,13], watershed-scale ecological sensitivity [14], and to achieve toxicity extrapolation among congeners to assess the risk of environmental contaminants [15]. Among these model-building methods, the quantitative structure–activity relationship method is a modeling approach based on the correlation between biological activity and molecular structure, which is widely recognized in the field of toxicology and pharmaceutical research [16–18].

In previous studies, NOAEL prediction using chemical SMILES structures, considering only a single kind of descriptors, may ignore the role of certain dominant descriptors [19–21]. Toropova built prediction models for NOAEL by SMILES [20]. The R^2 of the six models ranged between 0.52~0.78. This indicates that these models have poor predictive performance. Moreover, the prediction models developed in some studies only describe the toxic effects of chemicals on some organs, which has some limitations in prediction effects [22]. In addition, when calculating RfD values indirectly by the predicted values of NOAEL and LOAEL, it is difficult to fix the values of uncertainty factors generated by exposure time and experimental animals [23], and the critical endpoints are difficult to define. Although the benchmark dose method makes specific improvements to the NOAEL-based method, it does not address the problems related to non-carcinogenic risk evaluation [24]. Therefore, in the present study, a non-experimental method was considered for the derivation of reference dose values for pesticide-class substances. The toxicity of organic pesticides is closely related to the type and number of functional groups carried by their molecules, in this case, quantitative structure–activity relationship methods may be more effective in the prediction of the physicochemical properties of such substances.

Pesticide poisoning poses a serious threat to aquatic ecosystems. Many of these organisms are highly toxic even at very low concentrations [25]. In China, the rapid development of agriculture and the massive production of pesticides has resulted in the release of large quantities of pesticides into the environment, which are very dangerous due to their extreme toxicity, persistence, and bioaccumulation, posing a major challenge to the safety of ecosystems [26]. Even at concentrations below established lethal thresholds, some pesticides can cause fish kills [27]. For some species, such as carp and salmon, exposure to sublethal concentrations of pesticides can lead to abnormal behavior [27]. In addition, aquatic plants can be endangered or even die under the action of high concentrations of herbicides [28]. Various types of pesticides are currently detected in various environmental media such as water, soil, air, and in animals and humans, and their effects on human health cannot be overlooked [29]. Therefore, pesticide risk assessment and control in China now appear warranted, and the development of local water quality standards for pesticides is urgent.

In order to address the limitations of traditional methods and avoid the interference of uncertainty factors and critical values, this study uses a non-experimental method for predicting the RfD of pesticides directly. The data were collected from a public database called Integrated Risk Information System (IRIS, https://cfpub.epa.gov/ncea/iris/search/ (accessed on 27 July 2022)) and molecular descriptors were calculated based on molecular similarity [30,31]. It fills the data gap of RfD values of chemicals and explores the potential association between toxicity and physicochemical characteristics and chemical structure of pesticides. In addition, the predictive model constructed in this study is used to calculate the RfD values of priority pesticides. The exposure parameters, bioaccumulation coefficients, and other relevant indigenous parameters used to derive the indigenous human health water quality criteria values were determined through the survey data.

2. Materials and Methods

2.1. Dataset

Quantitative structure–activity relationship models could establish a quantitative relationship between chemical structures and their properties [32]. These computational models are used to predict physicochemical properties of similar compounds that currently lack of experimental data. In this study, the negative log of the reference dose was chosen as the model response value. The source data sets used in this study originate from IRIS, which contains risk information on the cancer and noncancer effects of chemicals, including oral reference dose which depends on the exposure pathway. There are 109 species of pesticide class chemicals that have been included in IRIS which have defined RfD.

The key to obtaining an ideal prediction model is reasonable molecular descriptors. The molecular descriptors of these pesticides were calculated with T.E.S.T. software mentioned by EPA's official website. This has resulted in 797 descriptors corresponding to 12 descriptor classes. In addition, the screen of the molecular descriptors above was performed by following principles: (1) deleting the descriptors that a variance of 0; (2) deleting the descriptors that have a number of non-zero values less than 10%; (3) deleting one of two descriptors that the correlation coefficients greater than 0.90. After the above pre-processing, 372 descriptors remained for the prediction analysis.

2.2. Model Building

The preprocessed set of molecular descriptors was used as the independent variable X, and the negative log of the reference dose RfD value (−logRfD) was used as the dependent variable Y. The multivariate stepwise linear regression method in SPSS software (version 26.0, IBM Inc. Chicago, IL, USA) was applied to establish the regression model between molecular descriptors and −logRfD. Moreover, the variance inflation factor (VIP) was used to verify whether there was multicollinearity among the descriptors in the model, and Durbin–Watson values (D-W) were used to test the model autocorrelation. In the total data set, 70~80% were randomly selected for the training set and 20~30% for the test set. Internal validation and external validation were used to verify the predictive ability and robustness of the model. The model was also used to predict the RfD of EPA-released priority pesticides lacking RfD values.

2.3. Derivation of Human Health Water Quality Criteria

This study focuses on the non-carcinogenic effects of p-p'DDE and α-HCH, and the human health water quality criteria are derived according to the *Technical Guideline for Deriving Water Quality Criteria for the Protection of Human Health* [33]. The human health ambient water quality criteria (AWQC) for non-carcinogenic effect is calculated according to the following equation:

$$\mathrm{AWQC} = \mathrm{RfD} \cdot \mathrm{RSC} \cdot \left(\frac{\mathrm{BW}}{\mathrm{DI} + \sum_{i=2}^{4} (\mathrm{FI_i} \cdot \mathrm{BAF_i})} \right) \quad (1)$$

where RfD is the reference dose ($mg \cdot kg^{-1} \cdot d^{-1}$) for non-carcinogenic effects; RSC is the relative source contribution to account for non-source exposures; BW is body weight (kg); DI is drinking water intake ($L \cdot d^{-1}$); FI_i is intake of aquatic products ($kg \cdot d^{-1}$) for each trophic levels (i = 2, 3, 4); BAF_i is the bioaccumulation factor ($L \cdot kg^{-1}$) for each trophic level (i = 2, 3, 4).

and the model is good. As indicated by the external validation results, $R_{tes}^2 = 0.683$ and RMSEP = 0.434, which indicates that the model has good stability and good external prediction ability. The cross-validation could be used for describing the fitting effect on the training set, and cross-validation correlation coefficients (q^2) are expected to be greater than 0.5. The dataset modeled in this paper contains only 109 compounds, so it is not suitable to divide the independent validation set. Consequently, the hold-out cross-validation method was chosen to evaluate the validity of the model performance. To further verify the reliability of the model, the validation method proposed by Roy for external testers is used, and $k = 0.983 > 0.88$ and $k' = 1.016 < 1.15$ are obtained, which satisfy the corresponding validation requirements [37]. This indicates that there is no systematic error in the model itself that would cause the prediction results to deviate in a particular direction. The combination of the above results indicates that the predictive ability of the model is acceptable.

Table 2. The description and statistical information of the predictive model.

N	R_{tra}^2	R_{tes}^2	RMSEP	*p*	D-W	q^2	k	k′
12	0.762	0.683	0.434	<0.05	1.952	0.648	0.983	1.016

The relationship between the predicted and actual values of −log RfD obtained from the MLR model is shown in Figure 1. Comparing the actual RfD values with those predicted by the model (Figure 2), it can be seen that for the vast majority of pollutants (>95%), the difference between the true and predicted values is within a factor of 10, and for most (>85%) pollutants the difference between the actual and predicted values is within a factor of 5. Consequently, the consistency between predicted and actual values also proved the accuracy of the predictive models.

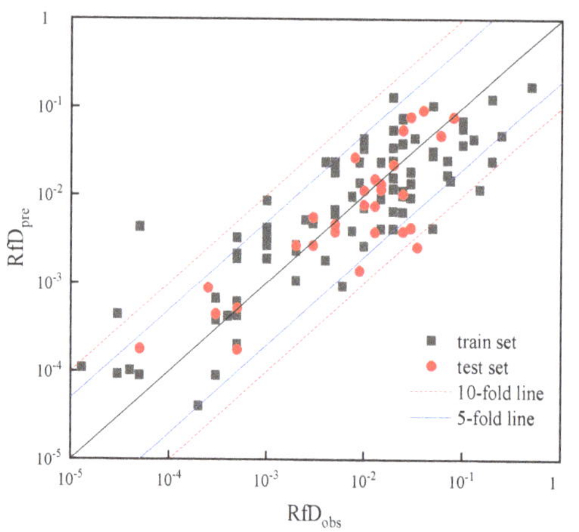

Figure 2. Comparison of observed and MLR-predicted RfD. The squares refer to data in training set and the dots are data in test set. The solid line represents the 1:1 line, while dot-dash lines and short-dashed lines represent a 5-fold and 10-fold difference, respectively, between these values.

The predicted RfD values of the two pesticides mentioned are 0.01271 $mg \cdot kg^{-1} \cdot d^{-1}$ and 0.0002124 $mg \cdot kg^{-1} \cdot d^{-1}$, respectively, which are obtained by the above equation with the corresponding molecular parameters. From the biological conception, the RfD values of these pesticides in this manuscript are identified as 0.01 $mg \cdot kg^{-1} \cdot d^{-1}$ and 0.0002 $mg \cdot kg^{-1} \cdot d^{-1}$.

3.2. Determination of Other Water Quality Criteria Parameters

The lipid fraction values were calculated using the average value of lipids of each species as the default value. The lgK_{ow} value of p-p'DDE is 6.76 and the lgK_{ow} value of α-HCH is 3.69, which obtained from the data disclosed on the official website of EPA. From Equation (4), the free dissolved state of p-p'DDE and α-HCH are 0.0319 and 0.9753, respectively.

According to the derivation steps of non-ionic organic compound bioaccumulation factors in the technical guideline, the baseline bioaccumulation factors and final trophic level bioaccumulation factors were calculated for different trophic levels, and the required parameter values and calculation results are shown in Table 3.

Table 3. Bioaccumulation factor parameters and calculated values.

Trophic Levels	f_l	Compounds	FCM	BL-BAF	F-BAF
2	0.019	p-p'DDE α-HCH	1.000 1.000	5.33×10^7 1.97×10^4	3.24×10^4 365
3	0.026	p-p'DDE α-HCH	13.30 24.70	5.18×10^8 3.55×10^5	4.30×10^5 9.00×10^3
4	0.030	p-p'DDE α-HCH	1.128 1.003	3.81×10^7 1.25×10^4	3.65×10^4 366

In the water quality criteria derivation formula, the human body weight BW and daily water intake DI refer to the handbook mentioned above published by the Ministry of Environmental Protection in 2013. The average body weight of adults over 18 years old in China is 60.6 kg, and the daily water intake is 1.85 $L \cdot d^{-1}$. Moreover, the intake of water products at each trophic level refers to the recommended values in the Nutrition and Dietary Guidelines for Chinese Residents [35]. The relative source contribution rate was taken with reference to the exposure decision tree method in the technical guidelines, and finally, the 20% default value was used as the RSC value in this study. The RfD value was the predicted value of the model constructed in this manuscript. The localized parameters required to calculate the human health water quality criteria were shown in Table 4. After calculating by Eq 1, the human health water quality criteria of p-p'DDE and α-HCH are 0.03 $\mu g \cdot L^{-1}$ and 0.02 $\mu g \cdot L^{-1}$, respectively.

Table 4. Statistical table of human health water quality parameters.

Compounds	RfD	BW	DI	FIi/kg · d^{-1}			BAF/L · kg^{-1}		
	$mg \cdot kg^{-1} \cdot d^{-1}$	kg	$L \cdot d^{-1}$	FI_2	FI_3	FI_4	2	3	4
p-p'DDE α-HCH	0.01 0.0002	60.60	1.850	0.0126	0.0100	0.0075	3.24×10^4 365	4.30×10^5 9.00×10^3	3.65×10^4 366

3.3. Health Risk Assessment

Organochlorine pesticides are highly fat-soluble and can enter the human body and animals through the food chain and can accumulate in the visceral tissues. Therefore, the health risk caused by organochlorine pesticides is of concern. In this study, the risk assessment of p-p'DDE and α-HCH in a domestic water environment was performed by the Hazard Quotient method (HQ) of Equation (5). The exposure concentrations used are the publicly released survey data in recent years, involving 192 and 254 sampling locations, respectively. In addition, the specific information is shown in Supplementary Table S1. The exposure concentration of p-p'DDE at the sampling sites ranged from 0.002 to 139 ng/L, with 49% of the sites having HQ values less than 0.1000; 18% of the sites had HQ values between 0.1000 and 1.000; and 33% of the sites had HQ values between

1.000 and 10.00. The exposure concentration of α-HCH at each sampling site ranged from 0.0151 to 297 ng/L, 60% of the sites had HQ values less than 0.1000; 36% of the sites had HQ values between 0.1000 and 1.000; and 4% of the sites had HQ values between 1.000 and 10.00. The median values of the monitored concentrations were used to represent the exposure levels of pollutants in the domestic water environment, and the HQ values of the two pollutants were calculated to be 0.13 and 0.08, respectively. The results show that at the current exposure levels, α-HCH basically poses no health risk to the residents around the watershed, and p-p'DDE poses a lower health risk to the residents. Based on the potential human health risks of organochlorine pesticides, it is necessary to control the pollution problems in the corresponding areas to ensure the water safety of the residents in these areas.

4. Conclusions

Currently, the development of water quality criteria and risk assessment of pesticide compounds is an issue of concern. Moreover, RfD is a key parameter of water quality criteria derivation. In this paper, through a non-experimental method, the RfD prediction model was constructed using molecular descriptors for the derivation of human health water quality criteria values. In the absence of experimental data, the predicted value of the model has a certain reference value, which is conducive to the advancement of pollutant health risk assessment.

The model obtained in this paper has good model stability in terms of statistics (Rtra2 = 0.762, Rtes2 = 0.683, RMSEP = 0.434). In a previous study, Mazzatorta used MLR to build prediction models for LOAEL [21], which had 15 descriptors (R2 = 0.50, RMSE = 0.727). Consequently, the predictive model for RfD inheres has higher reliability. In addition, some researchers used the QSAR approach to model the extrapolation of toxicity between BETX [15], and 81% of the species had a prediction error of 10 times or less. In this study, for the vast majority of pollutants (>95%), the difference between the true and predicted values is within a factor of 10, and for most (>85%) pollutants the difference between the actual and predicted values is within a factor of 5. In summary, the RfD prediction model in this paper has higher reliability.

The human health water quality criteria of p-p'DDE and α-HCH based on localized parameters are 0.03 $\mu g \cdot L^{-1}$ and 0.02 $\mu g \cdot L^{-1}$, respectively. Moreover, the quotient method was used to make a preliminary evaluation of the health risks of p-p'DDE and α-HCH in the water environment. The results demonstrated that, under the current exposure level, p-p'DDE is basically no health risk to the residents around the watershed, and α-HCH produces a lower health risk to the residents. This result may be due to the fact that p-p'DDE has a higher bioaccumulation factor and is therefore potentially more hazardous to human health.

Although there are important discoveries revealed by these studies, there are also limitations. Since there are only about 300 compounds with clearly defined RfD values in the IRIS system, even fewer pollutants meet the requirements of this study. Consequently, the dataset used in this paper contains only 109 compounds, which is a limited sample size. Therefore, if more compounds are added to the IRIS system in the future, this study should increase the sample size to optimize the model.

Supplementary Materials: The following supporting information can be downloaded at: https://www.mdpi.com/article/10.3390/toxics11040318/s1, Figure S1: The relationship between R_{adj}^2 and the number of molecular descriptors; Table S1: Parameters of health risk assessment in water environmental. References [38–43] are cited in the Supplementary Materials.

Author Contributions: Conceptualization, S.-H.M. and Z.-G.Y.; methodology, S.-H.M.; validation, S.-H.M.; investigation, S.-H.M.; data curation, Q.Z. and X.X.; writing—original draft preparation, S.-H.M.; writing—review and editing, Z.-G.Y. and C.-Y.J.; visualization, S.-H.M.; supervision, Z.-G.Y., X.Z. and C.-Y.J.; project administration, J.-Y.C. and X.Z.; funding acquisition, Z.-G.Y. All authors have read and agreed to the published version of the manuscript.

Funding: This work was financially supported by the National Natural Science Foundation of China (Grant No. 42277271) and the National Key Research and Development Program of China (Grant No. 2021YFC3201005).

Data Availability Statement: No new data were created or analyzed in this study. Data sharing is not applicable to this article.

Conflicts of Interest: The authors declare no conflict of interest.

References

1. Yan, Z.; Zheng, X.; Fan, J.; Zhang, Y.; Wang, S.; Zhang, T.; Sun, Q.; Huang, Y. China national water quality criteria for the protection of freshwater life: Ammonia. *Chemosphere* **2020**, *251*, 126379. [CrossRef] [PubMed]
2. Pham, L.L.; Borghoff, S.J.; Thompson, C.M. Comparison of threshold of toxicological concern (TTC) values to oral reference dose (RfD) values. *Regul. Toxicol. Pharmacol.* **2020**, *113*, 104651. [CrossRef] [PubMed]
3. Barnes, D.G.; Dourson, M. Reference dose (RfD): Description and use in health risk assessments. *Regul. Toxicol. Pharmacol. RTP* **1988**, *8*, 471–486. [CrossRef]
4. European Food Safety Authority and World Health Organization. *Review of the Threshold of Toxicological Concern (TTC) Approach and Development of New TTC Decision Tree*; European Food Safety Authority: Parma, Italy; World Health Organization: Geneva, Switzerland, 2016.
5. Kadry, A.M.; Skowronski, G.A.; Abdel-Rahman, M.S. Evaluation of the use of uncertainty factors in deriving RfDs for some chlorinated compounds. *J. Toxicol. Environ. Health* **1995**, *45*, 83–95. [CrossRef]
6. USEPA. *Methodology for Deriving Ambient Water Quality Criteria for the Protection of Human Health (2000)*; USEPA: Washington, DC, USA, 2000.
7. Alexeeff, G.V.; Broadwin, R.; Liaw, J.; Dawson, S.V. Characterization of the LOAEL-to-NOAEL uncertainty factor for mild adverse effects from acute inhalation exposures. *Regul. Toxicol. Pharmacol.* **2002**, *36*, 96–105. [CrossRef] [PubMed]
8. Hughes, B.J.; Cox, K.; Bhat, V. Derivation of an oral reference dose (RfD) for di 2-ethylhexyl cyclohexan-1,4-dicarboxylate (DEHCH), an alternative to phthalate plasticizers. *Regul Toxicol Pharm.* **2018**, *92*, 128–137. [CrossRef]
9. Katsnelson, B.A.; Chernyshov, I.N.; Solovyeva, S.N.; Minigalieva, I.A.; Gurvich, V.B.; Valamina, I.E.; Makeyev, O.H.; Sahautdinova, R.R.; Privalova, L.I.; Tsaregorodtseva, A.E.; et al. Looking for the LOAEL or NOAEL Concentration of Nickel-Oxide Nanoparticles in a Long-Term Inhalation Exposure of Rats. *Int. J. Mol. Sci.* **2021**, *22*, 416. [CrossRef] [PubMed]
10. Zhang, T.; Yan, Z.; Zheng, X.; Fan, J.; Wang, S.; Wei, Y.; Yang, L.; Wang, P.; Guo, S. Transcriptome analysis of response mechanism to ammonia stress in Asian clam (Corbicula fluminea). *Aquat. Toxicol.* **2019**, *214*, 105235. [CrossRef] [PubMed]
11. Council, N.R. *Toxicity Testing in the 21st Century: A Vision and a Strategy*; The National Academies Press: Washington, DC, USA, 2007; p. 216. [CrossRef]
12. Fan, J.; Yan, Z.; Zheng, X.; Wu, J.; Wang, S.; Wang, P.; Zhang, Q. Development of interspecies correlation estimation (ICE) models to predict the reproduction toxicity of EDCs to aquatic species. *Chemosphere* **2019**, *224*, 833–839. [CrossRef] [PubMed]
13. Men, S.; Xu, J.; Zhou, Q.; Liu, X.; Yan, Z. Reference dose prediction by using CDK molecular descriptors: A non-experimental method. *Chemosphere* **2022**, *305*, 135460. [CrossRef]
14. Fan, J.; Wang, S.; Li, H.; Yan, Z.; Zhang, Y.; Zheng, X.; Wang, P. Modeling the ecological status response of rivers to multiple stressors using machine learning: A comparison of environmental DNA metabarcoding and morphological data. *Water Res.* **2020**, *183*, 116004. [CrossRef]
15. Xu, J.; Zheng, L.; Yan, Z.; Huang, Y.; Feng, C.; Li, L.; Ling, J. Effective extrapolation models for ecotoxicity of benzene, toluene, ethylbenzene, and xylene (BTEX). *Chemosphere* **2020**, *240*, 124906. [CrossRef] [PubMed]
16. Myint, K.Z.; Xie, X.Q. Recent Advances in Fragment-Based QSAR and Multi-Dimensional QSAR Methods. *Int. J. Mol. Sci.* **2010**, *11*, 3846–3866. [CrossRef] [PubMed]
17. Hisaki, T.; Kaneko, M.A.n.; Hirota, M.; Matsuoka, M.; Kouzuki, H. Integration of read-across and artificial neural network-based QSAR models for predicting systemic toxicity: A case study for valproic acid. *J. Toxicol. Sci.* **2020**, *45*, 95–108. [CrossRef] [PubMed]
18. Wei, X.; Yang, M.; Zhu, Q.; Wagner, E.D.; Plewa, M.J. Comparative Quantitative Toxicology and QSAR Modeling of the Haloacetonitriles: Forcing Agents of Water Disinfection Byproduct Toxicity. *Env. Sci. Technol.* **2020**, *54*, 8909–8918. [CrossRef] [PubMed]
19. Veselinovic, J.B.; Veselinovic, A.M.; Toropova, A.P.; Toropov, A.A. The Monte Carlo technique as a tool to predict LOAEL. *Eur. J. Med. Chem.* **2016**, *116*, 71–75. [CrossRef] [PubMed]
20. Toropova, A.P.; Toropov, A.A.; Marzo, M.; Escher, S.E.; Dome, J.L.; Georgiadis, N.; Benfenati, E. The application of new HARD-descriptor available from the CORAL software to building up NOAEL models. *Food Chem. Toxicol.* **2018**, *112*, 544–550. [CrossRef]
21. Tilaoui, L.; Schilter, B.; Tran, L.-A.; Mazzatorta, P.; Grigorov, M. Integrated computational methods for prediction of the lowest observable adverse effect level of food-borne molecules. *Qsar Comb. Sci.* **2007**, *26*, 102–108. [CrossRef]
22. Jakubowski, M.; Czerczak, S. A proposal for calculating the no-observed-adverse-effect level (NOAEL) for organic compounds responsible for liver toxicity based on their physicochemical properties. *Int. J. Occup. Med. Environ. Health* **2014**, *27*, 627–640. [CrossRef]

23. Zarn, J.A.; Engeli, B.E.; Schlatter, J.R. Study parameters influencing NOAEL and LOAEL in toxicity feeding studies for pesticides: Exposure duration versus dose decrement, dose spacing, group size and chemical class. *Regul. Toxicol. Pharmacol.* **2011**, *61*, 243–250. [CrossRef] [PubMed]
24. Rabovsky, J.; Fowles, J.; Hill, M.D.; Lewis, D.C. A health risk benchmark for the neurologic effects of styrene: Comparison with NOAEL/LOAEL approach. *Risk Anal.* **2001**, *21*, 117–126. [CrossRef] [PubMed]
25. Dyer, S.D.; Versteeg, D.J.; Belanger, S.E.; Chaney, J.G.; Raimondo, S.; Barron, M.G. Comparison of species sensitivity distributions derived from interspecies correlation models to distributions used to derive water quality criteria. *Environ. Sci. Technol* **2008**, *42*, 3076–3083. [CrossRef] [PubMed]
26. Qian, X.; Lee, P.W.; Cao, S. China: Forward to the green pesticides via a basic research program. *J. Agric. Food Chem.* **2010**, *58*, 2613–2623. [CrossRef] [PubMed]
27. Cook, E.; Moore, P. The effects of the herbicide metolachlor on agonistic behaviorin the crayfish, Orconectes rusticus. *Arch. Environ. Contam. Toxicol.* **2007**, *55*, 94–102. [CrossRef]
28. Yan, Z.; Wang, w.; Zhou, J.; Yi, X.; Zhang, J.; Wang, X.; Liu, Z. Screening of high phytotoxicity priority pollutants and their ecological risk assessment in China's surface waters. *Chemosphere* **2015**, *128*, 28–35. [CrossRef]
29. He, J.; He, H.; Yan, Z.; Gao, F.; Zheng, X.; Fan, J.; Wang, Y. Comparative analysis of freshwater species sensitivity distributions and ecotoxicity for priority pesticides: Implications for water quality criteria. *Ecotoxicol Environ. Saf.* **2019**, *176*, 119–124. [CrossRef]
30. Bender, A.; Glen, R.C. Molecular similarity: A key technique in molecular informatics. *Org. Biomol. Chem.* **2004**, *2*, 3204–3218. [CrossRef]
31. Garcia-Domenech, R.; de Julian-Ortiz, J.V.; Besalu, E. True prediction of lowest observed adverse effect levels. *Mol. Divers.* **2006**, *10*, 159–168. [CrossRef]
32. Hansch, C.; Steward, A.R. The use of substituent constants in the analysis of the structure–activity relationship in penicillin derivatives. *J. Med. Chem.* **1964**, *7*, 691–694. [CrossRef]
33. Ministry of Ecology and Environment the People's Republic of China. *Technical Guideline for Deriving Water Quality Criteria for the Protection of Human Health*; Ministry of Ecology and Environment the People's Republic of China: Beijing, China, 2017.
34. Ministry of Ecology and Environment the People's Republic of China. *Exposure Factors Handbook of Chinese Population (Adult Volume)*; China Evironmental Science Press: Beijing, China, 2013.
35. Society, N. *Nutrition and Dietary Guidelines for Chinese Residents*; People's Medical Publishing House: Beijing, China, 2016.
36. Singh, J.; Singh, S.; Thakur, S.; Lakhwani, M.; Khadikar, P.V.; Agrawal, V.K.; Supuran, C.T. QSAR study on murine recombinant isozyme mCAXIII: Topological vs structural descriptors. *Arkivoc* **2006**, *14*, 103–118. [CrossRef]
37. Roy, P.P.; Paul, S.; Mitra, I.; Roy, K. On Two Novel Parameters for Validation of Predictive QSAR Models. *Molecules* **2009**, *14*, 1660–1701. [CrossRef]
38. Zhang, M.; Hua, R.M.; Li, X.D.; Zhou, T.T.; Yang, F.; Cao, H.Q.; Wu, X.W.; Tang, J. Residual Characteristic and Assessment of Organochlorine Pesticides in Water of Chaohu Lake Tributaries. In Proceedings of the Third National Symposium on Agricultural and Environmental Sciences, Tianjin, China, 1 October 2009; pp. 558–564.
39. Zhang, Z.; Chen, W.; Khalid, M.; Zhou, J.; Xu, L.; Hong, H. Evaluation and Fate of the Organic Chlorine Pesticides at the Waters in Jiulong River Estuary. *Environ. Sci.* **2001**, *22*, 88–92.
40. Wang, Y.H.; Qi, S.H.; Wang, W. The characteristic distribution of organic chlorine pesticides in the water and sediment of Diaocha Lake in Hubei Province. *Environ. Pollut. Control* **2007**, *29*, 415–418.
41. Yang, Q.S.; Mai, B.X.; Fu, J.M.; Sheng, G.Y.; Hu, X. Studies on organochlorine pesticites (OCPs) in waters of Pearl River artery estuary water. *China Environ. Sci.* **2005**, *25*, 47–51.
42. Gao, G.S.; Teng, M.D.; Xu, C.M. Determination of Fourteen Organochlorine Pesticides in the Middle of Chishui River by Gas Chromatography. *Arid. Environ. Monit.* **2011**, *25*, 193–202.
43. Tai, C.; Zhang, K.; Zhou, T.; Zhao, T.; Xiao, C.; Wu, L. Distribution characteristics and risk evaluation of organoclorine pesticides in runoff from typical area of Danjiangkou Reservoir. *China Environ. Sci.* **2012**, *32*, 1046–1053.

toxics

MDPI

Article

Risk Assessment of Phthalate Esters in Baiyangdian Lake and Typical Rivers in China

Yin Hou [1,2], Mengchen Tu [1], Cheng Li [3], Xinyu Liu [1], Jing Wang [1,2], Chao Wei [1], Xin Zheng [1,*] and Yihong Wu [3,*]

[1] State Key Laboratory of Environmental Criteria and Risk Assessment, Chinese Research Academy of Environmental Sciences, Beijing 100012, China
[2] College of Marine Ecology and Environment, Shanghai Ocean University, Shanghai 201306, China
[3] Institute of Green Development, Hebei Provincial Academy of Environmental Sciences, Shijiazhuang 050037, China
* Correspondence: zhengxin@craes.org.cn (X.Z.); yh_wuhb@126.com (Y.W.)

Abstract: Phthalate esters (PAEs) are frequently tracked in water environments worldwide. As a typical class of endocrine disruptor chemicals (EDCs), PAEs posed adverse effects on aquatic organisms at low concentration. Thus, they have attracted wide attention in recent years. In the present study, the concentrations of seven typical PAEs from 30 sampling sites in Baiyangdian Lake were measured, and the environmental exposure data of PAEs were gathered in typical rivers in China. Then, based on the aquatic life criteria (ALCs) derived from the reproductive toxicity data of aquatic organisms, two risk assessment methods, including hazard quotient (HQ) and probabilistic ecological risk assessment (PERA), were adopted to evaluate the ecological risks of PAEs in water. The sediment quality criteria (SQCs) of DEHP, DBP, BBP, DIBP and DEP were deduced based on the equilibrium partitioning method. Combined with the gathered environmental exposure data of seven PAEs in sediments from typical rivers in China, the ecological risk assessments of five PAEs in sediment were conducted only by the HQ method. The results of ecological risk assessment showed that in terms of water, DBP and DIBP posed low risk, while the risk of DEHP in Baiyangdian Lake cannot be ignored and should receive attention. In typical rivers in China, BBP and DEP posed no risk, while DIBP and DBP posed potential risk. Meanwhile, DEHP posed a high ecological risk. As far as sediment is concerned, DBP posed a high risk in some typical rivers in China, and the other rivers had medium risk. DEHP posed a high risk only in a few rivers and low to medium risk in others. This study provides an important reference for the protection of aquatic organisms and the risk management of PAEs in China.

Keywords: phthalate esters; reproductive toxicity; aquatic life criteria; sediment quality criteria; ecological risk assessment

Citation: Hou, Y.; Tu, M.; Li, C.; Liu, X.; Wang, J.; Wei, C.; Zheng, X.; Wu, Y. Risk Assessment of Phthalate Esters in Baiyangdian Lake and Typical Rivers in China. *Toxics* **2023**, *11*, 180. https://doi.org/10.3390/toxics11020180

Academic Editor: Paolo Montuori

Received: 4 January 2023
Revised: 5 February 2023
Accepted: 10 February 2023
Published: 15 February 2023

1. Introduction

Phthalate esters (PAEs), also known as phenolic acid ester, are intensively applied in plastic products as plasticizers to improve their elasticity, toughness and durability. In addition, PAEs are widely used in the production of paints, pesticides, fertilizers, herbicides, pesticides, solvents and cosmetics. It is reported that in 2014, PAEs accounted for 70% of the 8.4 million tons of plasticizers produced worldwide, and they will continue to grow at a high rate of 3.9% over the next 5 years [1]. Li et al. found that the total amount of PAEs consumed in China reached 2.2 million tons in 2011 alone [2]. Since PAEs do not chemically bond with polymer molecules of plastic products, they are easy to dissociate from the above products and leach into the environment under certain conditions [3,4]. After the mass production, use and disposal of substances containing PAEs, their pollution is distributed in various environments around the world, including air, soil, sediment, landfill leachate, urban runoff and natural water [5–7]. PAEs mainly enter the water environment through the discharge of domestic sewage and industrial wastewater, the input of surface runoff from agricultural and urban areas, and the dry and wet settlement of the atmosphere [8–10]. Then, they settle in the bottom sediment with particulate matter and accumulate continuously. Therefore, the concentrations of PAEs are relatively high in the sediments [11]. With the migration and transformation of PAEs in different environments and the long-distance transport in the global scale, PAEs has become a kind of global organic pollutant that is widely detected in rivers, lakes, reservoirs and their sediments around the world. For example, Vietnamese scholar Le et al. conducted a survey in six lakes including Tien Quang Lake in Vietnam and found that the highest concentration of PAEs reached 127 μg/L [12]. The highest concentration of PAEs was 4.64 μg/L in the Kaveri River in India [13]. Liu et al. investigated di (2-ethylhexyl) phthalate (DEHP) in 31 surface waters of seven major river basins in China, and the results showed that the concentration of DEHP ranged from 0.01000 to 2634 μg/L, among which the pollution of PAEs in Xuanwu Lake and Anshan Urban Rivers were the most serious, and the concentration was as high as 1.3×10^3 and 1.3×10^4 μg/L, respectively [14]. Li et al. studied the PAEs in the sediments of urban rivers in northeast China and found that the total concentration of PAEs in the sediment of Xi River, a tributary of Liao River, ranged from 22.4 to 369 μg/g dw, and the pollution of PAEs in the sediment was more serious [2]. Since PAEs are difficult to be degraded and have a high degree of bioaccumulation, they can produce reproductive, developmental and neurotoxic effects on aquatic organisms after entering the aquatic environment [15]. Otherwise, they can enter the human body through different exposure pathways, causing reproductive and developmental toxicity and even carcinogenesis [16,17]. The United States Environmental Protection Agency (EPA) has listed six PAEs including dimethyl phthalate (DMP), diethyl phthalate (DEP), di-n-butyl phthalate (DBP), butyl-benzyl phthalate (BBP), di-n-octyl phthalate (DOP) and DEHP in the 129 environmental priority pollutants field [18]. China also listed DBP, DMP and DOP as environmental priority pollutants [19].

ALC refers to the maximum concentration of pollutants that do not cause short-term or long-term adverse effects and harm to aquatic organisms [20]. Species sensitivity distribution (SSD) is an important extrapolation method, which can be used to derive ALCs from toxicological data of pollutants and extrapolate the corresponding pollutants concentration (HCx) for a target percentage of species affected [21]. Lethal effect is usually taken as the toxicity endpoint to construct SSD when deriving ALCs of conventional pollutants such as ammonia nitrogen and heavy metals, while PAEs as a typical class of EDCs that generally affect the reproduction of organisms at low concentrations [22,23]. Thus, ALCs derived from PAEs based on a lethal toxicity endpoint cannot provide sufficient protection for aquatic organisms. Previous studies have shown that reproductive toxicity endpoints were the most sensitive for EDCs [24,25], and reproductive toxicity includes fertility, fertilization rate, hatchability, gonadal index that lasts for multiple generations, and the synthesis of vitellogenin (VTG) [26]. Therefore, reproduction was the most suitable endpoint for deriving the ALCs of PAEs.

For the derivation of PAEs sediment quality criteria (SQC), considering that there are few studies on sediment benthic and the toxicity data are not enough to construct SSD, the equilibrium partitioning method recommended in the European Union Technical Guidelines for Risk Assessment (TGD) is referred. The equilibrium distribution method is applicable to nonionic organic compounds with $\lg K_{ow}$ (logarithm of octanol–water partition coefficient) > 3. This method is based on the following assumptions: (1) the organisms living in the sediment environment and in water have the same sensitivity to pollutants; (2) the concentrations of pollutants in sediment, interstitial water and benthic organisms are in thermodynamic equilibrium, and the equilibrium partition coefficient can be used to predict the concentration of pollutants in any phase.

Ecological risk assessment (ERA) refers to the assessment of the possibility of adverse ecological consequences after the ecosystem is affected by one or more stress factors [27]. The hazard quotient (HQ) is a point estimate method of ecological risk, which has the advantages of simplicity and low data requirement. However, the magnitude and probability of occurrence of ecological risk of pollutants cannot be evaluated by the HQ method, and it is only applicable to preliminary risk assessment [28,29]. Probabilistic ecological risk assessment (PERA) is a higher level ecological risk assessment method [28]. Joint probability curve (JPC) is one of the commonly used methods of probabilistic risk assessment. This curve is fitted based on the toxicity data and exposure data, which reflects the probability that the exposure concentration exceeds the corresponding critical concentration at different damage levels, that is, the risk degree of pollutants in the environment to aquatic organisms at different damage levels, and the probability of adverse effects concentration (HCx) of pollutants in water on a target percentage of aquatic organisms can be obtained [25,30]. The closer the joint probability curve is to the x axis, the less the aquatic organisms are affected by pollutants, and the less the occurrence probability of ecological risk of water. Each point on the JPC represents the probability that a target percentage of organisms being affected (events) will occur in the target water (the evaluation object).

As a global organic pollutant, PAEs are clearly harmful to the environment. However, a large number of PAEs are still used every year and are constantly released into the environment from production and living activities, affecting and endangering biosecurity and even ecosystem stability. Baiyangdian Lake is the largest freshwater lake in the North China Plain and Xiongan; the new area attaches great importance to the water environment. This study conducted a comprehensive field investigation on Baiyangdian Lake and assessed the ecological risk of PAEs, which can effectively control the pollution of PAEs and provide data support and a theoretical basis for the formulation of water quality standards and the prevention and control of PAEs pollution in the future.

2. Material and Methods

2.1. Solvents and Chemical Standards

DEHP, DBP, BBP, DEP, DMP, DOP and DIBP were investigated in this work, and their characteristics are summarized in Table S1. Methanol (pesticide grade) was purchased from J.T. Baker Co. USA. Hydrochloric acid, ethyl acetate and dichloromethane (pesticide grade) were purchased from Bailingwei Company (Beijing, China). A mixed standard solution of the 6 PAEs was used. Benzyl benzoate (BBZ) was used as the internal standard, and these substances were all obtained from Sigma-Aldrich (St. Louis, MO, USA).

2.2. Sample Collection and Preparation

Sampling sites were set based on the Baiyangdian Lake entrance, the interchange, the farmland area, the living area, etc., and a total of 15 sites were arranged on 8 rivers which entering the Baiyangdian Lake. Then, according to the national control monitoring sites and 40 lakes in Baiyangdian Lake, 15 sites were arranged, so 30 sites in total were arranged in Baiyangdian Lake. The sites distribution and the concentration of PAEs at each site are shown in Figure 1. In April 2019, water samples were collected in 2 L brown glass bottles. The sample bottles were cleaned with tap water, distilled water and methanol, respectively, in the laboratory for 3 times in advance, and then, they were moistened and washed with on-site water 3 times. After the samples were collected, the pH was adjusted to 2.0 with 4 $mol \cdot L^{-1}$ hydrochloric acid (to inhibit microbial activity), and the samples were stored in a refrigerator and transported back to the laboratory for pretreatment within 24 h.

The preparation of the water samples prepared for gas chromatography mass spectrometry (GC-MS) was performed. Firstly, 0.45 μm glass fiber filters (GF/F, Whatman, UK) were pre-burned for 4 h in a muffle furnace at 400 °C, and then, 1 L water samples were filtered using the filters. Secondly, solid phase extraction was performed. Before concentrating and enriching the sample, C18 solid phase extraction columns (BOJIN SPE Column C18) were activated with 5 mL dichloromethane, 5 mL ethyl acetate, 1 mL methanol and 10 mL ultrapure water, respectively. The controlled flow rate of the C18 SPE Columns was 3 $mL \cdot min^{-1}$. The sample bottles were cleaned with 5 mL ethyl acetate and entered into the collection bottle through the C18 SPE columns; then, they were cleaned with 5 mL dichloromethane and entered into the same collection bottle through the C18 SPE columns. Finally, the sample extracts were blown to nearly dry with nitrogen at 45 °C and reconstituted to 1 mL by adding 5 μL BBZ and ethyl acetate, after which they were sealed and stored at 4 °C before GC/MS analysis. After extraction, the C18 SPE columns were rinsed with 10 mL ultrapure water and then blown with nitrogen for 5 min to remove water.

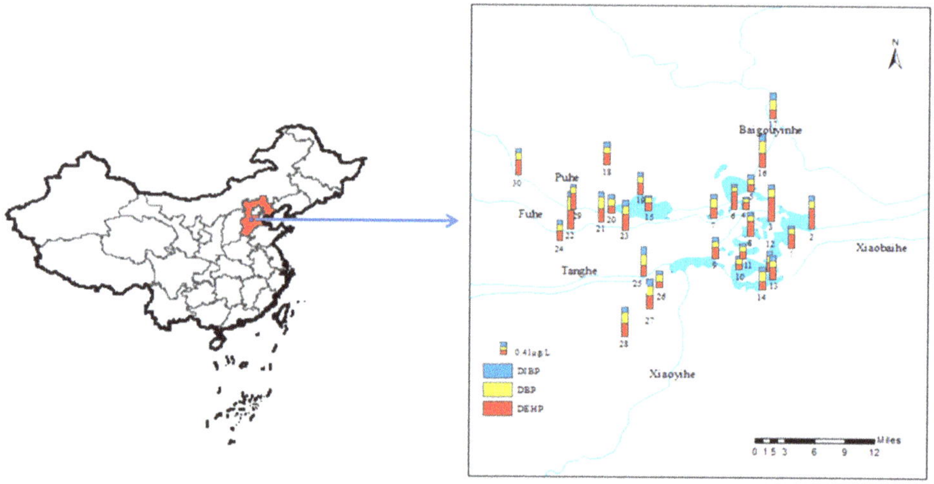

Figure 1. Location of sampling sites and PAEs concentration in Baiyangdian Lake.

2.3. Chemical Analysis

The analysis of sample extracts was conducted by a GC-MS system (Aglient7890-5975 C) with a DB-5 capillary column (30 m × 0.25 mm × 0.25 μm) (Agilent Technologies, Santa Clara, CA, USA). The oven temperature starts at 70 °C and was maintained for 2 min; then, it increased to 180 °C at 40 °C/min and was maintained for 2 min, and finally, it increased to 280 °C at 10 °C/min for 2 min. The samples were injected in splitless mode with helium as the carrier gas at a flow rate of 1 mL/min. The transfer line, quadrupole and ion source were 290 °C, 150 °C and 230 °C, respectively. The system was operated in electron impact (EI) and scan modes.

The instrument required for testing should be recalibrated daily according to the calibration standard before use, and the sample should also be processed with the program blank in order to improve the accuracy of data [31]. The recoveries of the PAEs in the water samples were 85.60–116.7%. The PAEs were calibrated with BBZ and recoveries of 94.8% in the water. The detection limits (signal to noise ratio = 3) for the PAEs were 0.060–0.84 ng/L for water.

2.4. Collection of Data

(1) Toxicity data

The toxicity data of PAEs were obtained from the ECOTOX database (https://cfpub.epa.gov/ecotox, accessed on 1 September 2021), the Web of Science database, the CNKI database and other published literature. However, the screening of toxicity data was only based on reproductive-related endpoints, such as fertility, fertilization rate, hatchability, gonadal index that lasts for multiple generations, and the synthesis of vitellogenin (VTG). The screening principles for toxicity data must be followed. No observed effect concentration (NOEC) was selected as the preferred toxicity endpoint data. The maximum acceptable toxicant concentration (MATC) was used when NOEC was not available. If neither NOEC nor MATC were available, the lowest observed effect concentration (LOEC) or 10% effective concentration (EC_{10}) value was used. The toxicity data of four PAEs are shown in Tables S2 and S3. In order to avoid possible data bias due to similar species and observed duration, the geometric mean value was adopted [32].

(2) Exposure data

In order to compare with the PAEs pollution in Baiyangdian Lake, the exposure data of PAEs in freshwater such as rivers, lakes and reservoirs and their sediments in China were also collected from the Web of Science database, the CNKI database and other published literature from 2007 to 2019.

2.5. Deriving of ALCs and SQCs for PAEs

(1) Deriving of ALCs

To construct SSDs based on the method of technical guideline [33].

For the chronic toxicity data collected, the MATC of reproductive toxicity of a species was calculated by the following formula:

$$MATC_{i,z} = \sqrt{NOEC_{i,z} \times LOEC_{i,z}} \tag{1}$$

where:

MATC = maximum acceptable toxicant concentration, mg/L or μg/L;

NOEC = no observed effect concentration, mg/L or μg/L;

LOEC = lowest observed effect concentration, mg/L or μg/L;

i = a certain species, dimensionless;

z = a toxic effect, dimensionless.

Then, chronic toxicity data (MATC, EC10, EC_{20}, NOEC, LOEC, EC_{50} and LC_{50}) were used as growth or reproductive CTV, and LC50 was used as survival CTV, which were substituted into Formula (3) to calculate the growth CVE, reproductive CVE and survival CVE of each species. This study calculated the reproductive CVE of each species.

$$CVE_{i,j} = \sqrt[n]{CTV_{i,j,1} \times CTV_{i,j,2} \times \ldots \times CTV_{i,j,n}} \tag{2}$$

where

CVE = chronic value for the same effect, μg/L or mg/L;

i = a certain species, dimensionless;

j = types of chronic toxic effects, generally classified as growth, survival and reproduction, dimensionless;

CTV = chronic toxicity value, mg/L or μg/L;

n = the number of CTV.

Arrange lgCVE from small to large, determine its rank R, and adopt Formula (3) to calculate the chronic cumulative frequency F_R of species.

$$F_R = \frac{\sum_1^R f}{N+1} \times 100\% \tag{3}$$

where

F_R = cumulative frequency, %;

R = the rank of toxicity value, dimensionless;

f = frequency, refers to the number of species corresponding to the rank of toxicity value;

N = the sum of all the frequencies.

In addition, with lgCVE as independent variable x, and the corresponding cumulative frequency F_R as dependent variable y, the SSD model was fitted by the log-logistic distribution, which was a good-fitting model for SSD (Figure 2) [21]. The hazardous concentration for 5% species affected (HC_5) was calculated from the SSD curves.

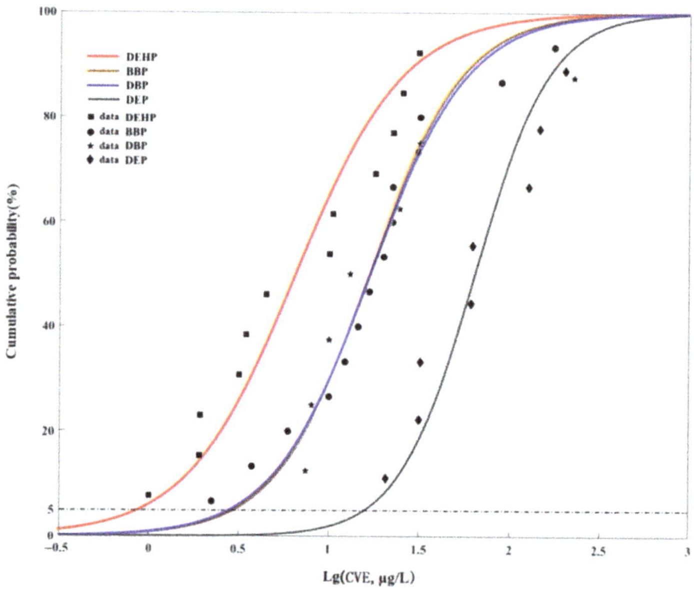

Figure 2. Species sensitivity distribution curves of PAEs.

Then, ALC is calculated as the HC_5 divided by an assessment factor of 3, because the number of species included in effective toxicity data was less than or equal to 15. The results are shown in Table 1. Since the amount of toxicity data of DIBP does not meet the minimum data requirements of SSD, the ALC of DIBP obtained by the assessment factor method was 0.90000 μg/L, and the AF value of 1000 was taken into account as the "worst case" in the ecological risk assessment [34].

Table 1. Parameters of SSDs for four PAEs (exclude DIBP).

PAEs	N	Mean	HC_5 (μg/L)	ALC (μg/L)	SQC (μg/g dw)
DEHP	12	223.44	0.87000	0.29000	0.77604
DBP	14	3009.4	2.8100	0.93667	0.76115
BBP	7	7427.7	2.9700	0.99000	0.88773
DEP	8	10,862	15.830	5.2767	4.1376
DIBP	-	-	-	0.90000	0.050000

-: No data. N: Number.

(2) Deriving of SQCs

The SQCs measured by wet mass were derived using the equilibrium partitioning method [35], and the formulas are as follows:

$$SQC_{wet\ mass} = \frac{K_{susp\text{-}water}}{RHO_{susp}} \times ALC \times 1000 \quad (4)$$

$$RHO_{susp} = Fsolid_{susp} \times RHO_{solid} + Fwater_{susp} \times RHO_{water} \quad (5)$$

$$K_{susp\text{-}water} = Fwater_{susp} + Fsolid_{susp} \times \frac{Foc_{susp} \times K_{oc}}{1000} \times RHO_{solid} \quad (6)$$

where

$SQC_{wet\ mass}$ = sediment quality criteria measured by wet mass, mg/kg;
RHO_{susp} = bulk density of (wet) suspended matter, kg/m^3;
$K_{susp\text{-}water}$ = partitioning coefficient in suspended matter and water, m^3/m^3;

$Fsolid_{susp}$ = φ (solid matter) in suspended matter, which was defined as 0.1 m^3/m^3;
RHO_{solid} = bulk density of solid matter, which was defined as 2500 kg/m^3;
$Fwater_{susp}$ = φ(H_2O) in suspended matter, which was defined as 0.9 m^3/m^3;
RHO_{water} = water density, which was defined as 1000 kg/m^3;
Foc_{susp} = organic carbon fraction of solid matter in suspended matter, 0.1 kg/kg in this study;
Koc = pollutant organic carbon—water partitioning coefficient, L/kg, that is, the ratio of the concentration of PAEs in sediment organic carbon and water.

2.6. Ecological Risk Assessment

In this study, two ecological risk assessment methods were used to evaluate the ecological risks of PAEs in Baiyangdian Lake and typical rivers in China, including the low-level HQ method and the high-level PERA method. However, only the HQ method was used to evaluate the ecological risks of PAEs in sediments from typical rivers in China.

(1) Hazard quotient (HQ)

HQ was calculated by the Equation (7):

$$HQ = \frac{EEC}{ALC} \text{ or } \frac{EEC}{SQC} \tag{7}$$

where EEC = environmental exposure concentration, μg/L or mg/kg.

The value obtained by the HQ method can be classified into the following 4 levels to evaluate the ecological risk [36]:
$HQ \leq 0.1$, no risk;
HQ = 0.1–1.0, there is low risk;
HQ = 1.1–10, medium risk;
$HQ \geq 10$, high risk.

The ecological risk assessments of PAEs in sediment also adopt the HQ method, but the risk classification is different from water: there was high risk when HQ > 1 (the lgK_{ow} of PAE congener was between 3 and 5) and high risk when HQ < 10 (lgK_{ow} > 5) [37]. For PAEs with lgK_{ow} < 3, their risk to aquatic organisms was not considered, because they are not easily adsorbed in sediment [34].

(2) Probabilistic ecological risk assessment (PERA)

Compared with the HQ method, PERA is an improved and higher ecological risk assessment method, because it can better describe the possibility that the concentration of a certain pollutant in water exceeds the toxic effect threshold and the risk of adverse effects [28]. Matlab software was used to draw JPCs to evaluate the ecological risks of PAEs. Firstly, log-normal distribution test was conducted on the exposure data of some PAEs in Baiyangdian Lake and typical rivers in China and the chronic toxicity data of aquatic organisms based on the endpoint of reproductive toxicity test. After that, the cumulative function of chronic toxicity data and the anti-cumulative function of PAEs exposure data were plotted to obtain the JPCs of PAEs [38,39].

3. Results

3.1. Occurrence and Composition of PAEs in Baiyangdian Lake

Through investigation and previous research [40], Baiyangdian Lake was divided into five functional areas: primitive area, tourism area, living area, breeding area and inflow area; each functional area had its own characteristics. The sampling sites were set in five functional areas and the concentrations of PAEs in the water at each sampling site are shown in Figure 1 and Table S4. The concentrations of seven typical PAEs were detected, including DMP, DEP, BBP, DBP, DEHP, DOP and DIBP. The results showed that DMP, DEP, BBP and DOP were not detected at all sampling sites, while DBP, DEHP and DIBP were 100% detected. It can be seen from Table S4 that the exposure concentrations of DBP and

DEHP were relatively high. The highest concentration of $\sum_3$PAEs was at sampling site 22 at a concentration of 1.3 μg/L. In addition, the concentration of $\sum_3$PAEs at sampling sites in the inflow rivers and inflow area were relatively high. The main reason might be that a large number of high-water-consuming and heavily polluting industrial enterprises such as chemical fiber, papermaking, and batteries were gathered in the upper reaches of the rivers, resulting in a lot of industrial sewage. The inflow rivers mixed with the water in the inflow area, which reduced the pollution concentration of $\sum_3$PAEs. Due to the influence of human activities, including tourism, domestic sewage discharge, and unreasonable application of pesticides and fertilizers, the $\sum_3$PAEs concentration of aquaculture, living areas and tourist areas was secondary. The lowest concentration was located in the primitive area, because it is an undeveloped area with less human intervention. Therefore, in general, the pollution of PAEs in the inflow area was the most serious. Meanwhile, DBP and DEHP were the most widely used PAEs, which had higher concentration in water [41].

3.2. The Ecological Risk Assessments of PAEs in Baiyangdian Lake

The ALCs of typical PAEs in this study are listed in Table 1. These values were obtained through SSD, which were constructed based on reproductive toxicity endpoints. Because the reproductive toxicity endpoints were more sensitive than the lethal toxicity effect endpoint, the obtained ALCs of PAEs can avoid the adverse effect to aquatic organisms due to long-term exposure at low concentration and can better protect aquatic organisms.

The HQs of DIBP, DBP and DEHP are shown in Table S4 and Figure 3. Obviously, the HQs of DBP at all sampling sites were in the range of 0.1–1.0, indicating that DBP posed a low ecological risk in Baiyangdian Lake. Moreover, in view of the DIBP, the HQs at about 86.7% of the sampling sites were also in the range of 0.1–1.0, showing that 86.7% of the sampling sites in Baiyangdian Lake had low risk, and the HQs at other sampling sites were less than 0.1, so the rest of the sampling had no risk. However, the HQs of DEHP were greater than 1.1 at most sampling sites, and the HQs of the remaining sampling sites were also in the range of 0.1–1.0, manifesting that DEHP had low or medium ecological risk. In contrast, the ecological risks of DBP and DIBP were much lower than DEHP, but their potential ecological risk cannot be ignored.

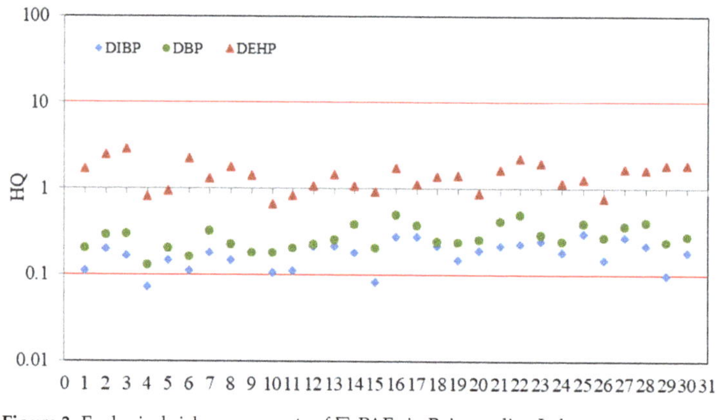

Figure 3. Ecological risk assessments of $\sum_3$PAEs in Baiyangdian Lake.

Compared with the HQ method, the PERA method can better describe the ecological risks of PAEs. JPC is one of the commonly used risk assessment methods in the process of the PERA method. The results of the PERA method in Baiyangdian Lake are shown in Figure 4a,b. It can be seen from the figure that the JPC of DBP was closer to the x-axis than that of DEHP, indicating that the potential ecological risk of DBP was higher than that of DEHP. The results of PERAs in Baiyangdian Lake showed that the probabilities of DEHP and DBP affecting 5% aquatic organisms by JPCs were 3.97% and 0.20%, respectively.

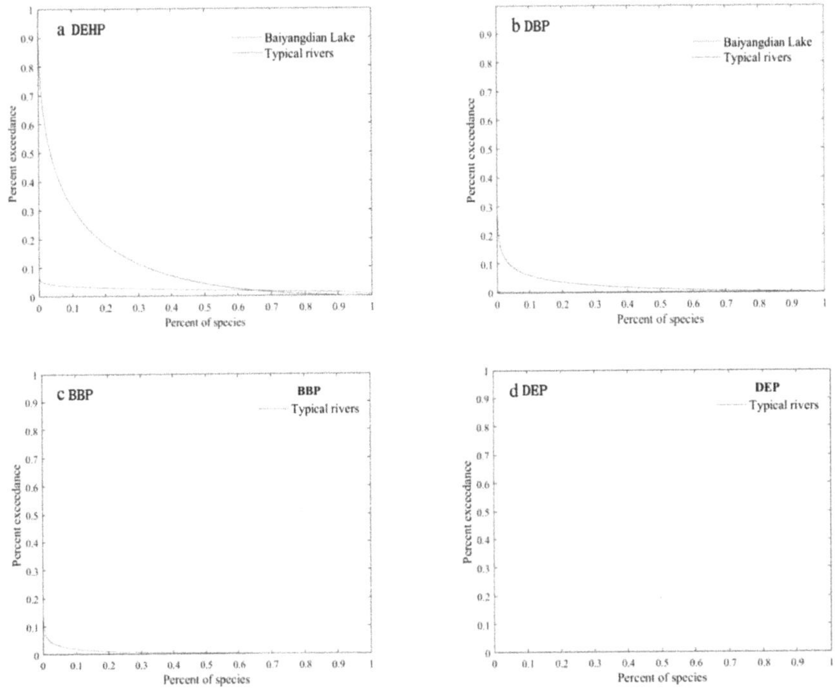

Figure 4. Joint probability curves for ecological risks of PAEs. (**a**) Joint probability curve of DEHP; (**b**) Joint probability curve of DBP; (**c**) Joint probability curve of BBP; (**d**) Joint probability curve of DEP).

3.3. The Ecological Risk Assessments of PAEs in Typical Rivers

In this study, the exposure concentrations of DBP, DEHP, DMP, DEP, BBP, DOP and DIBP in different typical rivers in China were collected from the published literature and listed in Table 2. The data were the average values of PAEs exposure concentrations except for the Zhenjiang section of the Yangtze River, which were the maximum exposure concentrations of PAEs.

The ecological risk assessments of PAEs in typical rivers in China were carried out by the HQ method. The results are shown in Table S5 and Figure 5. For the BBP, the HQs of Taihu Lake, Guanting Reservoir, Shichahai and Songhua River were in the range of 0.1–1.0, so these rivers had low risk, and the rest had no risk because their HQs were less than 0.1. Similarly, DEP had low risk in Taihu Lake and Songhua River; in addition, it also had a low risk in Zhenjiang, and its HQ was slightly higher than that of Taihu Lake and Songhua River. The risk posed by DBP was high at the Zhenjiang section of the Yangtze River and the middle and lower reaches of the Yellow River, while the rest had medium or low risk. In view of DEHP, about 21.1% of rivers had high risk, 10.5% had no risk, and others had medium or low risk. DIBP posed high risk in the Zhenjiang section of the Yangtze River and Pu River, a tributary of Liao River, and it posed medium risk in the Xi River and Jiulong River, while the rest had medium or low risk. DEHP mainly came from plastics and heavy chemical industry as well as domestic waste. DBP was widely used in cosmetics and personal care products, while DIBP has been widely used as a substitute for DBP in recent years. Therefore, with the extensive use of plastics and the development of urban industrialization, a large number of DEHP, DBP and DIBP were produced. The exposure concentrations of DEHP, DBP and DIBP in freshwater in China were higher due to surface runoff or atmospheric wet deposition.

Table 2. Exposure data of PAEs in typical rivers in China (μg/L).

Rivers	Sites	Concentration (μg/L)							
		DBP	**DEHP**	**DMP**	**DEP**	**BBP**	**DOP**	**DIBP**	**Reference**
Baiyangdian Lake		0.26	0.42	ND	ND	ND	ND	0.16	This study
Pearl River		8.5	5.6	2.4	0.046	ND	ND	ND	[42]
Jiulong River		0.67	1.7	0.088	0.085	ND	ND	3.4	[34]
Songhua River		5.1	0.20	2.5	2.4	2.5	2.4	ND	[43]
Chaohu Lake		3.2	0.20	0.42	0.14	0.071	0.035	ND	[44]
Yangtze River	Middle reaches of Jiangsu section	0.21	0.010	0.025	0.057	0.010	0.010	ND	[45]
	Lower jiangsu section	0.11	0.010	0.013	0.060	0.010	0.010	ND	[45]
	Zhenjiang	13	10	1.5	2.6	ND	1.1	13	[46]
	Nanjing	0.19	1.1	0.010	0.14	0.010	0.020	ND	[45]
	Suzhou	ND	ND	0.015	0.012	ND	0.034	ND	[47]
	Yangcheng Lake	7.2	17	0.13	0.086	0.072	0.34	ND	[47]
	Taihu Lake	1.6	1.3	0.71	0.72	0.50	0.16	ND	[48]
Yellow River	Lanzhou	0.80	0.83	0.64	0.46	ND	0.0020	0.48	[49]
	Middle and lower reaches	14	17	0.24	0.36	ND	1.9	ND	[9]
Haihe River	Summer Palace	0.34	0.26	0.062	0.0060	0.0060	0.019	0.26	[50]
	Guanting reservoir	0.30	0.087	0.056	ND	0.48	0.017	0.31	[50]
	Shichahai	0.066	0.24	0.081	0.0090	0.19	0.019	0.18	[50]
Liao River	Pu River	1.2	1.0	0.66	0.16	ND	ND	11	[51]
	Xi River	2.2	0.70	0.46	0.49	ND	ND	3.1	[51]

ND: Not Detected.

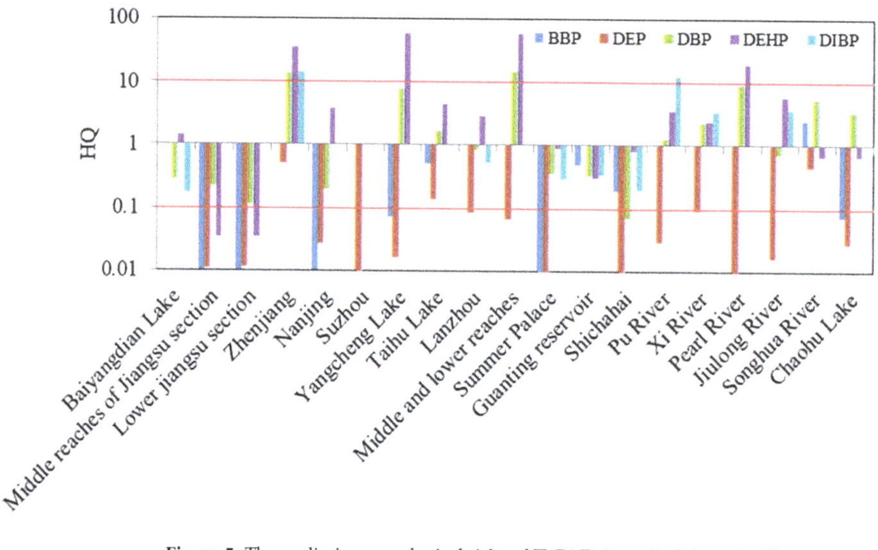

Figure 5. The preliminary ecological risks of $\sum_5$PAEs in typical rivers in China (in water).

The PERA method was used to evaluate the ecological risks of PAEs in typical rivers, and the JPCs are shown in Figure 4. It can be seen from the figure that the JPCs of BBP and DEP were closer to the *x*-axis than the DEHP and DBP, indicating that the potential ecological risks of DEHP and DBP were higher than BBP and DEP, which also indicated that aquatic organisms were affected by DEHP and DBP obviously, and they led to a higher probability of ecological risk. The results of PERAs in typical rivers in China showed that the probabilities of DEHP, DBP and BBP affecting 5% aquatic organisms by JPCs were

44.2%, 8.6% and 3.0%, respectively, while DEP was 0, meaning they affect less than 5% of the aquatic organisms.

3.4. The Ecological Risk Assessments of PAEs in Sediments

SQCs derived by the equilibrium partitioning method based on the ALCs of PAEs and the exposure data of DBP, DEHP, DMP, DEP, BBP, DOP and DIBP in sediments are listed in Tables 1 and 3 respectively. The exposure data were all the maximum values of PAEs.

Table 3. Part of PAEs in sediments from typical rivers in China (μg/gdw).

Rivers	Concentration (μg/L)							
	DMP	DEP	DBP	BBP	DEHP	DOP	DIBP	Reference
Songhua River	0.00300	0.0170	0.852	0.00500	6.56	0.0420	ND	[32]
Jiulong River	0.00400	0.00600	0.230	ND	1.28	ND	0.140	[34]
Taihu Lake	3.50	2.29	1.75	1.30	4.77	16.2	ND	[45]
Tongding River	0.0210	0.0540	0.165	ND	1.09	0.0200	0.750	[52]
Yellow River	1.04	1.12	72.2	ND	258	ND	ND	[9]
Yangtze River	2.24	1.24	246	ND	221	ND	ND	[53]
JiangHan Plain	0.238	1.87	0.290	0.0820	0.596	ND	0.639	[54]
Qiantang River	0.179	0.218	0.241	0.0210	6.24	0.0190	0.769	[55]
Pearl River	1.75	0.180	4.66	0.160	8.53	0.310	ND	[49]
Taiwan river	ND	1.10	30.3	1.80	23.9	ND	ND	[56]
Taiwan's rivers	ND	ND	1.30	3.10	46.5	ND	ND	[57]
Kaohsiung Harbor, Taiwan	ND	ND	1.31	ND	34.8	0.600	ND	[58]
Xi river	0.266	0.197	2.43	ND	8.30	4.35	11.2	[2]
Pu river	0.0530	0.0600	0.304	ND	44.5	1.47	0.404	[2]

ND: Not Detected.

For example, the content of DMP in sediments from Taihu Lake Basin was as high as 3.50 μg/gdw, which was hundreds of times higher compared with Baiyangdian Lake. The maximum content of DMP, DEP and DOP appeared in Taihu Lake. The concentrations of DMP and DEP were similar and in an order of magnitude, while the concentration of DOP is relatively high, reaching 16.2 μg/gdw.

The ecological risk assessments of PAEs in sediments from typical rivers in China were carried out only by the HQ method. The HQs of each river are shown in Table S6. Among them, BBP pose low risk in sediments from Taihu Lake, and its HQ was 1.47, which was inconsistent with previous study [45]. The reason may be that the ALC obtained in this study based on the reproductive toxicity endpoint was stricter than that based on the non-lethal toxicity endpoint in previous studies. Meanwhile, there was a medium risk in sediments from Taiwan's rivers, and the other rivers had low or no risk. As a result of the lgK_{OW} of DEP being 2.38, less than 3, and its hydrophobicity being weak, it is not easy to be adsorbed in sediment, so its risk to benthic organisms is not considered. For the DBP, about 35.7% of the rivers posed low risk, 42.9% of the rivers posed medium risk, and only 21.4% posed high risk. Risk posed by DEHP was high at 57.1% rivers, except for the JiangHan Plain, while other rivers posed a medium risk. DIBP posed a low risk in most rivers, and their HQs were within the range of 0.1–1, and only the Kaohsiung Harbor in Taiwan posed a medium risk. The preliminary ecological risks of $\sum_5$PAEs in sediments from typical rivers in China by HQ method are shown in Figure 6.

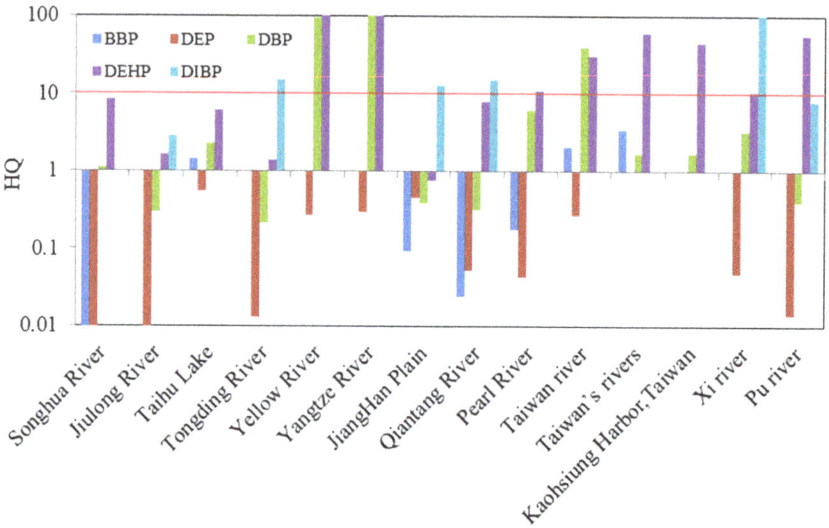

Figure 6. The preliminary ecological risks of $\sum_5$PAEs in typical rivers in China (in sediments).

4. Discussion

Compared with other typical rivers in Table 2, only DBP, DEHP and DIBP were detected in Baiyangdian Lake. Among them, the exposure concentration of DBP in Baiyangdian Lake was much lower than that in the Zhenjiang section of the Yangtze River basin, Yangcheng Lake, the middle and lower reaches of the Yellow River and the Pearl River, and there was an order of magnitude difference. It was equivalent to the exposure concentrations in the middle and lower reaches of the Jiangsu section of the Yangtze River, Nanjing section of the Yangtze River, Summer Palace and Guanting Reservoir. Only Shichahai exposure concentration was lower than the DBP in Baiyangdian Lake. In conclusion, the exposure concentration of DBP in Baiyangdian Lake was at a medium to low level compared with other typical rivers in China. Similar to DBP, the exposure concentration of DEHP in Baiyangdian Lake was also much lower than that in the Zhenjiang section of the Yangtze River, Yangcheng Lake and the middle and lower reaches of the Yellow River. In addition, the exposure concentration was similar to those in the Lanzhou section of the Yellow River, Summer Palace, Shichahai and Xi River. However, different from DBP, the exposure concentrations of DEHP in the middle and lower reaches of the Jiangsu section of the Yangtze River, Guanting Reservoir and Chao Lake were lower than that in Baiyangdian Lake, indicating that the concentration of DEHP in Baiyangdian Lake was at a medium to low level. Therefore, the pollution degree of PAEs is at a lower level in Baiyangdian Lake compared with other domestic water bodies, and the main pollutant is DEHP, which is followed by DBP. This result is consistent with the study of Taihu, which shows that DBP and DEHP were the most abundant PAE congeners in surface water [45], and DBP and DEHP have a certain homology. Although the concentration of DIBP was detected in Baiyangdian Lake, the concentration was very low.

For the rivers near or passing through industrial cities, such as the Zhenjiang section of the Yangtze River and Pu River, a tributary of the Liao River, the exposure concentrations of DIBP were relatively high, while in other rivers they were relatively low, and it was even not detected in a few rivers. A large number of DBP, DEHP and DIBP entered the water with surface runoff or atmospheric wet deposition. Because of their strong hydrophobicity, they often combine with particulate matter, settle in the bottom sediment and accumulate continuously, resulting in the high content of DBP, DEHP and DIBP in typical rivers sediments in China. The highest concentrations of DBP and DEHP occurred in the Yangtze

River and the Yellow River, respectively, and their concentrations can be as high as several hundred μg/gdw. DIBP has also been widely used as a substitute for DBP in recent years, which had increased the content in sediment. According to previous studies, the highest concentration appears in the Xihe River [59]. Although the production and consumption of these PAEs in daily life were very low, due to the developed agriculture around Taihu Lake, a large amount of agricultural runoff flowed into the lake, resulting in the high content of PAEs in sediment.

The ecological risk assessments of DEHP and DBP in Baiyangdian Lake were carried out by HQ and PERA. The results of the two methods were consistent, indicating that DEHP and DBP in Baiyangdian Lake water had potential ecological risk to aquatic organisms and regional ecosystem. However, the ecological risks of DEHP and DBP are evaluated based on Chinese environmental quality standards for surface water (the WQS of DEHP was 8 μg/L and DBP was 3 μg/L), the range of HQs were 0.0240–0.104 and 0.0400–0.153, respectively, indicating that there were no ecological risks of DEHP and DBP in most sampling sites of Baiyangdian Lake, and only a few sampling sites had low risk. Therefore, according to the Chinese current environmental quality standards, the ecological risks of PAEs in some rivers would be underestimated. The HQ method and PERA method were adopted for DEHP and DBP, and the results were consistent. In other words, both DEHP and DBP posed potential ecological risk in typical rivers. It is consistent with the results of the Liao River, and the study show that the ecological risk of DEHP in Liao River should be paid more attention [59]. So, it was necessary to strengthen the management and control of these two PAEs. DEHP values of 80.95 of the sediment samples of Songhua river exceeded the low effects range [32], and DEHP posed high ecological risk in typical river sediments, which was followed by DBP and DIBP. These studies have similar results, indicating that PAEs are potentially harmful to the aquatic environment.

5. Conclusions

The exposure concentrations of typical PAEs in Baiyangdian Lake were measured by field sampling in this study, and the results showed that the detection rates of DBP, DEHP and DIBP were higher. The PAEs exposure data in typical rivers and their sediments in China were obtained through published literature. Subsequently, according to the ALCs of typical PAEs derived from reproduction toxicity endpoints, the ecological risks of typical PAEs in Baiyangdian Lake water and typical rivers were evaluated by the HQ method and PERA method. Based on SQCs, which were derived using the equilibrium partitioning method by the ALCs of PAEs, only the HQ method was used to evaluate the ecological risks of typical PAEs in sediments from typical rivers. The results of the HQ method and PERA method showed that in terms of water, DBP and DIBP posed low risk in Baiyangdian Lake, but the risk of DEHP was slightly higher and cannot be ignored. Different from Baiyangdian Lake, in addition to the high ecological risk of DEHP in typical rivers in China, the risks of DBP and DIBP also cannot be ignored. Furthermore, the HQ method was used to evaluate the sediment risk about PAEs, and the results showed that DEHP posed a high ecological risk in typical rivers sediments, which was followed by DBP and DIBP. In conclusion, the ALCs of DEHP and DBP obtained in this study can be used to evaluate the ecological risks of Baiyangdian Lake and typical rivers more accurately. Meanwhile, in this study, the pollution status of PAEs in Baiyangdian was investigated, the pollution level, spatial distribution characteristics and sources of PAEs in typical lake were explored, and the ecological risks under their exposure levels were measured, so as to provide technical support for pollution prevention and environmental risk management in typical lakes worldwide.

Supplementary Materials: The following supporting information can be downloaded at: https://www.mdpi.com/article/10.3390/toxics11020180/s1, Table S1: Characterization of 7 PAEs investigated; Table S2: Reproductive toxicity data used to construct SSDs of DEHP, BBP, DEP, and DBP; Table S3: The average chronic toxicity value and cumulative frequency of DEHP, BBP, DEP, and

DBP; Table S4: Exposure concentrations and HQ values of ∑3 PAEs in Baiyangdian Lake; Table S5: Summary of HQ values in water; Table S6: Summary of HQ values in sediments.

Author Contributions: Conceptualization, Y.H. and X.L.; Methodology, Y.H.; Software, Y.H.; Validation, M.T., C.L. and J.W.; Writing–original draft, Y.H.; Writing–review & editing, M.T. and X.Z.; Visualization, M.T., X.L., C.W. and Y.W.; Supervision, X.Z. and Y.W.; Project administration, X.Z. and Y.W.; Funding acquisition, X.Z. and Y.W. All authors have read and agreed to the published version of the manuscript.

Funding: This research was funded by [Xin Zheng] grant number [Grant No. 42007378 and 41991315] and [Yihong Wu] grant number [20374204D]. And the APC was funded by [Grant No. 42007378 and 41991315].

Institutional Review Board Statement: Not applicable.

Informed Consent Statement: Not applicable.

Data Availability Statement: Not applicable.

Conflicts of Interest: All authors of this article declare that they have no conflict of interest regarding the publication of this study.

References

1. Zhang, Z.-M.; Zhang, H.-H.; Zhang, J.; Wang, Q.-W.; Yang, G.-P. Occurrence, distribution, and ecological risks of phthalate esters in the seawater and sediment of Changjiang River Estuary and its adjacent area. *Sci. Total Environ.* **2018**, *619–620*, 93–102. [CrossRef]
2. Li, B.; Liu, R.; Gao, H.; Tan, R.; Zeng, P.; Song, Y. Spatial distribution and ecological risk assessment of phthalic acid esters and phenols in surface sediment from urban rivers in Northeast China. *Environ. Pollut.* **2016**, *219*, 409–415. [CrossRef]
3. Janjua, N.R.; Mortensen, G.K.; Andersson, A.-M.; Kongshoj, B.; Skakkebæk, N.E.; Wulf, H.C. Systemic Uptake of Diethyl Phthalate, Dibutyl Phthalate, and Butyl Paraben Following Whole-Body Topical Application and Reproductive and Thyroid Hormone Levels in Humans. *Environ. Sci. Technol.* **2007**, *41*, 5564–5570. [CrossRef]
4. Prasad, B.; Prasad, K.S.; Dave, H.; Das, A.; Asodariya, G.; Talati, N.; Swain, S.; Kapse, S. Cumulative human exposure and environmental occurrence of phthalate esters: A global perspective. *Environ. Res.* **2022**, *210*, 112987. [CrossRef]
5. Hajiouni, S.; Mohammadi, A.; Ramavandi, B.; Arfaeinia, H.; De-La-Torre, G.E.; Tekle-Röttering, A.; Dobaradaran, S. Occurrence of microplastics and phthalate esters in urban runoff: A focus on the Persian Gulf coastline. *Sci. Total Environ.* **2021**, *806*, 150559. [CrossRef] [PubMed]
6. Nikolopoulou, V.; Alygizakis, N.A.; Nika, M.C.; Oswaldova, M.; Oswald, P.; Kostakis, M.; Slobodnik, J. Screening of legacy and emerging substances in surface water, sediment, biota and groundwater samples collected in the Siverskyi Donets River Basin employing wide-scope target and suspect screening. *Sci. Total Environ.* **2022**, *805*, 150253. [CrossRef] [PubMed]
7. Zhang, D.; Liu, W.; Wang, S.; Zhao, J.; Xu, S.; Yao, H.; Wang, H.; Bai, L.; Wang, Y.; Gu, H.; et al. Risk assessments of emerging contaminants in various waters and changes of microbial diversity in sediments from Yangtze River chemical contiguous zone, Eastern China. *Sci. Total Environ.* **2021**, *803*, 149982. [CrossRef] [PubMed]
8. Chai, X.L.; Ji, R.; Wu, J.; Tong, H.H.; Zhao, Y.C. Abiotic association of PAEswith humic substances and its influence on the fate of PAEs in landfillleachate. *Chemosphere* **2010**, *78*, 1362–1367.
9. Sha, Y.J.; Xia, X.H.; Yang, Z.F.; Huang, G.H. Distribution of PAEs in the middle and lower reaches of the YellowRiver, China. *Environ. Monit. Assess.* **2007**, *124*, 277–287. [CrossRef] [PubMed]
10. Wang, P.; Wang, S.L.; Fan, C.Q. Atmospheric distribution of particulate and gasphase phthalic esters (PAEs) in a Metropolitan City, Nanjing, East China. *Chemosphere* **2008**, *72*, 1567–1572. [CrossRef] [PubMed]
11. Qiu, Y.-W.; Wang, D.-X.; Zhang, G. Assessment of persistent organic pollutants (POPs) in sediments of the Eastern Indian Ocean. *Sci. Total Environ.* **2020**, *710*, 136335. [CrossRef] [PubMed]
12. Le, T.M.; Nguyen, H.M.N.; Nguyen, V.K.; Nguyen, A.V.; Vu, N.D.; Yen, N.T.H.; Hoang, A.Q.; Minh, T.B.; Kannan, K.; Tran, T.M. Profiles of phthalic acid esters (PAEs) in bottled water, tap water, lake water, and wastewater samples collected from Hanoi, Vietnam. *Sci. Total Environ.* **2021**, *788*, 147831. [CrossRef] [PubMed]
13. Selvaraj, K.K.; Sundaramoorthy, G.; Ravichandran, P.K.; Girijan, G.K.; Sampath, S.; Ramaswamy, B.R. Phthalate esters in water and sediments of the Kaveri River, India: Environmental levels and ecotoxicological evaluations. *Environ. Geochem. Health* **2015**, *37*, 83–96. [CrossRef] [PubMed]
14. Liu, N.; Wang, Y.; Yang, Q.; Lv, Y.; Jin, X.; Giesy, J.P.; Johnson, A.C. Probabilistic assessment of risks of diethylhexyl phthalate (DEHP) in surface waters of China on reproduction of fish. *Environ. Pollut.* **2016**, *213*, 482–488. [CrossRef]
15. Zhang, Q.Q.; Ying, G.G.; Pan, C.G.; Liu, Y.S.; Zhao, J.L. Comprehensive Evaluation of Antibiotics Emi-ssion and Fate in the River Basins of China: Source Analysis, Multimedia Modeling, and Linkage to Bacterial Resistance. *Environ. Sci. Technol.* **2015**, *49*, 6772–6782. [CrossRef]

16. Kamrin, M.A. Phthalate risks, phthalate regulation, and public health: A review. *Toxicol. Environ. Health B Crit. Rev.* **2009**, *12*, 157–174. [CrossRef] [PubMed]
17. Yuan, S.-Y.; Huang, I.-C.; Chang, B.-V. Biodegradation of dibutyl phthalate and di-(2-ethylhexyl) phthalate and microbial community changes in mangrove sediment. *J. Hazard. Mater.* **2010**, *184*, 826–831. [CrossRef]
18. US EPA. Priority Pollutants. 2014. Available online: http://water.epa.gov/scitech/methods/cwa/pollutants.cfm (accessed on 1 September 2021).
19. *GB 3838-2002; PRC-NS*; Environmental Quality Standard for Surface Water. Ministry of Environmental Protection of the People's Repubic of China and General Administration of Quality Supervision: Beijing, China; China Environmental Science Press: Beijing, China, 2022.
20. Wu, F.; Meng, W.; Zhao, X.; Li, H.; Zhang, R.; Cao, Y.; Liao, H. China Embarking on Development of its Own National Water Quality Criteria System. *Environ. Sci. Technol.* **2010**, *44*, 7992–7993. [CrossRef] [PubMed]
21. Wheeler, J.R.; Grist, E.P.; Leung, K.M.; Morritt, D.; Crane, M. Species sensitivity distributions: Data and model choice. *Mar. Pollut. Bull.* **2002**, *45*, 192–202. [CrossRef] [PubMed]
22. Jiang, M.; Li, Y.; Zhang, B.; Zhou, A.; Zhu, Y.; Li, J.; Xu, S. Urinary concentrations of phthalate metab-olites associated with changes in clinical hemostatic and hematologic parameters in pr-egnant women. *Environ. Int.* **2018**, *120*, 34–42. [CrossRef]
23. Zhang, B.; Zhang, T.; Duan, Y.; Zhao, Z.; Huang, X.; Bai, X.; Xie, L.; He, Y.; Ouyang, J.; Yang, Y.; et al. Human exposure to phthalate esters associated with e-waste dismantling: Exposure levels, sources, and risk assessment. *Environ. Int.* **2019**, *124*, 1–9. [CrossRef]
24. Caldwell, D.J.; Mastrocco, F.; Hutchinson, T.H.; Länge, R.; Heijerick, D.; Janssen, C.; Anderson, P.D.; Sumpter, J.P. Derivation of an Aquatic Predicted No-Effect Concentration for the Synthetic Hormone, 17α-Ethinyl Estradiol. *Environ. Sci. Technol.* **2008**, *42*, 7046–7054. [CrossRef] [PubMed]
25. Jin, X.; Wang, Y.; Jin, W.; Rao, K.; Giesy, J.P.; Hollert, H.; Richardson, K.L.; Wang, Z. Ecological Risk of Nonylphenol in China Surface Waters Based on Reproductive Fitness. *Environ. Sci. Technol.* **2013**, *48*, 1256–1262. [CrossRef] [PubMed]
26. Martino-Andrade, A.J.; Chahoud, I. Reproductive toxicity of phthalate esters. *Mol. Nutr. Food Res.* **2010**, *54*, 148–157. [CrossRef] [PubMed]
27. US EPA. *Guidelines for Ecological Risk assessment.Ecological Risk Assessment Step 2*; U.S. EPA: Washington, DC, USA, 1998.
28. Solomon, K.; Giesy, J.; Jones, P. Probabilistic risk assessment of agrochemicals in the environment. *Crop. Prot.* **2000**, *19*, 649–655. [CrossRef]
29. Wang, X.; Tao, S.; Dawson, R.; Xu, F. Characterizing and comparing risks of polycyclic aromatic hydrocarbons in a Tianjin wastewater-irrigated area. *Environ. Res.* **2002**, *90*, 201–206. [CrossRef]
30. Giesy, J.P.; Solomon, K.R.; Coats, J.R.; Dixon, K.R.; Giddings, J.M.; Kenaga, E.E. Chlorpyrifos: Ecological risk assessment in North American aquatic environments. *Rev. Environ. Contam. Toxicol.* **1999**, *160*, 1–129. [PubMed]
31. Paluselli, A.; Kim, S.K. Horizontal and vertical distribution of phthalates acid ester (PAEs) in seawater and sediment of East China Sea and Korean South Sea: Traces of plastic debris? *Mar. Pollut. Bull.* **2020**, *151*, 110831. [CrossRef] [PubMed]
32. Yan, Z.; Pan, J.; Gao, F.; An, Z.; Liu, H.; Huang, Y.; Wang, X. Seawater quality criteria derivation and ecological risk assessment for oil pollution in China. *Mar. Pollut. Bull.* **2019**, *142*, 25–30. [CrossRef]
33. MEP. *Ministry of Ecology and Environment of the People's Republic of China, HJ 831- 2022: Technical Guideline for Deriving Water Quality Criteria for the Protection of Freshwater Aquatic Organisms*; China Environmental Press: Beijing, China, 2022.
34. Li, R.; Liang, J.; Gong, Z.; Zhang, N.; Duan, H. Occurrence, spatial distribution, historical trend and ecological risk of phthalate esters in the Jiulong River, Southeast China. *Sci. Total Environ.* **2017**, *580*, 388–397. [CrossRef]
35. Jin, X.; Wang, Y.; Giesy, J.P.; Richardson, K.L.; Wang, Z. Development of aquatic life criteria in China: Viewpoint on the challenge. *Environ. Sci. Pollut. Res.* **2014**, *21*, 61–66. [CrossRef]
36. Lemly, A. Evaluation of the Hazard Quotient Method for Risk Assessment of Selenium. *Ecotoxicol. Environ. Saf.* **1996**, *35*, 156–162. [CrossRef]
37. *European Commission Technical Guidance on Risk Assessment in Support of Commission Directive 93/67/EEC on Risk Assessment for New Notified Substances Commission Regulation (EC) No 1488/94[R]*; Office for Official Publications of the European Communities: Luxembourg, 2003; pp. 1–337.
38. Shi, R.; Yang, C.; Su, R.; Jin, J.; Chen, Y.; Liu, H.; Giesy, J.P.; Yu, H. Weighted species sensitivity distribution method to derive site-specific quality criteria for copper in Tai Lake, China. *Environ. Sci. Pollut. Res.* **2014**, *21*, 12968–12978. [CrossRef]
39. Wang, Y.; Zhang, L.; Meng, F.; Zhou, Y.; Jin, X.; Giesy, J.P.; Liu, F. Improvement on species sensitivity distribution methods for deriving site-specific water quality criteria. *Environ. Sci. Pollut. Res.* **2014**, *22*, 5271–5282. [CrossRef] [PubMed]
40. Xia, L.L.; Liu, R.Z.; Zao, Y.W. Correlation analysis of landscape pattern and water quality in Baiyangdian watershed. *Procedia Environ. Sci.* **2012**, *13*, 2188–2196. [CrossRef]
41. Jin, D.; Kong, X.; Li, Y.; Bai, Z.; Zhuang, G.; Zhuang, X.; Deng, Y. Biodegradatio-n of di-n-Butyl Phthalate by Achromobacter sp. Isolated from Rural Domestic Wastewater. *Int. J. Environ. Res. Health* **2015**, *12*, 13510–13522. [CrossRef]
42. Li, X.; Yin, P.; Zhao, L. Phthalate esters in water and surface sediments of the Pearl River Estuary: Distribution, ecological, and human health risks. *Environ. Sci. Pollut. Res.* **2016**, *23*, 19341–19349. [CrossRef] [PubMed]
43. Gao, D.; Li, Z.; Wen, Z.; Ren, N. Occurrence and fate of phthalate esters in full-scale domestic wastewater treatment plants and their impact on receiving waters along the Songhua River in China. *Chemosphere* **2014**, *95*, 24–32. [CrossRef]

44. He, W.; Qin, N.; Kong, X.; Liu, W.; He, Q.; Ouyang, H.; Yang, C.; Jiang, Y.; Wang, Q.; Yang, B.; et al. Spatio-temporal distributions and the ecological and health risks of phthalate esters (PAEs) in the surface water of a large, shallow Chinese lake. *Sci. Total Environ.* **2013**, *461–462*, 672–680. [CrossRef] [PubMed]
45. He, H.; Hu, G.J.; Sun, C.; Chen, S.L.; Yang, M.N.; Li, J.; Zhao, Y.; Wang, H. Trace analysis of persistent toxic substances in the main stream of Jiangsu section of the Yangtze River, China. *Environ. Sci. Pollut. Res.* **2011**, *18*, 638–648. [CrossRef]
46. Chen, H.; Mao, W.; Shen, Y.; Feng, W.; Mao, G.; Zhao, T.; Wu, X. Distribution, source, and environmental risk assessment of phthalate esters (PAEs) in water, suspended particulate matter, and sediment of a typical Yangtze River Delta City, China. *Environ. Sci. Pollut. Res. Int.* **2019**, *26*, 24609–24619. [CrossRef]
47. Zhang, L.; Dong, L.; Ren, L.; Shi, S.; Zhou, L.; Zhang, T.; Huang, Y. Concentration and source identification of polycyclic aromatic hydrocarbons and phthalic acid esters in the surface water of the Yangtze River Delta, China. *J. Environ. Sci.* **2012**, *24*, 335–342. [CrossRef] [PubMed]
48. Gao, X.; Li, J.; Wang, X.; Zhou, J.; Fan, B.; Li, W.; Liu, Z. Exposure and ecological risk of phthalate esters in the Taihu Lake basin, China. *Ecotoxicol. Environ. Saf.* **2019**, *171*, 564–570. [CrossRef] [PubMed]
49. Zhao, X.; Shen, J.M.; Zhang, H.; Li, X.; Chen, Z.L.; Wang, X.C. The occurrence and spatial distribution of phthalate esters (PAEs) in the Lanzhou section of the Yellow River. *Environ. Sci. Pollut. Res.* **2020**, *27*, 19724–19735. [CrossRef]
50. Zheng, X.; Zhang, B.-T.; Teng, Y. Distribution of phthalate acid esters in lakes of Beijing and its relationship with anthropogenic activities. *Sci. Total Environ.* **2014**, *476–477*, 107–113. [CrossRef]
51. Li, B.; Hu, X.; Liu, R.; Zeng, P.; Song, Y. Occurrence and distribution of pht-halic acid esters and phenols in Hun River Watersheds. *Environ. Earth Sci.* **2015**, *73*, 5095–5106. [CrossRef]
52. Wang, X.T.; Ma, L.L.; Sun, Y.Z.; Xu, X.B. Phthalate esters in sediments from Guanting Reservoir and the Yongding River, Beijing, People's Republic of China. *Bull. Environ. Contam. Toxicol.* **2006**, *76*, 799–806. [CrossRef]
53. Fan, W.; Xinghui, X.; Yujuan, S. Distribution of phthalic acid esters in Wuhan section of the Yangtze River, China. *J. Hazard. Mater.* **2008**, *154*, 317–324.
54. Liu, H.; Liang, H.; Liang, Y.; Zhang, D.; Wang, C.; Cai, H.; Shvartsev, S.L. Distribution of phthalate esters in alluvial sediment: A case study at JiangHan Plain, Central China. *Chemosphere* **2010**, *78*, 382–388. [CrossRef]
55. Sun, J.; Huang, J.; Zhang, A.; Liu, W.; Cheng, W. Occurrence of phthalate esters in sediments in Qiantang River, China and inference with urbanization and river flow regime. *J. Hazard. Mater.* **2013**, *248–249*, 142–149. [CrossRef]
56. Yuan, S.Y.; Liu, C.; Liao, C.S.; Chang, B.V. Occurrence and microbial degrada-tion of phthalate esters in Taiwan river sediments. *Chemosphere* **2002**, *49*, 1295–1299. [CrossRef]
57. Huang, P.-C.; Tien, C.-J.; Sun, Y.-M.; Hsieh, C.-Y.; Lee, C.-C. Occurrence of phthalates in sediment and biota: Relationship to aquatic factors and the biota-sediment accumulation factor. *Chemosphere* **2008**, *73*, 539–544. [CrossRef] [PubMed]
58. Chen, C.W.; Chen, C.F.; Dong, C.D. Distribution of Phthalate Esters in Sediments of Kaohsiung Harbor, Taiwan. *Taylor Fr. Group.* **2013**, *22*, 119–131. [CrossRef]
59. Zheng, X.; Yan, Z.; Liu, P.; Li, H.; Zhou, J.; Wang, Y.; Fan, J.; Liu, Z. Derivation of aquatic life criteria for four phthalate esters and their ecological risk assessment in Liao River. *Chemosphere* **2018**, *220*, 802–810. [CrossRef] [PubMed]

toxics

MDPI

Article

Substantial Changes in Selected Volatile Organic Compounds (VOCs) and Associations with Health Risk Assessments in Industrial Areas during the COVID-19 Pandemic

Bhupendra Pratap Singh [1,2,*], Sayed Sartaj Sohrab [3,4], Mohammad Athar [5,6], Thamir A. Alandijany [3,4], Saumya Kumari [7], Arathi Nair [7], Sweety Kumari [7], Kriti Mehra [8], Khyati Chowdhary [8], Shakilur Rahman [9] and Esam Ibraheem Azhar [3,4]

1 Department of Environmental Studies, Deshbadhu College, University of Delhi, New Delhi 110019, India
2 Delhi School of Climate Change and Sustainability (Institute of Eminence), University of Delhi, New Delhi 110007, India
3 Special Infectious Agents Unit, King Fahd Medical Research Center, King Abdulaziz University, Jeddah 21589, Saudi Arabia
4 Department of Medical Laboratory Sciences, Faculty of Applied Medical Sciences, King Abdulaziz University, Jeddah 21589, Saudi Arabia
5 Science and Technology Unit, Umm Al-Qura University, Makkah 21955, Saudi Arabia
6 Department of Medical Genetics, Faculty of Medicine, Umm Al-Qura University, Makkah 21955, Saudi Arabia
7 Department of Zoology, Deshbandhu College, University of Delhi, New Delhi 110019, India
8 Department of Life Science, Deshbadhu College, University of Delhi, New Delhi 110019, India
9 Department of Medical Elementology and Toxicology, School of Chemical and Life Sciences, Jamia Hamdard, New Delhi 110019, India
* Correspondence: bpsingh@gmail.com

Abstract: During the COVID-19 pandemic, governments in many countries worldwide, including India, imposed several restriction measures, including lockdowns, to prevent the spread of the infection. COVID-19 lockdowns led to a reduction in gaseous and particulate pollutants in ambient air. In the present study, we investigated the substantial changes in selected volatile organic compounds (VOCs) after the outbreak of the coronavirus pandemic and associations with health risk assessments in industrial areas. VOC data from 1 January 2019 to 31 December 2021 were collected from the Central Pollution Control Board (CPCB) website, to identify percentage changes in VOC levels before, during, and after COVID-19. The mean TVOC levels at all monitoring stations were 47.22 ± 30.15, 37.19 ± 37.19, and 32.81 ± 32.81 μg/m^3 for 2019, 2020, and 2021, respectively. As a result, the TVOC levels gradually declined in consecutive years due to the pandemic in India. The mean TVOC levels at all monitoring stations declined from 9 to 61% during the pandemic period as compared with the pre-pandemic period. In the current study, the T/B ratio values ranged from 2.16 (PG) to 26.38 (NL), which indicated that the major pollutant contributors were traffic and non-traffic sources during the pre-pandemic period. The present findings indicated that TVOC levels had positive but low correlations with SR, BP, RF, and WD, with correlation coefficients (r) of 0.034, 0.118, 0.012, and 0.007, respectively, whereas negative correlations were observed with AT and WS, with correlation coefficients (r) of −0.168 and −0.150, respectively. The lifetime cancer risk (LCR) value for benzene was reported to be higher in children, followed by females and males, for the pre-pandemic, pandemic, and post-pandemic periods. A nationwide scale-up of this study's findings might be useful in formulating future air pollution reduction policies associated with a reduction in health risk factors. Furthermore, the present study provides baseline data for future studies on the impacts of anthropogenic activities on the air quality of a region.

Keywords: TVOCs; pandemic; T/B ratio; meteorological parameters; LCR

Citation: Singh, B.P.; Sohrab, S.S.; Athar, M.; Alandijany, T.A.; Kumari, S.; Nair, A.; Kumari, S.; Mehra, K.; Chowdhary, K.; Rahman, S.; et al. Substantial Changes in Selected Volatile Organic Compounds (VOCs) and Associations with Health Risk Assessments in Industrial Areas during the COVID-19 Pandemic. *Toxics* **2023**, *11*, 165. https://doi.org/10.3390/toxics11020165

Academic Editors: Zhen-Guang Yan, Zhi-Gang Li and Jinzhe Du

Received: 6 January 2023
Revised: 1 February 2023
Accepted: 6 February 2023
Published: 9 February 2023

1. Introduction

Coronavirus disease (COVID-19) is a disease caused by severe acute respiratory syndrome coronavirus 2 (SARS-CoV-2), and was first reported in Wuhan, China, in late December 2019 [1]. It rapidly spread across the world in a short time span, and the World Health Organization declared it a pandemic on 11 March 2020 [2]. The WHO and other agencies reported that, as of 5 January 2023, COVID-19 infected more than 66 million people, with more than 6.7 million deaths globally (including in India) [3]. The United States of America (USA) was the most adversely affected country, followed by India, the second most afflicted country in the world, with more than 10 million COVID cases and more than 1.1 million deaths reported as of 5 January 2023 [4].

Few studies have reported a relationship between air pollution and infectious disease transmission [5,6]. Several early pieces of evidence suggest that the link between prolonged exposure to air pollution and the impact of COVID-19 might increase the probability of severe outcomes [7–10]. Many researchers suggested that air pollutants may influence the severity of COVID-19 associated with respiratory infection [11], cardiovascular disease [12], as well as morbidity and mortality [13,14]. Among the air pollutants, volatile organic compounds (VOCs) are considered principal components and are often designated as specific hazardous or toxic air pollutants [15,16]. These VOCs also play a crucial role in forming tropospheric ozone and secondary pollutants through photochemical smog [17–20]. In many cities worldwide, significant reductions in atmospheric pollutant concentrations were observed during the lockdown periods of COVID-19 due to the complete or partial closures of industries, as well as transport and construction works [21–23]. It is very difficult to assess the air quality with respect to the contributions of different pollutants, and changes in individual pollutant levels are difficult to link to overall air pollution; therefore, it is difficult to compare their impacts on human health associated with concentrations of different pollutants [24,25]. Volatile organic compounds (VOCs) and nitrogen oxides (NOx) are the two main forerunners to tropospheric O_3 formation, with complex chemical mechanisms found in them [26,27]. The photochemical process depends upon the VOC to NOx ratio in the atmosphere, which has a pivotal role in O_3 formation [28].

Up to 60% of non-methane VOCs (NMVOCs) released into the atmosphere are BTEX [28]), and changes in BTEX ratios can be used as effective tools for investigating the causes of different photochemical processes that occur in the environment [17,29]. Traffic-related VOCs and VOCs released by industries as well as changes in VOC levels from many individual sources have been investigated to estimate the impact of the COVID-19 lockdown on the environment. Additionally, in some particular compounds, such as benzene, toluene, ethyl benzene, and xylene (represented by the acronym BTEX), some harmful effects of VOCs have been shown via short- and long-term adverse health effects. Furthermore, VOCs that are released into the ambient environment from various sources, such as oil and gas, play very crucial and important roles in petrochemical activities [16]. BTEX compounds are considered to be the main components of gasoline, and, due to their high evaporation rate, they can enter the ambient air environment from outer exhausts, vehicle carburetor engines, and petroleum product distribution stations [30]. Emission intensities of pollution sources and meteorological conditions play important roles in varying VOC levels, while in the present scenario, meteorological conditions significantly influence the chemical transformations involved in the production of O_3 concentrations [27]. During COVID-19, there were many studies under highly unusual conditions of partial or total internment and, therefore, vital information could be acquired for designing policies and strategies to prevent and control air pollution through evaluations of the effects of reduced emission sources on the local urban air quality [21–23,31].

Changes in atmospheric pollutants during the COVID-19 lockdown periods have been widely investigated; researchers have reported significant reductions in nitrogen dioxide (NO_2), particulate matter (PM), and carbon monoxide (CO) levels in many different cities across the world [10,23,24]. Subali et al. (2021) revealed a potential VOC-based breath analysis associated with high sensitivity and promising specificity for COVID-

19 screening [32]. Another study conducted in Maharashtra (India) reported that total VOC levels decreased during the lockdown periods in the corresponding year, 2019 [33]. However, due to shallower boundary layer depths, higher concentrations of aromatic volatile organic compounds (VOCs) and CO were found in the wintertime and transported from the polluted Indo-Gangetic Plain region. Relatively high loadings of benzene (~30%), toluene (45%), and CO (32%), respectively, were observed in vehicle exhaust by using the positive matrix factorization analysis method [34].

According to Ghaffari et al. (2021), the most toxic BTEX compound is benzene, which has been categorized as a Group 1 and class A human carcinogenic by the International Agency for Research on Cancer (IARC) and the United States Environmental Protection Agency (U.S. EPA), respectively [35]. Several studies have reported that individual VOCs are significantly associated with the adverse effects of cardiovascular and respiratory diseases [36,37] asthma [38], and chronic obstructive pulmonary disorder [39], and could possibly increase the chances of leukemia and aplastic [40]. Several studies have claimed that a considerably high concentration of benzene was associated with a high cancer risk for lifetime exposure in an ambient environment [41,42].

Changes in VOC levels due to particle and gaseous contaminants, in particular, have received more attention and several studies of various cities in India have revealed that the air quality improved significantly during the pandemic period [23–25]. Nevertheless, only a few studies have discussed links between BTEX compounds and health. In addition, there is a paucity of thorough research on BTEX compounds as well as the pandemic's health risks, and BTEX compounds during lockdown periods in North India have not been examined in any prior study. The key aims of the present study are: (i) to evaluate the spatiotemporal variations in TVOC levels, (ii) to identify the sources of BTEX, and (iii) to calculate the health risks associated with BTEX across various age groups.

2. Methodology

Study Area

The National Capital Territory of Delhi, India, has coordinates of 28.70° N and 77.10° E. It is situated on the Indo-Gangetic Plain of the northern region of India [24,25]. One report suggests that the National Capital Territory has subtropical and semi-arid climatic conditions. The area experiences all seasons, including summer, monsoon, and winter from April to June, from July to October, and from November to February, respectively. The climate of Delhi is humid and is greatly impacted by the annual monsoon. The average temperature from May to June is 35–40 °C, and from November to February it is 5–7 °C. The humidity is mostly felt during the months of July and August. Usually, there is a northeastern breeze in Delhi, but during the late summers, it is replaced by a southeastern wind. The sampling locations for all monitoring sites are presented in Figure 1.

Delhi is among the most polluted cities in India according to the index of global pollution [43]. Delhi is also one of the most populated cities in India, with 13.4 million registered vehicles on the roads [44]. In addition, Delhi's Metropolitan area has a huge number of public and private transportation vehicles compared with other Indian cities. Therefore, industrial areas play crucial roles in enhancing the level of pollution. In addition, automobiles, construction, and other anthropogenic activities are important key factors. All of these factors lead to the emissions of various compounds, including carcinogenic volatile organic compounds (VOCs). In the present study, selected monitoring stations in Delhi are presented in Table 1.

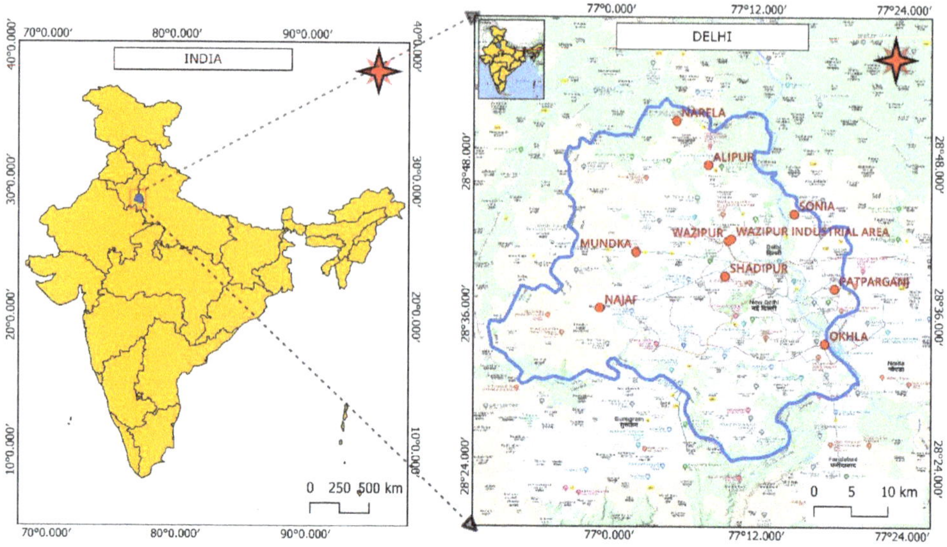

Figure 1. Sampling stations for all monitoring sites at Delhi.

Table 1. Different monitoring stations with latitude, longitude, and population census at Delhi.

S. No.	Monitoring Stations	Latitude (0 E)	Longitude (0 N)	Population Density (2021 Census) (Km^{-2})
1.	Alipur (AL)	77.1331	28.7972	6369
2	Bawana (BW)	77.0483	28.7932	6660
3	Mundaka (MD)	77.0349	28.6823	10,275
4	Najafgarh (NG)	76.9855	28.6090	5213
5	Narela (NL)	77.0892	28.8549	3071
6	Okhala (OKH)	77.2914	28.5626	31,087
7	Patparganj (PG)	77.3046	28.6347	22,088
8	Shadipur (SP)	77.1582	28.6516	23,942
9	SoniaVihar (SON)	77.2496	28.7332	5662
10	Wazirpur (WA)	77.1604	28.6975	24,908

3. Data and Sources

In the present study, hourly and daily data of volatile organic compounds, especially benzene, toluene, ethylene, and xylene (BTEX), were collected from the Central Pollution Control Board (CPCB) website (https://app.cpcbccr.com/ccr/#/caaqm-dashboard-all/caaqm-landing, accessed on 5 January 2022). The measurements and technical specifications of the instruments can be found elsewhere [45].

Several previous studies have reported that the CPCB provides data quality assurance and quality control (QA/QC) programs and detection limits of each BTEX compound through rigorous sampling, analysis, and calibration procedures [35].

In the present study, data were procured from 1 January 2019 to 31 December 2021, to identify percentage changes in the VOC levels (https://app.cpcbccr.com/ccr/#/caaqm-dashboard-all/caaqm-landing, accessed on 5 January 2023). The data were procured in three time periods before, during, and after COVID-19. To examine the relative and temporal changes in VOC levels in the ambient atmosphere, the time period between 1 January 2019 and 31 December 2019 represented the pre-pandemic period, the time period between 1 January 2020 and 31 December 2020 represented, the pandemic period, and the time period between 1 January 2021, and 31 December 2021, represented the post-pandemic period. All 10 industrial air quality monitoring stations in Delhi that were selected for this study, with their latitudes, longitudes, and population census data, are presented in Table 1. The monitoring stations are Alipur (AL), Bawana (BW), Mundka (MD), Najafgarh (NG), Narela (NL), Okhla (OKH), Patparganj (PG), Shadipur (SP), Sonia Vihar (SON), and Wazirpur (WA). Meteorological parameters, such as solar radiation (SR in kWh/m^2), pressure (BP in kg/ms^2), atmospheric temperature (AT in Celsius), rainfall (RF in mm), wind speed (WS in km/h), and wind direction (WD in degree/cardinal direction) were observed on an hourly basis at all 10 monitoring sites.

4. Human Health Risk Assessment

A human health risk assessment (HHRA) can be performed to assess the nature and probability of different pollutants in a population based on acute and chronic exposure.

4.1. Hazard Identification

The pollutants that cause major impacts on human health are considered hazardous. In this current study, VOCs such as BTEX are hazardous to human health and can cause cancer.

4.2. Exposure Assessment

An exposure assessment (EA) was performed to examine the duration and magnitude of the pollutants based on different parameters. In the present study, inhalation was the major route of exposure for the identified pollutants. We estimated the daily and annual readings of normal and acute exposure periods for different age groups, namely males (70 years), females (60 years), and children (36 years) [46]. The values of the parameters used in the health risk assessment model are presented in Table 2.

$$EC\left(\frac{\mu}{m^3}\right) = \frac{CA\left(\frac{\mu}{m^3}\right) \times ET\left(\frac{h}{day}\right) \times EFyear \times ED\left(\frac{day}{year}\right)}{AT\ (year) \times 365\left(\frac{day}{year}\right) \times 24\left(\frac{h}{day}\right)} \tag{1}$$

$$EDI\left(\frac{mg}{kg.day}\right) = \frac{CA\left(\frac{\mu}{m^3}\right) \times \left(\frac{1}{1000}\right)\left(\frac{mg}{\mu g}\right) \times IR\left(\frac{m3}{day}\right) \times ET\left(\frac{h}{day}\right) \times EFyear \times ED\left(\frac{day}{year}\right)}{AT\ (year) \times 365\left(\frac{day}{year}\right) \times 24\left(\frac{h}{day}\right) \times BW(kg)} \tag{2}$$

$$HQ = \frac{EC}{RfC} \tag{3}$$

$$ILCR = CDI \times SF \tag{4}$$

where EC ($\mu g/m^3$) represents the exposure concentration, defined as the number of TVOCs present per cubic meter; CA ($\mu g/m^3$) = VOC, the average concentrations of benzene, toluene, ethylene, and xylene; ET (h/d) is the exposure time, the total time duration per day in which exposure to TVOCs takes place; EF (d/y) represents exposure frequency, defined as the number of exposures taking place in a day;

Table 2. Correlation between the different monitoring stations for the pre-pandemic period.

	AL	BW	MD	NG	NL	OKH	PG	SP	SON	WA
AL	1									
BW	0.003	1								
MW	0.306 **	0.306 **	1							
NG	0.234 **	0.302 **	0.734 **	1						
NL	−0.006	0.434 **	0.286 **	0.448 **	1					
OKH	0.062	0.153 **	0.296 **	0.351 **	0.178 **	1				
PG	−0.140 **	0.081	0.116 *	0.019	0.116 *	0.094	1			
SP	0.192 **	0.206 **	0.441 **	0.317 **	0.325 **	0.139 **	0.310 **	1		
SON	0.227 **	0.148 **	0.545 **	0.337 **	0.036	0.167 **	0.222 **	0.106 *	1	
WA	−0.081	0.044	−0.005	−0.207 **	0.022	−0.109 *	0.236 **	0.301 **	0.048	1

** Correlation is significant at the 0.01 level (2-tailed). * Correlation is significant at the 0.05 level (2-tailed).

ED (y) represents the exposed length of working, the difference of the average age of exposure and the average age at the beginning; AT (h) is the average exposure time, during the carcinogenic assessment, the average lifetime (per capita life expectancy $\times$ 365 d/y $\times$ 24 h/d) was adopted, and during the non-carcinogenic assessment, the average period of exposure cycle (ED $\times$ 365 d/y $\times$ 24 h/d) was adopted; HQ ($\mu g/m^3$) is the hazard quotient, the ratio of exposure to chemicals and the measure at which no defined results can occur; RfC is the reference concentration of inhalation toxicity, which refers to continuous exposure to the human population without any cancerous health risks; SF (kg d mg^{-1}) represents the carcinogenic slope factor, defined as an upper bound, approximating a 95% confidence limit in the escalated cancer crisis from the lifetime exposure to a chemical [23]; ILCR (incremental lifetime cancer risk) refers to increasing the chances of any person having cancer due to exposure to a pollutant during his/her lifetime.

5. Results and Discussion

5.1. Total VOC Levels for 2019–2021

The current study focused on establishing the significant changes in air pollutants in different industrial zones, especially the total volatile organic compound (TVOC) levels, from 2019 to 2021, in Delhi, India. At all monitoring stations, the mean TVOC levels were 47.22 $\pm$ 30.15, 37.19 $\pm$ 37.19, and 32.81 $\pm$ 32.81 $\mu g/m^3$ for 2019, 2020, and 2021, respectively (Figure 2). The results show that the TVOC levels gradually deteriorated over successive years due to the pandemic in India. The main aspects behind the significant decrease in TVOC levels during the lockdown were complete and partial restrictions on transport, industrial activities, and marketplace openings. The annual mean TVOC levels at all monitoring stations ranged from 6.70 $\pm$ 4.71 to 103.86 $\pm$ 80.37, from 3.65 $\pm$ 7.36 to 97.57 $\pm$ 68.39, and from 5.21 $\pm$ 5.12 to 128.56 $\pm$ 74.43 $\mu g/m^3$, for 2019, 2020, and 2021, respectively. The trend of annual mean TVOC levels was observed to be BW > MD > NL > SON > WA > OKH > SP > AL > NG > PG for 2019, whereas the trend was NL > BW > OKH > SP > MD > SON > NG > AL > PG for 2020, and NL > OKH > SP > MD > BW > AL SON > WA > PG > NG for 2021.

The maximum TVOC levels were 347.27 $\mu g/m^3$ (BW), 408.91 $\mu g/m^3$ (NL), and 467.30 $\mu g/m^3$ (NL) for 2019, 2020, and 2021, respectively. The maximum TVOC levels at all stations varied from 23.40 $\mu g/m^3$ (PG) to 347.27 $\mu g/m^3$ (BW), from 75.06 $\mu g/m^3$ (SON) to 408.91 $\mu g/m^3$ (NL), and from 37.18 $\mu g/m^3$ (PG) to 467.30 $\mu g/m^3$ (NL) for 2019, 2020, and 2021, respectively.

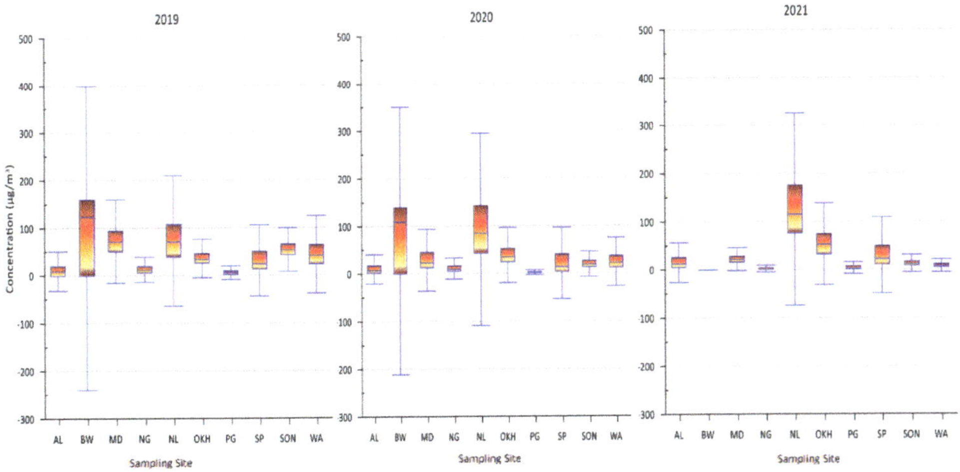

Figure 2. Box plot for TVOCs for different monitoring stations in Delhi for the years 2019, 2020, and 2021.

5.2. Identification of VOC Characteristic Pollutants for 2019

The BTEX concentrations for 2019–2021 at all selected monitoring regions are presented in Figure 2. The mean values for individual VOCs were 3.48 ± 2.43, 48.33 ± 22.86, 3.68 ± 12.29, and 3.7 ± 3.64 µg/m^3 for benzene, toluene, eth-benzene, and mp-xylene, respectively, for 2019. The average concentration values for benzene at all monitoring stations varied from 1.26 (AL) to 5.43 (SP) µg/m^3, whereas, for toluene, the average concentration values varied from 4.58 (PG) to 74.22 (MD) µg/m^3. The trend of the mean benzene value was observed to be SP > OKH > MD > SON > WA > NL > PG > NG > AL; the trend of the mean toluene value was NL > MD > SON > WA > OKH > SP > AL > NG > PG.

5.3. Identification of VOC Characteristic Pollutants for 2020

The mean values of benzene, toluene, eth-benzene, and mp-xylene at all monitoring stations were calculated to be 12.85 ± 9.42, 30.42 ± 19.06, 4.06 ± 7.06, and 8.60 ± 13.71 µg/m^3, respectively, for 2020–2021. The standard deviation value was high, indicating large variations in emission sources at various monitoring stations. The average benzene values at the monitoring stations ranged from 1.47 (AL) to 98.42 (NL) µg/m^3, whereas the average toluene values ranged from 2.39 (NG) to 145.22 (BW) µg/m^3, respectively. The maximum mean values among all monitoring stations were reported at Narela (benzene) and Bawana (toluene) monitoring stations, which are hubs of industrial zones in Delhi. Most plastic industries in BW operated during the pandemic period due to their association with the production of sanitizing bottles. These industries might have contributed more benzene and toluene compound emissions. Considering all of the selected monitoring stations, the trend for the average benzene values was NL > NG > SP > WA > SON > OKH > MD > AL > PG > BW, and the trend for the average toluene values was BW > OKH > MD > WA > SP > SON > AL > PG > NL > NG.

5.4. Identification of VOC Characteristic Pollutants for 2021

The mean values for individual VOCs at all monitoring stations were 2.89 ± 2.66, 43.01 ± 22.26, 505 ± 5.01, and 6.23 ± 8.14 µg/m^3 for benzene, toluene, eth-benzene, and mp-xylene, respectively, for 2021–2022. The average values for benzene at the monitoring stations ranged from 5.26 (SP) to 0.55 (PG), whereas the average values for toluene ranged between 164.91 (BW) and 0.91 (NG) µg/m^3. Considering all of the selected monitoring stations, the trend of average benzene values was SP > OKH > MD > SON > NL > WA >

AL NG > PG > PG, and the trend for average toluene values was BW > NL > OKH > SP > MD > AL > SON > WA PG > NG.

5.5. Comparative Analysis of Pre-Pandemic, Pandemic, and Post-Pandemic Periods

The average TVOC values at all of the monitoring stations declined from 9 to 61% during the pandemic period compared with those of the pre-pandemic period. The highest decline was observed at the SON monitoring station (−61%) and the lowest decline was at the NG monitoring station (−9%); the reason could be that the SON monitoring station was observed to restrict measures during the pandemic period, which caused significant changes in TVOC values compared with the pre-pandemic period (Figure 3). However, the NG monitoring station is considered to be India's second most pollutant cluster, with air and water in the critical category. On the one hand, most industrial activities were performed during the pandemic period, and there were insignificant changes in TVOC values compared with those of the pre-pandemic period. On the other hand, increased TVOC values were reported at OKH (24%) and NL (15%) monitoring stations during the pandemic period. The location of the OKH monitoring station is considered to be an industrial zone (waste-to-energy plant) where municipal solid wastes (generated from households) are used as fuel, which continued during the pandemic period [24,25]. Therefore, the increase in the level of VOCs reported by the OKH monitoring station was attributed to waste burning. Several previous studies have reported that higher source emissions could be attributed to local source emissions from burning waste and construction activities near a monitoring station [47].

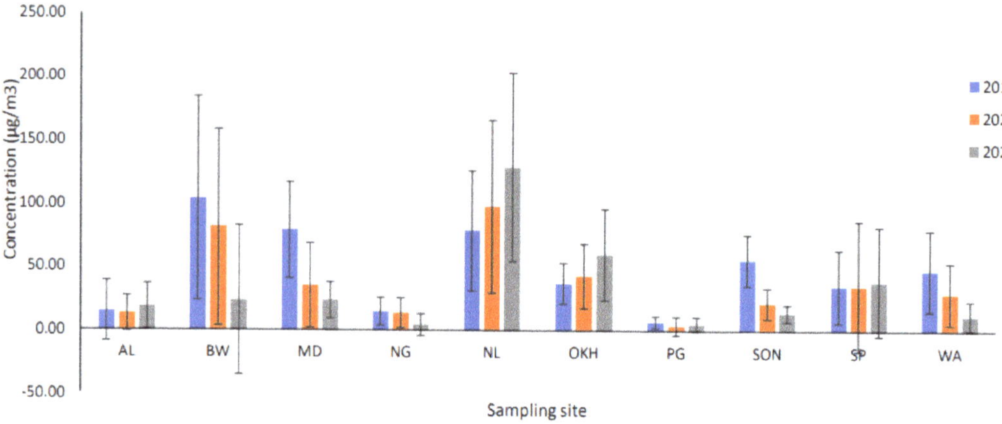

Figure 3. A comparative analysis of TVOCs at all sampling sites for the years 2019, 2020, and 2021.

The average TVOC values continued to decline even during the post-pandemic period due to restricted measures in a few of the monitoring stations. The government declared restricted measures for the post-pandemic period, under which schools, colleges, cinemas, and gyms were to remain closed, shops dealing in non-essential items were opened only on an odd–even basis, and metro trains and buses in the city ran at 50% seating capacity.

The decreases in TVOC levels varied from −77% (BW) to −22% (PG), whereas there were significant increases in TVOC levels with changes at 64%, 62%, and 11% at the NL, OKH, and SP monitoring stations. The highest increase in TVOC levels was reported at the NL monitoring station because the NL monitoring station is located near plastic industries for making shoe soles and other plastic goods, such as Rexine, adhesives, and other highly inflammable items, which could be a significant source of emissions during the reopening of industrial activities. Similarly, the OKH monitoring station witnessed a further increase in the amount of waste generated from domestic and industrial sectors, from BTEX pollutant sources that included plastics, paints, resins, rubber, adhesives, lubricants, and detergents [30].

A study in Maharashtra reported that TVOC levels declined by 84% during the lockdown period as compared with those of the previous year [23]. The average TOVC value at the PG monitoring station declined by 46% and 22% during the pandemic and post-pandemic periods, respectively, as compared with that of the pre-pandemic period. Various industrial sectors, such as paper, scraps of leather, and polythene, were located near the PG monitoring station, which attributed to source emissions. The decline in the average TVOC levels was higher during the pandemic (46%) compared with the post-pandemic period due to the reopening of these industrial activities. Additionally, the PG monitoring station is located near the Ghazipur landfill site, which further contributed to higher VOC levels in this monitoring station.

5.6. Source Identification

Identification and estimation of VOC emission sources can be assessed using diagnostic ratios. The toluene/benzene ratio can be used to evaluate the impact of traffic and non-traffic sources [23]. The ratio of toluene/benzene (T/B) is frequently used to inspect the relative importance of vehicular exhaust, industrial emissions, and combustion sources and to provide crucial insight into the vicinity of vehicular discharge sources and photochemically aged air masses [48]. A T/B ratio that is less than 2 indicates that vehicular emissions have a significant influence on aromatic VOC emissions. Several studies have reported that T/B ratios close to or more than 2 refer to non-traffic sources, and ratios higher than 10 indicate industrial activity as a considerable factor [23,48,49].

In the current study, the T/B ratio values ranged from 2.16 (PG) to 26.38 (NL), indicating that the major VOC contributors were the traffic and non-traffic sources during the pre-pandemic period. The calculated T/B ratios at some monitoring stations, such as the WZ, SON, AL, MD, and NL monitoring stations, were reported to be 10.44, 11.39, 12.00, 16.49, and 26.38, respectively, indicating the activities of industries and factories were the primary causes of VOC emissions (Figure 4). During 2020, the T/B ratios ranged from 0.03 (NL) to 13.47 (OKH), indicating that the major VOC contributors were traffic and non-traffic sources. Narela and Najafgarh had ratios of 0.03 and 0.21, respectively, stipulating that vehicular emissions were the main sources of VOC emissions. The T/B ratio during 2021 ranged from 3.54 (NG) to 42.10 (NL), where high ratios indicated non-traffic sources and much higher ratios indicated industries, factories, and petrol pumps as the predominant VOC contributors. The calculated T/B ratios at OKH and NL monitoring stations were 13.81 and 42.10, respectively, indicating that the activities of industries and factories were the predominant sources of VOC emissions.

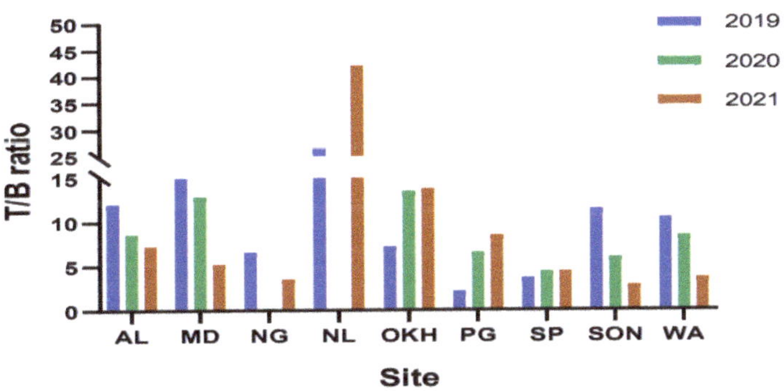

Figure 4. T/B ratio at all industrial monitoring stations for years 2019, 2020, and 2021.

5.7. Correlations among the Monitoring Stations

Correlations among VOC levels can help one to understand the source of origin of the different constituents. If the correlation between different pollutants is similar, it depicts that their source of origin might be the same. The correlation value was found by using the Pearson correlation coefficient for the mean VOC levels on a daily basis. These correlation data were classified into various categories depending on the coefficient value (from −1 to 1). A coefficient value from 0.8 to 1.0 indicated a very strong correlation, a coefficient value from 0.6 to 0.8 indicated a strong correlation, a coefficient value from 0.4 to 0.6 indicated a moderate correlation, a coefficient value from 0.2 to 0.4 indicated a weak correlation, and a coefficient value from 0 to 0.2 indicated irrelevant data [50]. In the present study, a very strong positive correlation was observed between the NG and MD monitoring stations (0.734) during the pre-pandemic period, whereas other monitoring stations reported moderate to low correlations. In a similar study in Maharashtra, the authors reported a significantly strong correlation between the Thane and Bandra monitoring stations (0.73) in the pre-lockdown period [19]. The correlations among the different monitoring stations are presented in Tables 2–4.

Table 3. Correlation between the different monitoring stations during the pandemic period.

	AL	BW	MW	NG	NL	OKH	PG	SP	SON	WA
AL	1									
BW	0.255 **	1								
MW	0.300 **	0.360 **	1							
NG	0.343 **	0.440 **	0.575 **	1						
NL	0.454 **	0.499 **	0.413 **	0.665 **	1					
OKH	0.554 **	0.343 **	0.373 **	0.598 **	0.628 **	1				
PG	0.141 **	0.107 *	0.129 *	0.139 **	0.066	0.092	1			
SP	0.736 **	0.340 **	0.422 **	0.550 **	0.547 **	0.619 **	0.173 **	1		
SON	0.468 **	0.356 **	0.515 **	0.536 **	0.347 **	0.434 **	0.194 **	0.530 **	1	
WA	0.080	0.187 **	0.096	−0.013	0.077	0.023	−0.067	0.041	−0.038	1

** Correlation is significant at the 0.01 level (2-tailed). * Correlation is significant at the 0.05 level (2-tailed).

Table 4. Correlation between the different monitoring stations for the post-pandemic period.

	AL	BW	MW	NG	NL	OKH	PG	SP	SON	WA
AL	1									
BW	0.497 **	1								
MW	−0.021	−0.076	1							
NG	0.349 **	0.646 **	0.021	1						
NL	0.130 *	0.157 **	0.119 *	0.273 **	1					
OKH	0.219 **	0.176 **	0.197 **	0.309 **	0.571 **	1				
PG	0.240 **	−0.004	0.094	0.094	0.354 **	0.725 **	1			
SP	0.654 **	0.553 **	0.027	0.426 **	0.254 **	0.287 **	0.292 **	1		
SON	0.565 **	0.513 **	0.054	0.546 **	0.377 **	0.612 **	0.534 **	0.617 **	1	
WA	0.572 **	0.813 **	0.016	0.648 **	0.112 *	0.255 **	0.105 *	0.587 **	0.618 **	1

** Correlation is significant at the 0.01 level (2-tailed). * Correlation is significant at the 0.05 level (2-tailed).

During the pandemic period, there was a strong positive correlation between the SP and AL monitoring stations (0.736) and between the SP and OKH monitoring stations (0.619). For the post-pandemic period, strong positive correlations were shown for the BW and WA (0.813), BW and NG (0.646), PG and OKH (0.725), SP and AL (0.654), OKH and SON (0.612), SON and SP (0.617), and WZ and NG (0.648) monitoring stations. Apart from this, various monitoring stations showed moderate correlations, such as WA and AL (0.572), and WA and SP (0.587) monitoring stations.

5.8. Correlations between the TVOC Levels and Meteorological Parameters

In the present study, we investigated the correlations between total VOC (TVOC) levels and meteorological parameters during the period from 2019 to 2021. Table 5 shows the correlation statistics of the TVOC significant levels ($p = 0.05$) with solar radiation, pressure, atmospheric temperature, rainfall, wind speed, and wind direction.

Table 5. The correlation coefficient between TVOCs and meteorological parameters for the year 2019–2021.

Parameters	TVOCs	SR	BP	AT	RF	WS	WD
TVOCs	1						
SR	0.034	1					
BP	0.118	−0.176	1				
AT	−0.168	0.146	0.169	1			
RF	0.012	0.059	−0.992 **	−0.070	1		
WS	−0.150	0.123	0.077	−0.137	−0.047	1	
WD	0.007	−0.355	0.102	0.100	−0.061	−0.308	1

** Correlation is significant at the 0.01 level (2-tailed).

Our results indicated that TVOC level had positive but low correlations with SR, BP, RF, and WD with correlation coefficients (r) of 0.034, 0.118, 0.012, and 0.007, respectively, whereas negative correlations were observed with AT and WS with correlation coefficients (r) of −0.168 and −0.150, respectively. These observations indicated that VOC levels were lower during high AT and WS, possibly due to photodegradation and wind dispersion, which played crucial roles in the VOC levels. Similarly, RF showed a strong negative correlation with BP (−0.992) and a lesser correlation with AT (−0.070). A similar study in Delhi reported variations in pollutant concentrations associated with meteorological parameters [51–53].

5.9. Health Risk Assessment

Hazard Quotient (HQ)

The hazard quotient (HQ) defines the ratio of the exposure concentration for the specific VOC species to an acute reference concentration (RfC) of non-carcinogenic compounds [54,55]. An HQ value of less than 1 indicates a minor or insignificant non-carcinogenic effect, whereas higher values indicate greater non-carcinogenic risks resulting in significant adverse effects on human health [56–58]. The current study estimated the total HQ values for benzene for the years 2019, 2020, and 2021 to be 0.11, 0.43, and 0.09 for males, 0.13, 0.51, and 0.11 for females, and 0.23, 0.85, and 0.19 for children at all industrial sites, respectively. All HQ values were reported to be below 1, indicating negligible human health risks [59]. A similar study conducted in industrial regions reported the HQ values to be less than 0.1 in Tehran, Iran [59] and Rayong Province, East Thailand [58]. According to Baberi et al. (2022), during the lockdown, people spent more than 80% of their schedules in enclosed areas that were associated with hazardous pollutants (benzene), which helped lower the load of disease and thereby reduced national healthcare costs [60]. For the pre-pandemic, pandemic, and post-pandemic periods, for all monitoring regions, in the present study, we estimated the LCR values for benzene in all age groups for males, females, and children, as shown in Figure 5.

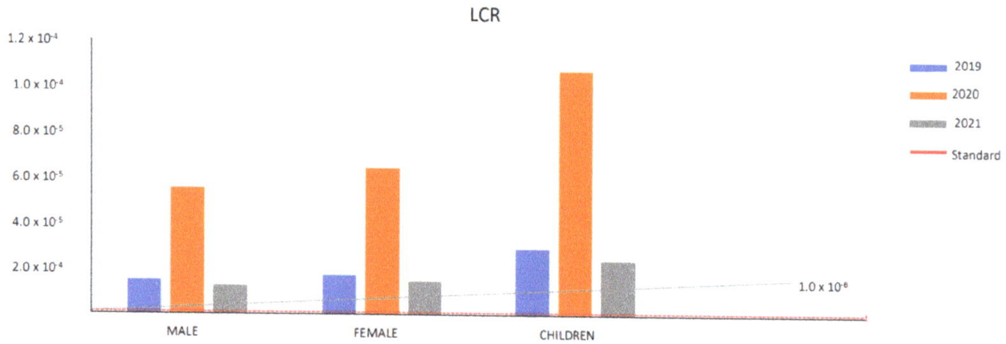

Figure 5. Lifetime cancer risk value for different gender for the years 2019, 2020, and 2021.

For males, the total LCR values for benzene at all monitoring stations were calculated as 1.49×10^{-5}, 5.51×10^{-5}, and 1.24×10^{-5}, whereas the values for toluene were calculated as 1.5×10^{-4}, 1.57×10^{-4}, and 1.5×10^{-4}, for 2019, 2020, and 2021, respectively. The LCR values for benzene at all monitoring stations varied from 5.41×10^{-6} to 2.32×10^{-5}, from 1.24×10^{-6} to 4.22×10^{-4}, and from 2.36×10^{-6} to 2.26×10^{-5} for 2019, 2020, and 2021, respectively. The LCR values for benzene at all monitoring stations were estimated as the lowest for AL and the highest for SP in 2019, while in 2020 they were the lowest for BW and the highest for NL, and in 2021 they were the lowest for PG and the highest for SP. The results indicated that LCR values were higher during the pandemic period than those in the pre- and post-pandemic periods. For benzene, some monitoring stations had LCR values that exceeded the standard LCR value as prescribed by the CPCB (1.0×10^{-6}), such as values of 1.23×10^{-5}, 1.78×10^{-5}, 1.92×10^{-5}, 1.93×10^{-5}, 1.93×10^{-5}, and 2.33×10^{-5} for NL, WZ, SON, MD, OKH, and SP, respectively, in 2019; 1.08×10^{-5}, 1.27×10^{-5}, 1.35×10^{-5}, 1.35×10^{-5}, and 1.99×10^{-5} for MD, OKH, SON, WZ, and SP, respectively, in 2020; 1.11×10^{-5}, 1.28×10^{-5}, 1.55×10^{-5}, 1.69×10^{-5}, 1.73×10^{-5}, and 2.26×10^{-5} for WZ, NL, SON, MD, OKH, and SP, respectively, in 2021.

For females, the totals of LCR values for benzene at all monitoring stations were calculated to be 1.73×10^{-5}, 6.42×10^{-5}, and 1.44×10^{-5}, whereas the values for toluene were calculated to be 1.82×10^{-4}, 1.82×10^{-4}, and 1.82×10^{-4} in 2019, 2020, and 2021, respectively. The LCR values for benzene at all of the monitoring stations varied from 6.31×10^{-6} to 2.71×10^{-5}, from 1.45×10^{-6} to 4.92×10^{-4}, and from 2.76×10^{-6} to 2.63×10^{-5} for 2019, 2020, and 2021, respectively. The LCR values for benzene, at all of the monitoring stations, were estimated as the lowest for AL and the highest for SP in 2019, while they were the lowest for BW and the highest for Narela in 2020, and the lowest for PG and the highest for SP in 2021.

For benzene, some monitoring stations had LCR values that exceeded the standard LCR value as prescribed by the CPCB, such as values of 1.06×10^{-5}, 1.43×10^{-5}, 2.07×10^{-5}, 2.24×10^{-5}, 2.25×10^{-5}, 2.25×10^{-5}, and 2.71×10^{-5} for PG, NL, WZ, SON, MD, OKH, and SP, respectively, in 2019; 1.26×10^{-5}, 1.48×10^{-5}, 1.57×10^{-5}, 1.58×10^{-5}, 2.32×10^{-5}. 5.69×10^{-5}, and 4.92×10^{-4} for MD, OKH, SON, WZ, SP, NG, and NL, respectively, in 2020; 1.06×10^{-5}, 1.29×10^{-5}, 1.49×10^{-5}, 1.81×10^{-5}, 1.98×10^{-5}, 2.02×10^{-5}, and 2.63×10^{-5} for AL, WZ, NL, SON, MD, OKH, and SP, respectively, in 2021. For the pre-lockdown period, a value of LCR was established to be similar to 2.15×10^{-5} and 2.05×10^{-5} for male and female residents, respectively, in China, which showed discernibly higher carcinogenic risks for male and female residents [61].

For children, the total LCR values for benzene at all of the monitoring stations were calculated to be 2.89×10^{-5}, 1.07×10^{-4}, and 2.41×10^{-5}, whereas the values for toluene were calculated to be 3.05×10^{-4}, 3.05×10^{-4}, and 3.05×10^{-4}, for 2019, 2020, and 2021, respectively. The LCR values for benzene for children ranged from 1.05×10^{-6} to

4.52×10^{-5}, from 2.42×10^{-6} to 8.21×10^{-4}, and from 4.59×10^{-6} to 4.39×10^{-5} in 2019, 2020, and 2021, respectively. The LCR values for benzene were estimated for all of the monitoring stations with the lowest LCR value for AL and the highest LCR value for SP in 2019, while in 2020, the lowest LCR value was for BW and the highest LCR value was for NL, and in 2021, the lowest LCR value was for PG and the highest LCR value was for SP. For benzene, some monitoring stations had LCR values that exceeded the standard LCR value as prescribed by the CPCB, such as in 2019; all stations exceeded the LCR value ranging from 1.05×10^{-5} for AL to 4.52×10^{-5} for SP; in 2020, the LCR values were 1.22×10^{-5}, 2.10×10^{-5}, 2.46×10^{-5}, 2.61×10^{-5}, 2.63×10^{-5}, 3.86×10^{-5}, 9.49×10^{-5}, and 8.20×10^{-4} for AL, MD, OKH, SON, WZ, SP, NG, and NL, respectively; in 2021, the LCR values were 1.77×10^{-5}, 2.16×10^{-5}, 2.49×10^{-5}, 3.01×10^{-5}, 3.29×10^{-5}, 3.36×10^{-5}, and 4.39×10^{-5} for AL, WZ, NL, SON, MD, OKH, and SP, respectively. For all monitoring stations, the LCR values for benzene, for the pre-lockdown, lockdown, and post-lockdown periods, were higher than the authorized value (1×10^{-6}), except during the lockdown period, which is a guideline limit value in some circumstances [62].

6. Conclusions

The quantifications of the selected volatile organic compounds (VOCs) were executed in various industrial areas in Delhi, India, from 2019 to 2021. The VOC data from 2019 to 2021 were acquired from the Central Pollution Control Board (CPCB) website, with reference to the pre-pandemic, pandemic, and post-pandemic periods. Using statistical analysis, the current study concluded that anthropogenic activities were considerable sources of emission for VOCs in industrial areas. At all monitoring stations, the mean VOC levels were 47.22 ± 30.15, 37.19 ± 37.19, and 32.81 ± 32.81 μg/m^3 for 2019, 2020, and 2021, respectively. As a result, the level of TVOCs gradually deteriorated over consecutive years due to the pandemic. During the lockdown, the major factors behind the crucial decrease in TVOC levels were complete and partial restrictions on industrial activities, transport, and marketplace openings. The average TVOC values at all the monitoring stations declined from 9 to 61% throughout the pandemic period in contrast to the pre-pandemic period. The change in TVOC levels was reported to be the highest in NL, because NL is renowned for plastic manufacturing industries that create shoe soles and other additional plastic goods, such as adhesive, Rexine, and other tremendously explosive items, which could be significant sources of emissions during the reopening of industrial activities. During 2020, the T/B ratio was estimated in the range of 0.03–13.47, indicating that the major contributors were traffic and non-traffic sources, whereas, during 2021, it ranged from 3.54 to 42.10, where high ratios stipulated non-traffic sources and much higher ratios indicated industries, factories, and petrol pumps as the predominant contributors. The correlation results revealed that TVOC levels had negative relationships with wind speed and atmospheric temperature, which might play a significant role in the dispersion of TVOCs. Comparatively, the lifetime cancer risk (LCR) value for males and females was estimated to be higher throughout the lockdown period than in the pre- and post-lockdown periods. The reason could be the longer exposure time to increase the production of plastic and resin manufacturing units during the pandemic period. Further, the present study aims to increase the scientific accuracy of research on VOCs.

Author Contributions: Data curation, B.P.S.; formal analysis, M.A., S.S.S. and T.A.A.; funding acquisition, S.S.S.; investigation, B.P.S. and S.K. (Saumya Kumari); methodology, S.K. (Sweety Kumari); project administration, S.S.S.; Figures, A.N. and K.M.; writing—review and editing, K.M., K.C., S.R. and E.I.A. All authors have read and agreed to the published version of the manuscript.

Funding: This research work was funded by the Institutional Fund Projects under grant no. (IFPIP:1544-141-1443). The authors gratefully acknowledge the technical and financial support provided by the Ministry of Education and King Abdulaziz University, DSR, Jeddah, Saudi Arabia.

Institutional Review Board Statement: Not applicable.

Informed Consent Statement: Not applicable.

Data Availability Statement: Not applicable.

Conflicts of Interest: The authors declare that there are no known competing financial interests or personal relationships that could have appeared to influence the work reported in this paper.

References

1. Martelletti, L.; Martelletti, P. Air pollution and the novel COVID-19 disease: A putative disease risk factor. *SN Compr. Clin. Med.* **2020**, *2*, 383–387. [CrossRef] [PubMed]
2. World Health Organisation. World Health Organization Coronavirus Disease (COVID-19) Pandemic, WHO. 2020. Available online: https://www.who.int/emergencies/diseases/novel-coronavirus-2019 (accessed on 31 March 2020).
3. World Health Organisation. World Health Organization, Coronavirus Disease. (COVID-19) India Situation Report-1. 2020. Available online: https://www.who.int/docs/default-source/wrindia/india-situation-report-1.pdf?sfvrsn5ca2a672_0 (accessed on 31 January 2020).
4. World Metros. 2022. Available online: https://www.worldometers.info/coronavirus/ (accessed on 5 January 2021).
5. Gorai, A.K.; Tchounwou, P.B.; Mitra, G. Spatial variation of ground level ozone concentrations and its health impacts in an urban area in India. *Aerosol Air Qual. Res.* **2017**, *17*, 951–964. [CrossRef] [PubMed]
6. Li, H.; Xu, X.-L.; Dai, D.-W.; Huang, Z.-Y.; Ma, Z.; Guan, Y.-J. Air pollution and temperature are associated with increased COVID-19 incidence: A time series study. *Int. J. Infect. Dis.* **2020**, *97*, 278–282. [CrossRef] [PubMed]
7. Ogen, Y. Assessing nitrogen dioxide (NO_2) levels as a contributing factor to the coronavirus (COVID-19) fatality rate. *Sci. Total Environ.* **2020**, *726*, 138605. [CrossRef]
8. Wu, X.; Nethery, R.C.; Sabath, B.M.; Braun, D.; Al, E. Exposure to air pollution and COVID-19 mortality in the United States. *medRxiv* **2020**. [CrossRef]
9. Zhu, Y.; Xie, J.; Huang, F.; Cao, L. Association between short-term exposure to air pollution and COVID-19 infection: Evidence from China. *Sci. Total Environ.* **2020**, *727*, 138704. [CrossRef] [PubMed]
10. Singh, B.P.; Kumar, P. Spatio-temporal variation in fine particulate matter and effect on air quality during the COVID-19 in New Delhi, India. *Urban Clim.* **2021**, *40*, 101013. [CrossRef]
11. Brandt, E.; Beck, A.; Mersha, T. Air pollution, racial disparities and COVID-19 mortality. *J. Allergy Clin. Immunol.* **2020**, *146*, 61–63. [CrossRef]
12. Comunian, S.; Dongo, D.; Milani, C.; Palestini, P. Air pollution and COVID-19: The role of particulate matter in the spread and Increase of COVID-19's morbidity and mortality. *Int. J. Environ. Res. Public Health* **2020**, *17*, 4487. [CrossRef]
13. Paital, B.; Agrawal, P.K. Air pollution by NO_2 and PM2.5 explains COVID-19 infection severity by overexpression of angiotensin-converting enzyme 2 in respiratory cells: A review. *Environ. Chem. Lett.* **2020**, *19*, 25–42. [CrossRef]
14. Pozzer, A.; Dominici, F.; Haines, A.; Witt, C.; Münzel, T.; Lelieveld, J. Regional and global contributions of air pollution to risk of death from COVID-19. *Cardiovasc. Res.* **2020**, *116*, 2247–2253. [CrossRef] [PubMed]
15. Kumar, A.; Singh, B.P.; Punia, M.; Singh, D.; Kumar, K.; Jain, V.K. Assessment of indoor air concentrations of VOCs and their associated health risks in the library of Jawaharlal Nehru University, New Delhi. *Environ. Sci. Pollut. Res.* **2014**, *21*, 2240–2248. [CrossRef] [PubMed]
16. Kumar, A.; Singh, B.P.; Punia, M.; Singh, D.; Kumar, K.; Jain, V.K. Determination of volatile organic compounds and associated health risk assessment in residential homes and hostels within an academic institute, New Delhi. *Indoor Air* **2014**, *24*, 474–483. [CrossRef] [PubMed]
17. Kumar, A.; Singh, D.; Singh, B.P.; Singh, M.; Anandam, K.; Kumar, K.; Jain, V.K. Spatial and temporal variability of surface ozone and nitrogen oxides in urban and rural ambient air of Delhi-NCR, India. *Air Qual. Atmos. Health* **2015**, *8*, 391–399. [CrossRef]
18. Pandey, N.; Singh, B.P.; Singh, M.; Tyagi, S. Diurnal Variation of Ozone Levels in Academic Hostel in Delhi-A Case Study of JNU campus. *Int. J. Appl. Environ. Sci.* **2017**, *12*, 6.
19. Shuai, J.; Kim, S.; Ryu, H.; Park, J.; Lee, C.K.; Kim, G.B.; Ultra, V.U.; Yang, W. Health risk assessment of volatile organic compounds exposure near Daegu dyeing industrial complex in South Korea. *BMC Public Health* **2018**, *18*, 528. [CrossRef]
20. Singh, B.P.; Zughaibi, T.A.; Alharthy, S.A.; Al-Asmari, A.I.; Rahman, S. Statistical analysis, source apportionment, and toxicity of particulate- and gaseous-phase PAHs in the urban atmosphere. *Front. Public Health* **2023**, *10*, 1070663. [CrossRef]
21. Singh, H.; Meraj, G.; Singh, S.; Shrivastava, V.; Sharma, V.; Farooq, M.; Kanga, S.; Singh, S.K.; Kumar, P. Status of Air Pollution during COVID-19-Induced Lockdown in Delhi, India. *Atmosphere* **2022**, *13*, 2090. [CrossRef]
22. Chen, K.; Wang, M.; Huang, C.; Kinney, P.L.; Anastas, P.T. Air pollution reduction and mortality benefit during the COVID-19 outbreak in China. *Lancet Planet. Health* **2020**, *4*, E210–E212. [CrossRef]
23. Dutheil, F.; Baker, J.S.; Navel, V. COVID-19 as a factor influencing air pollution? *Environ. Pollut.* **2020**, *263*, 114466. [CrossRef]
24. Singh, B.P.; Rana, P.; Mittal, N.; Kumar, S.; Athar, M.; Abduljaleel, Z.; Rahman, S. Variations in the Yamuna River Water Quality During the COVID-19 Lockdowns. *Front. Environ. Sci.* **2022**, *10*, 940640. [CrossRef]
25. Singh, B.P.; Kumar, K.; Jain, V.K. Source identification and health risk assessment associated with particulate- and gaseous-phase PAHs at residential sites in Delhi, India. *Air Qual. Atmos. Health* **2021**, *14*, 1505–1521. [CrossRef]
26. Singh, B.P.; Kumar, K.; Jain, V.K. Distribution of ring PAHs in particulate/gaseous phase in the urban city of Delhi, India: Seasonal variation and cancer risk assessment. *Urban Clim.* **2021**, *40*, 101010. [CrossRef]

27. Pollack, I.B.; Ryerson, T.B.; Trainer, M.; Parrish, D.D.; Andrews, A.E.; Atlas, E.L.; Blake, D.R.; Brown, S.S.; Commane, R.; Daube, B.C.; et al. Airborne and ground-based observations of a weekend effect in ozone, precursors, and oxidation products in the California South Coast Air Basin. *J. Geophys. Res.* **2012**, *117*, D00V05. [CrossRef]
28. Singh, D.; Kumar, A.; Kumar, K.; Singh, B.; Mina, U.; Singh, B.B.; Jain, V.K. Statistical modeling of O3, NOx, CO, PM2.5, VOCs and noise levels in commercial complex and associated health risk assessment in an academic institution. *Sci. Total Environ.* **2016**, *572*, 586–594. [CrossRef] [PubMed]
29. Luecken, D.; Napelenok, S.; Strum, M.; Scheffe, R.; Phillips, S. Sensitivity of ambient atmospheric formaldehyde and ozone to precursor species and source types across the United States. *Environ. Sci. Technol.* **2018**, *52*, 4668–4675. [CrossRef] [PubMed]
30. Pakkattil, A.; Muhsin, M.; Varma, M.K.R. COVID-19 lockdown: Effects on selected volatile organic compound (VOC) emissions over the major Indian metro cities. *Urban Clim.* **2021**, *37*, 100838. [CrossRef]
31. Allahabady, A.; Yousefi, Z.; Tahamtan RA, M.; Sharif, Z.P. Measurement of BTEX (benzene, toluene, ethylbenzene and xylene) concentration at gas stations. *Environ. Health Eng. Manag.* **2022**, *9*, 23–31. [CrossRef]
32. Molekoa, M.D.; Kumar, P.; Choudhary, B.K.; Yunus, A.P.; Kharrazi, A.; Khedher, K.M.; Alshayeb, M.J.; Singh, B.P.; Minh, H.V.T.; Kurniawan, T.A.; et al. Spatio-temporal variations in the water quality of the Doorndraai Dam, South Africa: An assessment of sustainable water resource management. *Curr. Res. Environ. Sustain.* **2022**, *4*, 100187. [CrossRef]
33. Subali, A.D.; Wiyono, L.; Yusuf, M.; Zaky, M.F.A. The potential of volatile organic compounds-based breath analysis for COVID-19 screening: A systematic review & meta-analysis. *Diagn. Microbiol. Infect. Dis.* **2022**, *102*, 115589. [CrossRef]
34. Singh, B.P.; Kumari, S.; Nair, A.; Kumari, S.; Wabaidur, S.M.; Avtar, R.; Rahman, S. Temporary reduction in VOCs associated with health risk during and after COVID-19 in Maharashtra, India. *J. Atmos. Chem.* **2022**, *17*, 1–24. [CrossRef] [PubMed]
35. Sahu, V.; Gurjar, B.R. Spatial and seasonal variation of air quality in different microenvironments of a technical university in India. *Build. Environ.* **2020**, *185*, 107310. [CrossRef]
36. Ghaffari, H.R.; Kamari, Z.; Hassanvand, M.S.; Fazlzadeh, M.; Heidari, M. Level of air BTEX in urban, rural and industrial regions of Bandar Abbas, Iran; indoor-outdoor relationships and probabilistic health risk assessment. *Environ. Res.* **2021**, *200*, 111745. [CrossRef] [PubMed]
37. Yousefian, F.; Mahvi, A.H.; Yunesian, M.; Hassanvand, M.S.; Kashani, H.; Amini, H. Long-term exposure to ambient air pollution and autism spectrum disorder in children: A case-control study in Tehran, Iran. *Sci. Total Environ.* **2018**, *643*, 1216–1222. [CrossRef] [PubMed]
38. Cerón-Bretón, J.G.; Cerón-Bretón, R.M.; Kahl, J.D.W.; Ramírez-Lara, E.; Guarnaccia, C.; Aguilar-Ucán, C.A.; López-Chuken, U.; Montalvo-Romero, C.; Anguebes-Franseschi, F. Diurnal and seasonal variation of BTEX in the air of Monterrey, Mexico: Preliminary study of sources and photochemical ozone pollution. *Air Qual. Atmos. Health* **2015**, *8*, 469–482. [CrossRef]
39. Paterson, C.A.; Sharpe, R.A.; Taylor, T.; Morrissey, K. Indoor PM2.5, VOCs and asthma outcomes: A systematic review in adults and their home environments. *Environ. Res.* **2021**, *202*, 111631. [CrossRef]
40. Alford, K.L.; Kumar, N. Pulmonary Health Effects of Indoor Volatile Organic Compounds—A Meta-Analysis. *Int. J. Environ. Res. Public Health* **2021**, *18*, 1578. [CrossRef]
41. Baker, E.L.; Smith, T.J.; Landrigan, P.J. The neurotoxicity of industrial solvents: A review of the literature. *Am. J. Ind. Med.* **1985**, *8*, 207–217. [CrossRef]
42. Hazrati, S.; Rostami, R.; Farjaminezhad, M.; Fazlzadeh, M. Preliminary assessment of BTEX concentrations in indoor air of residential buildings and atmospheric ambient air in Ardabil, Iran. *Atmos. Environ.* **2016**, *132*, 91–97. [CrossRef]
43. Tiwari, S.; Srivastava, A.K.; Bisht, D.S.; Parmita, P.; Srivastava, M.K.; Attri, S.D. Diurnal and seasonal variations of black carbon and PM2.5 over NewDelhi, India: Influence of meteorology. *Atmos. Res.* **2013**, *125*, 50–62. [CrossRef]
44. Srivastava, A.; Jain, V.K.; Srivastava, A. SEM EDX analysis of various sizes aerosols in Delhi India. *Environ. Monit. Assess.* **2009**, *150*, 405–416. [CrossRef] [PubMed]
45. Central Pollution Control Board (CPCB). Central Pollution Control Board (CPCB), Ministry of Environment, Forest and Climate Change. New Delhi: Government of India. 2022. Available online: https://app.cpcbccr.com/ccr/#/caaqm-dashboard-all/caaqm-landing (accessed on 5 January 2022).
46. Zhang, H.; Li, H.; Zhang, Q.; Zhang, Y.; Zhang, W.; Wang, X.; Bi, F.; Chai, F.; Gao, J.; Meng, L.; et al. Atmospheric volatile organic compounds in a typical urban area of Beijing: Pollution characterization, health risk. *Atmosphere* **2017**, *8*, 61. [CrossRef]
47. Zhang, Z.; Xu, J.; Ye, T.; Chen, L.; Chen, H.; Yao, J. Distributions and temporal changes of benzene, toluene, ethylbenzene, and xylene concentrations in newly decorated rooms in southeastern China, and the health risks posed. *Atmos. Environ.* **2020**, *246*, 118071. [CrossRef]
48. Srivastava, A.; Gupta, S.; Jain, V.K. Source apportionment of total suspended particulate matter in coarse and fine ranges over Delhi. *Aerosol. Air Qual. Res.* **2008**, *8*, 188–200. [CrossRef]
49. Sahu, L.K.; Saxena, P. High time and mass resolved PTR-TOF-MS measurements of VOCs at an urban site of India during winter: Role of anthropogenic, biomass burning, biogenic and photochemical sources. *Atmos. Res.* **2015**, *164–165*, 84–94. [CrossRef]
50. Raysoni, A.U.; Stock, T.H.; Sarnat, J.A.; Chavez, M.C.; Sarnat, S.E.; Montoya, T.; Holguin, F.; Li, W.W. Evaluation of VOC concentrations in indoor and outdoor microenvironments at near-road schools. *Environ. Pollut.* **2017**, *231*, 681–693. [CrossRef]
51. Akoglu, H. User's guide to the correlation coefficient. *Turk. J. Emerg. Med.* **2018**, *18*, 91–93. [CrossRef]
52. Punia, M.; Nain, S.; Kumar, A.; Singh, B.P.; Prakash, A.; Kumar, K.; Jain, V.K. Analysis of temperature variability over north-west part of India for the period 1970–2000. *Nat. Hazards* **2015**, *75*, 935–952. [CrossRef]

53. Bhardwaj, P.; Singh, B.P.; Pandey, A.K.; Jain, V.K.; Kumar, K. Characterization and Morphological Analysis of Summer and Wintertime PM 2.5 Aerosols Over Urban-Rural Locations in Delhi-NCR. *Int. J. Appl. Environ. Sci.* **2017**, *12*, 5.
54. Singh, B.P.; Singh, D.; Kumar, K.; Jain, V.K. Study of seasonal variation of PM2.5 concentration associated with meteorological parameters at residential sites in Delhi, India. *J. Atmos. Chem.* **2021**, *78*, 161–176. [CrossRef]
55. Su, F.C.; Jia, C.; Batterman, S. Extreme value analyses of VOC exposures and risks: A comparison of RIOPA and NHANES datasets. *Atmos. Environ.* **2012**, *62*, 97–106. [CrossRef] [PubMed]
56. Alsbou, E.M.; Omari, K.W. BTEX indoor air characteristic values in rural areas of Jordan: Heaters and health risk assessment consequences in winter season. *Environ. Pollut.* **2020**, *267*, 115464. [CrossRef] [PubMed]
57. Kim, H.; Lim, Y.; Shin, D.; Sohn, J.R.; Yang, J. Risk Assessment of Volatile Organic Compounds (VOCs) and Formaldehyde in Korean Public Facilities: Derivation of Health Protection Criteria Levels. *Asian J. Atmos. Environ.* **2011**, *5*, 121–133. [CrossRef]
58. Pinthong, N.; Thepanondh, S.; Kondo, A. Source Identification of VOCs and their Environmental Health Risk in a Petrochemical Industrial Area. *Aerosol Air Qual. Res.* **2022**, *22*, 210064. [CrossRef]
59. Yousefian, F.; Sadegh, M.; Nabizadeh, R. The concentration of BTEX compounds and health risk assessment in municipal solid waste facilities and urban areas. *Environ. Res.* **2020**, *191*, 110068. [CrossRef] [PubMed]
60. Baberi, Z.; Azhdarpoor, A.; Hoseini, M.; Baghapour, M. Monitoring Benzene, Toluene, Ethylbenzene, and Xylene (BTEX) Levels in Mixed-Use Residential-Commercial Buildings in Shiraz, Iran: Assessing the Carcinogenicity and Non-Carcinogenicity Risk of Their Inhabitants. *Int. J. Environ. Res. Public Health* **2022**, *19*, 723. [CrossRef]
61. Qin, N.; Zhu, Y.; Zhong, Y.; Tian, J.; Li, J.; Chen, L.; Fan, R.; Wei, F. External Exposure to BTEX, Internal Biomarker Response, and Health Risk Assessment of Nonoccupational Populations near a Coking Plant in Southwest China. *Int. J. Environ. Res. Pub. Health* **2022**, *19*, 847. [CrossRef]
62. Ramírez, N.; Cuadras, A.; Rovira, E.; Borrull, F.; Marcé, R.M. Chronic risk assessment of exposure to volatile organic compounds in the atmosphere near the largest Mediterranean industrial site. *Environ. Int.* **2012**, *39*, 200–209. [CrossRef]

MDPI
St. Alban-Anlage 66
4052 Basel
Switzerland
www.mdpi.com

Toxics Editorial Office
E-mail: toxics@mdpi.com
www.mdpi.com/journal/toxics

www.ingramcontent.com/pod-product-compliance
Lightning Source LLC
Chambersburg PA
CBHW061503210926
51850CB00008B/1805
9783725808342